T0143155

Poisonous Skies

Poisonous Skies
Acid Rain and the Globalization of Pollution

Rachel Emma Rothschild

The University of Chicago Press :: Chicago and London

The University of Chicago Press, Chicago 60637
The University of Chicago Press, Ltd., London
© 2019 by The University of Chicago
Published 2019
Printed in the United States of America

28 27 26 25 24 23 22 21 20 19   1 2 3 4 5

ISBN-13: 978-0-226-63471-5 (cloth)
ISBN-13: 978-0-226-63485-2 (e-book)
DOI: https://doi.org/10.7208/chicago/9780226634852.001.0001

Library of Congress Cataloging-in-Publication Data

Names: Rothschild, Rachel Emma, author.
Title: Poisonous skies : acid rain and the globalization of pollution /
    Rachel Emma Rothschild.
Description: Chicago ; London : The University of Chicago Press, 2019. |
    Includes bibliographical references and index.
Identifiers: LCCN 2018054763 | ISBN 9780226634715 (cloth :
    alk. paper) | ISBN 9780226634852 (e-book)
Subjects: LCSH: Acid rain—Research—History. | Acid rain—Political
    aspects. | Acid rain—Environmental aspects. | Air—Pollution—
    Research—History.
Classification: LCC TD195.42 .R67 2019 | DDC 363.738/6—dc23
LC record available at https://lccn.loc.gov/2018054763

♾ This paper meets the requirements of ANSI/NISO Z39.48-1992
(Permanence of Paper).

Contents

Acronyms

CEGB   (Central Electricity Generating Board)
CERL   (Central Electricity Research Laboratories)
CSCE   (Conference on Security and Cooperation in Europe)
DAFS   (Department of Agriculture and Fisheries in Scotland)
EACN   (European Air Chemistry Network)
EPA   (Environmental Protection Agency)
IBP   (International Biological Program)
IIASA (International Institute for Applied Systems Analysis)
IPCC   (Intergovernmental Panel on Climate Change)
ITE   (Institute for Terrestrial Ecology)
NAPAP (National Acid Precipitation Assessment Program)
NATO   (North Atlantic Treaty Organization)
NAVF   (Norwegian Research Council for Science and the Humanities, Norges almenvitenskapelige forskningsråd)
NILU   (Norwegian Institute for Air Research, Norsk institutt for luftforskning)
NTNF   (Royal Norwegian Council for Scientific and Industrial Research, Norges Teknisk-Naturvitenskapelige Forskningsråd)
OECD   (Organisation for Economic Cooperation and Development)
OEEC   (Organisation for European Economic Cooperation)
OPEC   (Organization of Petroleum Exporting Countries)
RAINS (Regional Air Pollution Information and Simulation)

SNSF   (Acid Rain's Effects on Forests and Fish Project, Sur nedbørs
        virkning på skog og fisk)
SWAP   (Surface Water Acidification Program)
UN   (United Nations)
WHO   (World Health Organization)

Introduction: A Rain of Ashes

> Direr visions, worse foreboding,
> Glare upon me through the gloom!
> Britain's smoke-cloud sinks corroding
> On the land in noisome fume,
> Smirches all its tender bloom,
> All its gracious verdure dashes,
> Sweeping low with breath of bane,
> Stealing sunlight from the plain,
> Showering down like rain of ashes
> On the city of God's doom.[1]

The earliest recorded reference to air pollution from one country harming the environment of its neighbors appeared in the above epigraph, written in 1865 by the Norwegian playwright Henrik Ibsen in his tragedy *Brand*.[2] While Ibsen lacked our modern-day understanding of air pollution and its ecological effects, he was responding to a dramatic change in human society and its relationship with nature. The smoke-clouds of Britain's industrialization, powered by the country's ample coal deposits, would soon spread throughout Europe and North America. Over the course of the nineteenth century, coal extraction grew ten-fold. By the end of the twentieth century, it had increased nearly seven-fold as developing countries like China and India sought to replicate the vast economic growth of industrialized nations. The environmental consequences have been severe, with emissions of pollutants like sulfur dioxide more than quadrupling in the last century.[3]

Yet the idea that fossil fuel pollution could travel across vast stretches to rain down on distant lands was not seen as a topic worthy of scientific investigation until nearly a century after Ibsen wrote those words.[4] Now in many countries it's simply assumed that even an educated layperson is aware of the environmental damages fossil fuels can inflict. There are entire fields devoted to examining the biological, geophysical, and atmospheric aspects of pollution. Environmental science is a fast-growing interdisciplinary area of research, with scientists from an array of backgrounds studying the earth's processes and in what ways pollution might be damaging the natural world. This book is an attempt to understand the history of our knowledge about fossil fuel pollutants and how scientists and policymakers came to grasp the global nature of their environmental threat. It does so by looking at the first air pollution problem identified as having damaging effects on areas far from the source of emissions: acid rain.

The problem of acid rain changed scientific and popular understandings of how fossil fuel pollution could cause regional, and perhaps global, environmental harms. The term has been used to describe any form of precipitation, including rain, snow, or fog, with high levels of sulfur dioxides or nitrogen oxides.[5] These chemicals are important components of many biological and physical processes on the earth, but after the Industrial Revolution of the nineteenth century they were produced in much greater quantities than occur in nature. The construction of electrical power systems and the invention of new technologies like steam engines and automobiles led to widespread use of coal and oil as energy sources and the release of progressively larger quantities of air pollutants, particularly after the Second World War.

In the late 1960s, Swedish scientists first suggested that growing amounts of fossil fuel pollution could be responsible for an increase in acid rain across the southern portions of Norway, Sweden, and Finland. While sulfur dioxide was initially identified as the likely culprit, nitrogen oxides were soon suspected of partly contributing to the problem. When American scientists claimed to have identified a similar phenomenon along the eastern part of the country in the early 1970s, acid rain quickly became a top environmental issue facing industrialized nations. Once emitted into the atmosphere, sulfur dioxides and nitrogen oxides undergo several chemical reactions that result in the production of sulfate and nitrate, which are acidic pollutants that lower the pH of precipitation. Upon discovering an increase in precipitation acidity across their countries, Scandinavian and American scientists became concerned that acid rain could cause severe environmental damage by contaminating soil, polluting water, and destroying flora and fauna in freshwater and terrestrial ecosystems. Throughout the next two decades, researchers across Europe and North America would seek to understand whether fossil fuel pollutants were in fact responsible for acid rain, how much each country contributed to the air pollution of neighboring states, and what the environmental impact of acid rain would be in susceptible regions.

However, there were more than a few challenges with answering these questions. The first was that a scientific apparatus to study acid rain did not yet exist. Atmospheric scientists had never examined the dispersion of fossil fuel pollutants across long distances, nor had biologists and soil scientists tried to track their deposition patterns and effects on organisms. Investigations into acid rain led to new interdisciplinary collaborations among physicists, biologists, foresters, and geologists on a vast scale, and helped to form an emerging field of "environmental science" equipped to tackle the challenges of global pollution problems.

These collaborative, interdisciplinary efforts contributed to a change in how scientists conceptualized environmental systems from discrete, isolated parts to an integrated whole of physical and biological components. Today, we take for granted that environmental studies are concerned with tracing the movement of fossil fuel pollutants from smokestacks through ecosystems, across continents and national borders. But it was studies of acid rain and other pollutants that brought about a reimagining of how to investigate the natural world as a complete entity. The success of this method was not preordained. Other models for environmental research put forward around the time of acid rain's discovery eschewed "big science" projects, envisioning more of a bottom-up method that pieced together small studies rather than a top-down, all encompassing approach.

Who should be involved in these efforts and who should pay for them were not easy questions to answer. Acid rain forced countries to grapple with the benefits and drawbacks of having governments, industry, or academic institutions fund environmental research. In addition, governments faced the prospect of trying to facilitate cooperation not only across different fields but across national boundaries as well. Yet acid rain emerged at a seemingly inconvenient moment for the international community, with the world divided by the ongoing Cold War and struggling to decide how to allocate diplomatic power to intergovern-mental groups like the United Nations (UN), Organisation for Economic Cooperation and Development (OECD), and European Communities.

When acid rain exploded to the forefront of scientific and political agendas, energy security was also becoming a prominent diplomatic con-cern as fossil fuel companies struggled with the impacts of national en-vironmental regulations. Frustrated by what they viewed as a misguided attack on their operations and the economic prosperity of the Western world, the coal industries of both Britain and the US mounted an immense effort to counter environmental research on acid rain with their own stud-ies of the problem. The research establishment they created influenced scientific, political, and public perceptions of acid rain, and one of the major goals of this book is to examine the extent of their involvement in environmental science and politics as well as some of its questionable results.

I have chosen to tell this story as a history of knowledge and its po-litical implications, focusing on the scientists and environmental officials involved in acid rain from its inception through attempts to regulate fos-sil fuel pollution in the late 1980s and 1990s. Because the bulk of acid rain science and diplomacy was led by citizens of Norway, Sweden, Brit-ain, the US, and Canada, with West Germany arriving on the scene a bit later, the book follows actors in these countries while incorporating the perspectives of other European nations where relevant, notably the Soviet Union. Writing about the history of acid rain in this way invari-ably leaves out certain topics. For example, if a particular subject did not prove important in international negotiations but had more local resonance, it was left on the cutting room floor. Or in an archival box, I should say. But I hope this work will open up new avenues and questions for other scholars to answer, whether on domestic aspects of acid rain, its connections to popular social movements of the period, or changing cultural perspectives about the natural world.

Researching this book took me to eight different countries and more than a dozen archives. In the process, I was able to slowly unravel how decisions were made that had an enormous impact on the course of scientific and political events concerning acid rain. Often documents of crucial importance to understanding the actions of individuals or gov-ernments from one country were found buried in another state's national archives or the holdings of an intergovernmental institution. It would not have been possible to piece together this narrative without under-taking an investigation that crossed national borders in the same ways that acid rain has done. To cite just a few examples, this multi-archival research revealed that the British government's concerns about the 1970s oil shocks strongly impacted its international position on acid rain, that Norwegian scientists were the driving force behind creating an air pollution monitoring network across the iron curtain with the Soviet Union, and that a US State Department ultimatum to its allies during UN negotiations was the key factor that led to the first successful treaty on acid rain. None of this is public information, and part of my aim in writing this book was to uncover what was going on behind the scenes in constructing acid rain research and policy.

I obtained some of the most valuable insights from personal interviews with scientists who worked on behalf of governments as well as the coal industry. Many of them granted me extensive swaths of their time, and I am grateful for their candor. To verify the accounts I received, I cross-referenced claims or stories as much as possible with other interview subjects or with the documentary record. In instances where I received conflicting accounts, I have acknowledged this in my notes to the text. Although indebted to these scientists for the detailed information I was able to obtain about the events I discuss in the book, all the opinions and arguments expressed in this work are my own and should not be assumed to represent their personal perspectives.

When one is talking to those who spent the better part of their professional lives studying acid rain, the significance of the problem is quite evident, whether for understanding how even "natural" pollutants can have harmful environmental consequences or for establishing norms in environmental diplomacy. But if you polled your average person on the street, acid rain would probably seem like an environmental problem that's already been solved. Why concern ourselves with the details of its history? We have far more pressing environmental problems, the argument might go, climate change being the most obvious contender for our attention.

There are two reasons that I have found persuasive in undertaking the task of working on this project for nearly a decade, and they are ones that I hope readers who care about the environment might find compelling as well. The first is that without past knowledge of how societies have dealt with environmental pollution, we have little chance of improving upon these efforts. An important caveat: this book is first and foremost a work of history, and as such, those interested in finding succinct prescriptions for tackling the environmental challenges ahead may be frustrated with its contents. No historical period will ever be a perfect laboratory to model how the future will unfold. There are always differences between the past and present, and it would be unwise to ignore them when trying to draw lessons from acid rain. With that said, we have no better options. Just as we cannot run experiments on the earth, neither can we test out human behavior on the enormous scales of science and politics. We are still wrestling with how to weigh expertise in political decisions, how to balance the interests of the environment against economic growth, and how to convince many nations that they should work together for the protection of a planet we all share. Acid rain can tell us how at least some people tried to resolve these issues, which will better position each of us in addressing the serious tasks to come.

The second reason is less directly utilitarian and speaks more to the interests of historians and other scholars. It is to obtain as accurate an account as possible of the events surrounding acid rain and venture several arguments about why they transpired as they did and their larger impact. This is the first book to try to tell its history in an international context. It will hopefully upend some commonly received wisdom about the acid rain story as well as science, the environment, and diplomacy more generally. For those interested in the history of science, acid rain's relationship with the burgeoning field of environmental science can illuminate how the legacy of military sponsorship of physics, chemistry, and other fields during wartime influenced the direction of research on the environment. Scholars of the Cold War will likely find the chapters on acid rain's relationship with the détente process especially revealing about how environmental issues intersected with Cold War politics, in addition to the importance of non–super power states in diplomacy. Economic and business historians may appreciate the portions of this work that address the role of the British and American coal industries in environmental science and discussions concerning how to balance economic growth with pollution regulation. Environmental historians will ideally glean the most from a study of one of the most significant pollution problems of the twentieth century, particularly its role in shaping ideas about environmental risk and the precautionary principle.

The question of what level of scientific proof is necessary before acting to address an environmental threat runs throughout this text. When acid rain was first discovered, scientific experts had, by and large, earned a prominent seat at the table in many Western governments thanks to the contributions they made in fighting the previous two world wars.[6] Many policymakers from across the political spectrum hoped that scientists could play an equally important part in deciding whether we needed to reduce fossil fuel pollution and, if so, the best way to do it. Scientists themselves who were involved early on in acid rain research shared a similar faith in their expertise and a desire to get involved in the diplomatic process.

Environmental scientists were certainly crucial to identifying acid rain as a threat and pinpointing fossil fuels as a possible cause. Without the tools of science and technology, it would have been practically impossible for humans to discover that invisible chemicals were the culprit behind the observed damages. However, when it came time to enact policy on acid rain at both the national and international levels, environmental scientists could not carry government officials across the finish line to achieve agreement on how to solve the problem. In part, this was because the coal industries of Britain and the US mounted an enormous campaign to discredit their work, continually raising the bar for how much scientific certainty was needed before society could act. But it was also because scientific experts could not answer fundamental questions about environmental values that were at the heart of disagreements over what do about acid rain.

There is a word in Norwegian, "friluftsliv," which roughly translates to "open air life" in English. First made famous by Henrik Ibsen, it connotes a deep, abiding connection with the natural world and access to an unspoiled environment. For Scandinavians, acid rain not only threatened to reduce fish populations or diminish forests. It endangered an entire way of life and culture that prized nature for its spiritual and restorative properties. There was no clear way to calculate the impact of acid rain on friluftsliv. Even if it were possible to tally all the salmon killed in a given year and the resulting loss of sports-fishing in a region, it was not so simple to calculate the value of sailing through pristine fjords or exploring the deep woods along towering mountain ranges.

It was comparatively straightforward to measure the cost of installing new pollution control technologies in coal-fired power plants along the Ruhr valley, the British Midlands, and the American Midwest. The financial burden was also immediate, with direct impacts on the price of energy for citizens of countries that might implement restrictions on fossil fuel emissions. Humankind has had a notoriously difficult history with anticipating crises far in the future and taking action in the present for potential long-term benefits, whether on an individual, national, or international level.

We are still grappling with how to value the environment and what sacrifices our societies are willing to make to protect it. In the case of acid rain, failure to examine the ethical issues involved created an unspoken gulf between the positions of polluters and recipients of acid rain.

Instead of discussing the benefits and drawbacks of tackling the problem, including who would be helped or harmed by various approaches, too often scientific and political debates concerned only whether we knew enough to act. This was a driving force behind the eventual turn toward precautionary approaches, which allowed governments to sidestep the issue of scientific certitude and place the burden of proof on industry. While a remarkable advance in policymaking, the history of acid rain exemplifies the dangers of not having these crucial conversations out in the open. If countries are to protect the planet and maintain the benefits of economic development, it is imperative to be clear about what is at stake for the natural world as well as what we risk, and have risked, by not acting.

1 Creating a Global Pollution Problem

Looking back, I would say we entered the acid age in 1952 although we didn't know it at the time. That is when a "killer smog"—a manmade soup of noxious chemicals—settled over London, England. Four thousand deaths were attributed to the smog. Due to that incident and a number of less dramatic ones around the world, a great deal was done to clean up air pollution. In many cases, the answer was simple. Build a tall smokestack so that industrial emissions could be widely dispersed in the atmosphere. It was a relatively cheap and effective solution. The trouble is that what goes up, must also come down.[1]

John Roberts, minister of the environment, Canada, 1981

Acid rain heralded a shift from local environmental problems to concern about the impact of man's activities on a regional or global scale.[2]

Gro Harlem Brundtland, minister of the environment, Norway, 1974

Air pollution has beset societies for centuries, but it increased noticeably after the Industrial Revolution brought about pervasive use of fossil fuels throughout Europe and North America. Black smoke from steam-engine coal fires darkened the skies and coated buildings with soot. Even the weather in certain areas began to change. Thick smog carrying pollution from factories and power plants blanketed cities for days, at times causing respiratory problems so severe that dozens to thousands of the very young or very old were killed.[3]

The smoke and fogs of industrialization were visible to any citizen walking the streets of London or traveling the coal rich region of the Ruhr valley in Western Germany.[4]

Public petitions to reduce soot, dirt, and dust in the air surfaced repeatedly from the late nineteenth century through the years after the Second
World War, when a series of severe air pollution incidents spurred governments in Europe and North America to act. Like the destruction of
forests and wild areas or the disposal of sewage into waterways, smoke
from power plants required no special equipment or knowledge to identify. During the Second World War pollution reached such high levels
in some cities that automobile headlights needed to be kept on during
the daytime.[5] Many governments began implementing air quality standards
to mitigate the harmful health effects of air pollution, and industries responded by installing new technologies to reduce smoke from burning
fuels or increasing the height of smokestacks to eject pollution higher
into the air.[6]

Although these changes reduced the amount of soot in emissions and
decreased the visible smoke from fossil fuel combustion, they did not
remove the chemical byproducts. These "invisible" pollutants included
sulfur dioxide, which scientists began to identify as the major respiratory
irritant in smog episodes. Sulfur dioxide subsequently became the first
fossil fuel pollutant suspected of posing a danger to public health during
the 1930s.[7] The shift toward a more chemical understanding of pollution
and its environmental impacts deepened in the years after the Second
World War as thousands of new manmade chemicals entered the market-
place. This new way of thinking about pollution raised a number of questions about how to identify dangerous chemicals in our air, water, soil,
and food. In contrast to the visibility of black soot in urban cities, sewage in waterways, or the destruction of forests and wild areas, chemical
pollutants like sulfur dioxide were not immediately or clearly detectable
without scientific documentation and analysis. As evidence accumulated
about the potential for these invisible chemicals to damage the environment during the 1950s and 1960s, scientists began to assume new roles as
interpreters of potential environmental threats for government officials
and the public. Research teams descended upon cities with sampling devices in hand, redefining air quality from the blackness of smoke to the
presence of certain chemicals in various concentrations. It was a crucial
step in understanding the mounting environmental crisis from fossil fuels, but it was also a transfer of power from everyday citizens and urban
residents to scientists and policymakers. The privileging of new kinds of
expert knowledge about pollutants would transform our understandings
of environmental health and the tradeoffs in relying upon ever increasing
amounts of fossil fuels for industrial production.

The identification of particular fossil fuel chemicals as agents of harm, rather than visible smoke, also suggested that their dispersion might prove far more widespread than previously thought. With the introduction of atomic weapons and nuclear testing after the Second World War, the scientific community had begun to document the spread of radioactive fallout to nearly every corner of the planet. The ability to trace radioactive particles through ecosystems and the human body led to several novel scientific insights about the processes of bioaccumulation and showed that it was possible for fallout to affect populations far from nuclear detonations. Yet it wasn't at all evident that other types of pollutants could travel such great distances and impact the environment in distant regions. The potential for other chemicals to similarly accumulate in locations far from their emission only received sustained scientific attention during the mid-1960s as evidence emerged that pesticides had reached areas as remote as the arctic. The discovery of "acid rain" subsequently played a crucial role in demonstrating that fossil fuel pollution was not only a local issue, but a global problem with the potential to cause long-term, serious harm to the environment.

Swedish scientists first identified acid rain as a continental phenomenon in Europe during the late 1960s, and their findings prompted international debates over whether countries should work together on pollution problems through supranational institutions. The increasing reliance on science and technology to investigate the "chemical" nature of pollution led many countries to first turn to the Organisation for Economic Cooperation and Development (OECD) to facilitate environmental cooperation. As the only intergovernmental body that included the majority of Western, capitalist countries at this time in history, the OECD initially served as the primary forum for collaborative acid rain research as well as projects on a host of other environmental pollution problems.

However, as evidence mounted that chemical pollutants had regional and global effects, many government officials began to question whether a more inclusive institution should lead environmental diplomacy. Acid rain became the galvanizing issue for many scientists, policymakers, and activists who wanted to see the United Nations (UN) serve as the world leader on environmental issues. The problem eventually came to serve as a case study for the famous 1972 UN Conference on the Human Environment, the first global meeting of its kind, as well as justification for the UN to assume responsibility for negotiations on international environmental problems.

However, a multitude of difficulties with managing the 1972 conference combined with the UN's limited experience in overseeing scientific research led to mounting objections to cooperating on environmental problems through the organization. Government officials in Western nations were faced with the difficult choice of either working through an organization with limited membership but a historically robust reliance on scientific experts or pressing on with a global institution that lacked experience in managing large research projects. Faced with ongoing Cold War tensions and disagreements with developing nations over which environmental issues should take precedence, Western governments turned to the OECD rather than face the bureaucracy and diplomatic morass of the UN. For work on acid rain, this meant an influx of support to environmental scientists during the 1970s, which united the emerging field and built close relationships with policymakers in the international arena. Scientists' newfound authority as decoders of nature for government officials and the public thus played a major part in determining how and where future research and policymaking would occur on acid rain, shaping a new era of pollution in environmental diplomacy.

Death-Dealing Fogs

On December 1, 1930, unusual meteorological conditions caused a thick fog to settle across the Meuse Valley in Belgium as factories and power plants released plumes of smoke into the air. Normally, air that is close to the earth's surface is the warmest and rises vertically, dispersing any pollutants. But with the sun at a low angle in the winter sky, the ground radiated more heat into the atmosphere than it received from the sun's rays, creating a pocket of cold air near the ground. As a mass of warm air moved across the valley, it reversed the normal temperature gradient of the atmosphere and produced a meteorological "inversion," which trapped cool ground air within the mountainous terrain. In just a few days, the small town of Liège in the valley was filled with the noxious smell of rotten eggs and more than three dozen fatalities were reported.[8]

It was the first documented case of a major air pollution disaster and attracted attention from government officials, the scientific community, and the general public across Europe and the US, making the front page of the *Sun,* the *New York Times,* the *Washington Post,* and the *Los Angeles Times.*[9] Victims described difficulty breathing and chest pains that would only abate upon leaving the town, leading to suspicion that they had been poisoned by something in the fog. The incident so frightened urban residents that Belgium's Queen Elizabeth visited the small city to persuade local health officials to investigate the cause of the fog.[10] At first, many scientists and doctors in Belgium as well as throughout Europe were skeptical about the role of air pollution in the disaster. Some eminent scientists, such as the British biologist J. B. S. Haldane, suggested the fatalities could have resulted from an illness like the Black Death, while others suspected that they were caused by an accidental leak of old, buried German chemical war gases.[11] Eventually, however, a scientific investigation launched by the Belgium government concluded that the deaths were the result of "sulphurous bodies, either in the form of sulphur dioxide or sulphuric acid," and recommended the implementation of pollution control policies to prevent such accidents from occurring in the future during similar atmospheric conditions.[12]

The possibility that chemicals in fossil fuel emissions could be hazardous to human health had not been extensively studied by scientists at the time of these events. Some research had been conducted on very high levels of occupational exposures, but few of these studies focused on sulfur oxides.[13] Although sulfur gases had been identified as a potential byproduct of burning coal since the nineteenth century, discussions of its harmful effects as a pollutant in urban areas were limited to its potentially corrosive effect on buildings.[14] Only after the Liège disaster did scientific surveys of air pollution begin to measure sulfur dioxide concentrations and identify the chemical as an important pollutant, and by the Second World War, some doctors had linked exposure to sulfur dioxide as a possible contributor to asthma attacks.[15] However, many scientists and public health officials remained unconvinced that sulfur dioxide was to blame for the 1930 catastrophe, and its concentrations in cities were still commonly believed to pose no risk to public health.[16] Since the discovery of steam power, most of the general public had viewed smoke as a sign of prosperity, a testament to the economic growth of industrialized nations and the promise of better lives through increasing energy consumption.[17]

This began to change over the next two decades as more lethal incidents occurred in the US and Britain. In Donora, Pennsylvania, dozens died in 1948 during a meteorological inversion that trapped air pollution around the city.[18] Just a few years later in 1952, the most severe pollution disaster to date hit London, Great Britain, resulting in thousands of deaths as well as numerous respiratory illnesses.[19] Transportation ground to a halt as the smog grew so thick it became too difficult to drive without flares lighting the streets. Londoners, who had experienced many such "pea soup" fogs since industrialization, reported that this fog was unique in thickness and intensity.[20] Some of the city's elderly population, having already lived through two wars, perished in the months thereafter. But the fog also struck down seemingly healthy

young people as well as cattle and other farm animals. A British atmospheric scientist who was five years old and living in London at the time
described the smog as being so thick that it permeated his home, leaving him extremely ill. He recalled lying in bed day in and day out trying
to breathe, unable to tell which of his parents was checking in on him
through the darkened air.[21]

Research into air pollution surged in the 1950s across Western, industrialized countries as a result of these events. Following the Liège
disaster, the number of scientific publications per year on atmospheric
pollution doubled, and then quadrupled after the next major disaster in
Donora, Pennsylvania, in 1948.[22] In the aftermath of the 1952 London
smog episode, even more scientists across Europe and North America
began devoting their research to smog and air pollution. In response to
the disaster, the British government created the National Smoke and Sulfur Dioxide Survey in 1953 to monitor air pollution and subsequently
enacted the British Clean Air Act of 1956, which introduced federal control over industrial emissions and mandated increased chimney heights
to disperse pollutants high enough to prevent them from becoming
trapped around cities.[23] Several research groups at US universities, such
as the California Institute of Technology and the University of Illinois,
undertook independent investigations of air pollution in cities deemed
vulnerable to a pollution disaster, notably Los Angeles.[24]

Public protests in areas at risk of experiencing a similar smog disaster also led to several new national programs to collect data on air pollution and to advise on possible industry regulations outside Britain. In
West Germany, the government sponsored its own "smog study" in the
Ruhr valley, an area heavily populated with coal and steel plants, after
years of community pressure following the London smog.[25] Additionally, in 1955 the West German Parliament put together its first scientific
committee, the Clean Air Commission, to review atmospheric pollution
in Germany and propose ways to reduce emissions, and in 1960 its civil
code was amended to require authorization by the government for any
industrial installations that might pollute the atmosphere.[26] Americans
also took to the streets to voice their concern about air quality. Many
were women anxious about the health implications for their children,
such as a group of housewives who donned gas masks and paraded
through Pasadena in 1954 to draw attention to smog problems in California. The group included a small child with a doll adorned in protective gear.[27] Shortly thereafter the state began an inquiry into air pollution, with other cities and states following California's lead in seeking to
improve air quality. Though most of the political actions in the US were

taken by state and local governments, in 1955 the US Congress passed legislation declaring air pollution a threat to public health and promising assistance to states in tackling the problem. In addition, the bill set aside $15 million in funding for scientific research to investigate smog formation and its impact on human health.[28] Comparable government efforts followed in France; its President ordered the Ministry of Health and Population to invest more resources in air pollution studies in 1960.

In the years after the London smog, these investigations revealed the potentially lethal chemical cocktail of fossil fuel byproducts in air pollution. The dirty, sooty emissions masked an underlying invisible mass of compounds, whose interactions with one another and the surrounding environment were still largely a mystery. Like the reports following the 1930 pollution disaster in Belgium, several of the scientific studies conducted after the pollution disasters of Donora and London on smog began to differentiate between the dangers of "smoke," consisting of condensed particles of carbon, dust, and soot, and the "invisible" chemicals in fossil fuels, such as sulfur dioxide, nitrogen oxides, and carbon monoxide.[29] Many researchers singled out sulfur dioxide as the "invisible and far more dangerous component of smoke" and the likely cause of death and illness in Donora and London.[30] As one scientist explained in a 1954 address for the American Association for the Advancement of Science's first symposium on air pollution:

> Air pollution is not simply a matter of coal smoke or other visible things; instead, the vast quantities of invisible gaseous pollutants constitute the major part of the problem.[31]

Based on these findings, more and more scientists and medical professionals argued for the importance of government regulation of air pollution, and specifically sulfur dioxide, in the interests of public health.[32]

But despite the mounting evidence for the potential harm from invisible chemicals in fossil fuel pollution as opposed to dust and soot, some scientists and public health officials were still unconvinced that there was clear proof of the damaging effects of sulfur dioxide on human health.[33] It was extremely difficult to attribute cause of death directly to a particular pollutant in the midst of confounding environmental factors and health conditions of the victims, many of whom were already ill or elderly. Levels of sulfur dioxide during the smog incidents also did not appear to rise above levels considered tolerable for factory workers.[34] Frustratingly for public health researchers, a British government committee investigating the alleged 4,000 deaths from the 1952 London smog claimed it could not definitively determine whether sulfur dioxide or smoke had been the primary culprit.[35]

Government regulations reflected this ambiguity in the scientific literature concerning what, precisely, was so harmful about smog. Rather than attempting to reduce the chemical components of pollution emissions, governments sought to lower levels of smoke, grit, and dust in the air.[36] Focusing on these aspects of smog was also attractive to governments because there were readily available technological solutions that were relatively inexpensive for industries to utilize.[37] Power plants and factories met these new regulations on smoke, grit, and dust by modifying the design of furnaces to more efficiently burn fuel and decreasing the density of emission sources in vulnerable geographic locations.[38] In addition to these strategies, countries in Europe and North America instituted regulations to increase the height of chimney stacks with the aim of dispersing chemical pollutants high enough into the atmosphere so that they could not become trapped during meteorological inversions.[39]

While these technological changes improved air quality and visibility on a day to day basis and may have reduced the threat of deadly smog around urban areas, industries continued to discharge chemical pollutants at an ever growing rate. They simply released sulfur dioxide and other pollutants higher into the atmosphere without regard for where they might end up. Neither the possibility that sulfur dioxide emissions could harm the environment nor the potential implications for public health were taken into account by industries, scientists, or government leaders in implementing air quality policies.

This would begin to change as scientists identified other "invisible" chemicals as potential environmental threats, notably radioactive fallout, nuclear waste, and pesticides. Scientific and public concerns about radioactive fallout became rampant in the 1950s after unexpected wind shifts during a US atomic test in the Marshall Islands caused radioactive ash to fall on a nearby Japanese fishing boat, haplessly named the "Lucky Dragon." The fallout sickened crewmembers and contaminated nearby tuna fish to such a high degree that they were deemed unfit for human consumption.[40] Fears about the potential environmental dangers of radioactivity erupted throughout the world in the following weeks and months.[41] With the London smog episode still fresh in the public's mind, many people voiced concerns about the possibility of a "death-dealing" fog imbued with radioactive ash.[42] A London doctor even refused to pay his taxes in 1957 until the British government checked coal for radiation in order to avoid "radioactive smog" that could cause lung cancer.[43]

Though smog reduction efforts may have allayed the menace of radioactive smog, two scientific studies released in the years after the Lucky Dragon accident resulted in escalating public ire over fallout from nuclear testing. The first, published in 1958, showed alarming levels of strontium-90 in baby teeth as a result of nuclear testing. The second, published in 1959, revealed that milk products in fifty different areas in the US contained strontium-90 as well.[44] Protests against nuclear testing eventually resulted in the US, Britain, and the Soviet Union signing the Limited Test Ban Treaty in 1963, which banned aboveground atomic weapons testing.[45]

The specter of radiation sensitized certain scientists, such as the biologist Rachel Carson, to the environmental hazards of invisible chemicals.[46] Carson's *Silent Spring*, published in 1962, utilized common understandings of radiation's effects to explain the risk to human health and the environment from pesticides that had entered into commercial use after the Second World War, particularly DDT.[47] Carson alerted the public to research conducted by the US Fish and Wildlife Bureau that demonstrated how DDT persisted in the environment, accumulating in animals through ecological food chains similarly to radioactive contamination and damaging birds and fish. Her work was one of the major catalysts for the modern environmental movement, which spread throughout North America and Europe over the course of the 1960s. In addition to selling over 2 million copies in the US, Carson's *Silent Spring* had a significant impact on Western Europe, spawning a host of similar books by European authors in France, Germany, and Sweden.[48] In response to massive public protests inspired by Carson's work, governments in Europe and North America introduced environmental legislation to regulate harmful chemicals and reduce air and water pollution; while four laws related to environmental protection were passed in these countries in the period from 1956 to 1960, this increased to ten from 1960 to 1965, to eighteen from 1966 to 1970, and to thirty-one from 1970 to 1975.[49]

As states formulated these national policies on pollution, many began to consider the desirability of sharing scientific and technical expertise with one another to obtain more rapid results and maximize their investments in research.[50] Collaborating on research into pollution's harmful effects and the development of technologies to monitor and control chemicals in the environment would allow governments to share the costs of large projects and learn from one another's efforts. As new findings began to suggest that sulfur dioxide and other chemicals could impact the environment far from sources of emissions, the question of how countries could work jointly to address pollution began to have greater salience for scientists, government officials, and a new cadre of professionals at intergovernmental institutions.

From the Local to the Global

The recognition of invisible chemical pollutants as a potential public health and environmental threat coincided with the increasing participation of intergovernmental organizations in international scientific cooperation. Their involvement in organizing research projects among politically allied governments expanded after the Second World War because of a growing emphasis on scientific research in foreign policy among countries in Western Europe and the US. This was particularly true for the Organisation for Economic Cooperation and Development (OECD) and its precursor, the Organisation for European Economic Cooperation (OEEC). The only intergovernmental group consisting of the majority of Western states and excluding Eastern Europe, it functioned as the primary venue for performing cooperative research on a number of environmental problems among its member governments throughout the 1950s and 1960s. Like many intergovernmental organizations at this time, the OECD "Secretariat" was staffed by a new breed of civil service professionals who worked at its Paris office full-time in support of representatives from its member countries.[51] It was subdivided into several committees, subcommittees, and working groups on topics such as trade, education, science, and economic policy, with academic experts and government officials from each country directing research and facilitating policy discussions.

Following the pollution disasters, the OECD began to assist with studies and information exchange on pollution monitoring and possible smoke abatement measures.[52] With rising concerns over invisible chemicals in the environment, this role expanded to include overseeing joint scientific projects and coordinating government policies on sulfur dioxide and other chemical products. In this way, the OECD's work brought about a distinct change in collaboration on air pollution research and government policies among Western governments, with profound ramifications for international environmental diplomacy in the late 1960s and early 1970s once states began to grapple with the possibility that pollutants could have impacts far from their point of origin.

While international scientific cooperation has a long history, after 1945, it became institutionalized in intergovernmental bodies to an extent never before seen.[53] In the immediate decade after the Second World War, 58 new intergovernmental and nongovernmental international scientific organizations were created, more than twice the number than had been founded in the previous twenty years. By the mid-1960s, there were over 250 nongovernmental and 50 intergovernmental groups engaged in scientific activities.[54] To a large extent, this transformation was the result of greater government involvement in financing scientific research and technological development to support the war effort. Innovations in atomic physics and the application of nuclear research to military purposes during the Second World War fundamentally changed the relationship between science and government in America and Europe as states began shouldering a large share of investment costs and control over scientific work.[55] The successes of international collaborations on wartime technologies, such as work on radar, atomic weapons, and penicillin, convinced many governments of the advantages in continuing joint efforts in research, particularly given rising budgetary investments in science and technology.[56]

Coordination among these many new institutions became a major objective of governments in Europe and North America during the 1950s and early 1960s to avoid duplication of work and wasting of resources.[57] This shifted the formulation of many scientific programs and policies to international forums from national governments and provided additional motivation for joint projects.[58] Cooperation was also rooted in the need to combine resources to pursue research in areas where the cost of equipment was very high, especially for smaller nations.[59] With the onset of the Cold War, the incorporation of scientific expertise into security considerations also assumed more importance as the US sought to bolster the military and economic strength of a beleaguered Western Europe. Devastated from the Second World War, governments in Western Europe were in dire need of US aid for economic reconstruction. Under President Truman, the US government had authorized $13 billion in aid to Western Europe in 1948 in order to provide short-term assistance in stabilizing these countries financially and politically in what became known as the "Marshall Plan."[60] While science was not initially part of the strategy for assisting Europe, CIA fears about the defection of scientists from Western Europe to the Soviet Union contributed to a shift in policy.[61] Collaborating with private philanthropies like the Rockefeller and Ford Foundations, the US government sought to finance the rehabilitation of European infrastructure and manpower in science and technology so that it would not have to shoulder the entire responsibility for balancing capabilities against the Soviet Union.[62]

The OEEC emerged as an important institution in implementing these policies after it was founded on April 16, 1948. Its purpose was to distribute the Marshall Plan aid from the US and Canada, as well as promote the economic integration and cooperation of Western Europe.[63] The emphasis on science intensified once the ravaged economies had fully recovered in the late 1950s. After the launch of Sputnik and renewed fears of Soviet superiority, the OEEC formed the Committee for Scientific and Technical Personnel to bolster Europe's scientific manpower.[64] The OEEC efforts focused on providing American expertise in atomic physics to assist in the creation of a nuclear power industry on the continent, while the North Atlantic Treaty Organization (NATO), with financing from the Rockefeller and Ford foundations, put programs in place to train additional scientists and engineers.[65]

The fruitfulness of these endeavors and advances in economic integration motivated US and European leaders to refashion the OEEC in order to continue coordinating scientific research with the common goal of economic expansion.[66] Its successor, the OECD, was first proposed by President Eisenhower late in 1959 and enthusiastically welcomed by many officials in Western Europe and Canada.[67] After Eisenhower discussed his plans with President de Gaulle of France, Chancellor Adenauer of Germany, and Prime Minister Macmillan of Britain, negotiations began among all 18 member countries and concluded in December 1960 with the signing of the OECD convention in Paris.[68] Significantly, the drafting committee proposed keeping the Scientific and Technical Personnel group together while also adding a specific committee to be focused on scientific research.[69] It recommended that member countries provide financial resources to further international cooperation in science and technology and earmark a certain sum each year for joint research endeavors.[70] The new focus on such projects reflected a growing belief among member countries that scientific discoveries and their technological applications could spur economic growth as much as, if not more than, the accepted classical factors of production.[71]

The OECD's burgeoning responsibilities in assisting governments with collectively forming science policies paved the way for its development as the primary agency for political and scientific collaboration on air pollution.[72] The OECD was one of the first intergovernmental organizations to exhibit an interest in air pollution studies, and soon became the major international forum for its member governments to work cooperatively on the problem after the escalating number of pollution disasters.[73] Following the 1952 London smog incident, in 1957 the OEEC established a working group of scientific representatives to study the best methods of measuring different components of air pollution. These included grit and dust, suspended matter, and sulfur oxides, and later hydrocarbons and fluorine compounds. With two leading British experts in air pollution studies as chairmen, the delegates met each year to review sampling techniques with some assistance from the World Health Organization (WHO), which was developing air quality standards at this time.[74] Throughout these meetings, scientists attempted to standardize procedures and undertake experiments to test various methods of monitoring pollution concentrations, as well as discuss the latest developments in the field. Upon the OEEC's refashioning into the OECD and expansion to include the US and Canada in 1961, air pollution became a major focus of the Committee for Scientific Research.[75]

Although several other intergovernmental and nongovernmental organizations also facilitated collaborative endeavors for a number of environmental issues, few paid attention to air pollution during the 1950s and 1960s.[76] For example, the United Nations Educational, Scientific, and Cultural Organization was largely focused on development and natural resources in the Third World, while scientific nongovernmental organizations, such as the International Association of Meteorology and Atmospheric Physics, were involved with less applied research.[77] Among the major intergovernmental organizations that did deal in various ways with air pollution, most did not look at fossil fuel pollution or focused narrowly on air pollution from particular industries. In the early 1960s, for instance, the International Atomic Energy Agency organized a series of panels on proper safety guidelines for radiation releases from atomic tests and nuclear power plants, but did not move beyond its focus on atomic energy.[78] Though the WHO arranged a series of government consultations and workshops on air quality criteria and the health effects of pollutant exposure after the pollution disasters, much of its work during the 1960s involved studies on radioactive wastes, clean water standards, the increasing use of pesticides, the ecology and biology of disease vectors, drug abuse, and noise.[79] The UN Economic Commission for Europe was much more focused on water pollution at this time, and its work on air pollution was limited to a few studies of smokeless fuels and air pollution by coking plants in the steel industry. The World Meteorological Organization did develop two types of air pollution networks in 1967, but these were intended to study long-term climatic changes and could not be used for regional studies of air pollution because of their location.[80]

One exception was the Council of Europe, which formed its first expert committee on air pollution in 1966 and set out to adopt, in general terms, principles to address the problem of air pollution from fossil fuels.[81] It worked regularly to facilitate discussions on the environmental impact of air pollutants, convening a scientific conference on the effects of air pollution on plants, both cultivated and wild.[82] Yet beyond these activities, its involvement was limited to political debates concerning the development of recommendations; while making use of ongoing scientific research, the Council of Europe did not coordinate technical studies.[83]

In contrast to these groups, the OECD became much more involved in facilitating joint scientific research and information exchange on air pollution from fossil fuels, especially sulfur dioxide. Their work grew considerably after 1966 as the environmental movement spread across Europe and North America. In response to the public's demand for government action, the OECD promoted its Committee for Cooperative Research from a subcommittee to a full Committee for Scientific Research and initiated new research efforts on environmental topics.[84] Some of these collaborations involved sharing technical expertise and scientific information about fossil fuel pollutants as governments struggled to improve their air quality.[85] To this end, the Committee worked to create a common glossary of terminology used in studies on air pollution to be published in English, French, German, and Italian, as well as facilitate the exchange of publications and data among its member states. It also hosted seminars, such as on the creation of "zero-stations" located far away from pollution sources, so scientific representatives could learn from other countries' research experiences.[86] This helped standardize methods of sulfur dioxide detection, which was vital in laying the groundwork for future scientific cooperation on air pollution problems.[87]

Beyond serving as a forum for brainstorming research strategies and environmental policies, the OECD was also instrumental in making data itself global by standardizing pollution measurement techniques. In addition to exchanging information among its members, the OECD began to oversee some of the first jointly run air pollution studies during the 1960s. It created its own network of pilot observatories to begin trials on atmospheric measurements of fossil fuel emissions, establishing stations with the German Research Foundation in the Federal Republic of Germany, the Research Institute of Applied Chemistry in France, the Institute for Water and Air Pollution Research in Sweden, and the Swiss Federal Laboratory for Testing of Materials in Switzerland. Through these activities, the OECD cultivated a group of experienced laboratories that could test sampling and analysis methods for future regional networks of air pollution monitoring stations.[88] Without synchronizing their testing procedures against one another, scientists in different countries could not be sure of the comparability of their results, a fundamental prerequisite for conducting a research study with different national groups.

These efforts began to assume greater importance within the OECD as concerns grew about the environmental impact of invisible chemical pollutants on areas remote from their emission source.[89] Since the 1950s, studies on nuclear fallout released into the atmosphere from atomic testing had made it clear that these pollutants could be transported worldwide. Radiation from atomic tests in the US and Soviet Union produced radioactive rainfall as far away as Germany, Japan, and Northern Canada.[90] However, it was unclear whether chemical pollutants released through other means could be similarly transported to distant areas.

The discovery of high levels of pesticides in the Antarctic in 1965 suggested that radioactive fallout was not unique in its ability to travel long distances in the atmosphere. Though *Silent Spring* had presented data showing DDT was accumulating in organisms far from locations where pesticides had been sprayed, measurements of pesticides in seals and penguins living in the Antarctic provided unmistakable evidence that large-scale atmospheric or oceanic transport of these chemicals was occurring.[91] In response to these concerns about pesticide accumulation in the environment, from 1966 through 1971 the OECD oversaw a cooperative study on chemical residues in wildlife. Scientists from thirteen of its member states participated, and it was the first international, joint research project undertaken on the fate of chemical pollutants in the environment.[92]

These findings on DDT, in concert with knowledge about the atmospheric transport of radioactive fallout, prompted scientists and policymakers at the OECD to question the degree to which fossil fuel pollutants could also travel long distances across the planet.[93] This shift in thinking about pollution as a regional and perhaps even global phenomenon compelled the OECD and other intergovernmental organizations to consider whether a common approach to studying and regulating sulfur dioxide pollution was needed, and if so, what institution would be best suited to lead scientific and diplomatic efforts.

The Discovery of Acid Rain

The recognition that pollutants did not respect international boundaries was part of a widespread transformation in foreign policy circles concerning the importance of environmental issues in international diplomacy. As more governments sought to enact legislation during the late 1960s in response to the environmental movement, many government officials, diplomats, scientists, and business leaders began to argue that environmental problems needed to be addressed on a global scale. Issues ranging from fears about population growth to natural resource conservation helped spur this shift in thinking and prompted extensive debate about how countries should engage in international cooperation on environmental problems.[94] Swedish scientists' detection of acid rain serendipitously coincided with the emergence of these conversations on global environmental cooperation. After first receiving scant attention from the Swedish government, acid rain went on to become the paramount focus of its efforts leading up to the famous 1972 UN Conference on the Human Environment, which was the first international conference to address the environmental crises facing nations around the world. While helping to generate attention to the acid rain problem, the UN meeting also underscored the challenges in tackling pollution at the supranational level. Paradoxically, it hardened divisions in the international community at the same time as the conference underscored the importance of environmental diplomacy.

Atmospheric and soil scientists first discovered an increase in acidic rainfall quite accidentally in the course of studies conducted through the European Air Chemistry Network (EACN) during the 1950s and 1960s. The EACN originated out of research by a Swedish soil scientist at the Ultuna Agricultural College in Uppsala, Hans Egnér, who set out to examine the relationship between chemicals in the atmosphere and agricultural productivity after the Second World War.[95] With the help of his assistant, the meteorologist Erik Eriksson, in 1948 Egnér fashioned precipitation samplers at farms throughout Sweden in order to study the input of nutrients like sodium, chloride, and calcium into the soil.[96] These were analyzed monthly alongside samples of air, which were tested for levels of chemical compounds such as ammonia and sulfur dioxide.[97]

In the spring of 1952, their data set came to the attention of Carl-Gustaf Arvid Rossby, a leading Swedish atmospheric physicist who had recently returned to Stockholm from the US after helping to train American meteorologists during the Second World War.[98] Rossby developed an interest in studying global atmospheric processes shortly before the war began, and his encounter with Egnér and Eriksson's work inspired him to consider the need for meteorology to investigate the role of the atmosphere in biogeochemistry.[99] Little research had been done previously in global atmospheric chemistry and nutrient cycles, and Rossby believed that it could have important practical applications for forestry and agriculture as well as understanding climatic change over long periods. In the hopes of pursuing work on these large-scale earth pro-

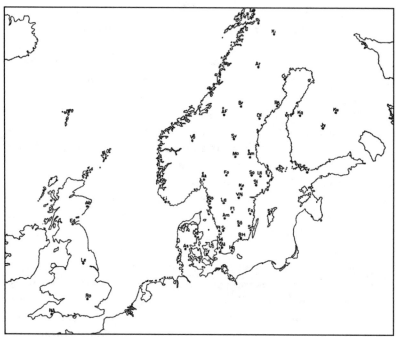

FIGURE 1.1 Map of stations participating in the European Air Chemistry Network in the 1950s. From Hans Egnér and Erik Eriksson, "Current Data on the Chemical Composition of Air and Precipitation," *Tellus 7*, no. 1 (February 1, 1955): 134–139.

cesses, he helped Egnér and Eriksson expand the number of station sites in their small network to Norway, Sweden, and Finland between 1952 and 1954. Rossby then recruited Eriksson to help him analyze the atmospheric data at the newly founded International Meteorological Institute in Stockholm.[100] In addition to increasing the geographic scope of the measurements, Rossby, Egnér, and Eriksson introduced other components to the analysis, notably pH.[101] A pH value indicates whether a chemical is acidic or basic, two extremes similar to whether a substance is hot or cold. It is made on a logarithmic scale ranging from 0 (acidic) to 14 (basic), which means that for every step on the scale a substance is 100 times more acidic or basic. Pure water is neutral and measures "7" on the pH scale, though rainfall has historically been slightly acidic with a range between 5.0 and 5.5. Originally conceived in the early twentieth century for use in chemistry experiments, pH measurements had not been used to determine rainfall acidity on any kind of regional basis. The pioneering EACN measurements of pH and other chemical properties were eventually extended to Germany and Belgium over the course of the 1950s, and by the end of the decade, the network had accumulated a considerable amount of data on changes in precipitation and air composition.[102]

One of the most striking findings for Rossby was the pattern of increasing concentrations of rainfall acidity throughout Scandinavia. Although having no direct data on its origins, Rossby correctly inferred that this increase in acid rain was likely caused by pollution from British and German industries.[103] But despite identifying the ominous trend, Rossby didn't recognize that a rise in precipitation acidity could damage the environment, and his investigations ended with his unexpected death in 1957.[104] Other biologists elsewhere in Scandinavia, however, were beginning to accumulate evidence indicating acid rain could in fact cause serious damages to organisms. As early as 1950, Norwegian biologists suggested that fish populations in the country might be declining from an increase in the acidity of precipitation.[105] Several other researchers published similar conjectures about local effects of acidic rainfall in Europe during the 1950s and 1960s, but their work was not widely disseminated within the scientific community.[106]

It was not until the late 1960s that a Swedish soil scientist, Svante Odén, recognized acid rain could have potentially serious environmental consequences across the region. Odén had agreed to manage the EACN after Rossby's untimely death and Eriksson's simultaneous decision to leave Sweden for a job with the International Atomic Energy Agency.[107] He had previously worked on research examining the heat properties and organic constituents of clay and humus at the Agricultural College in Uppsala and had little experience in atmospheric monitoring.[108] But as he managed the network stations over the ensuing decade, Odén began to suspect that atmospheric pollution was an ecological threat after beginning a project to map the precipitation data gathered each month. Through this work, he produced a cartographic image of changing pH across Scandinavia for the first time, which painted a striking picture of the intensification of acid rainfall since the network's creation. It was through this visualization of the EACN's data on acidity in precipitation that Odén realized both the magnitude of the decrease in pH and its geographic extent.[109]

Concerned about the environmental impact of this "acid rain," Odén took the unusual step of publishing his scientific findings in a popular daily newspaper in October 1967 before submitting them for peer-review in an academic journal. Writing in *Dagens Nyheter*, Sweden's most popular daily press, he argued that industrial emissions of sulfur dioxide were causing acid rain and damaging the country's ecosystems.

FIGURE 1.2 Photograph of Svante Odén. Erik Lotse, "NF Svante Odén," *Svenskt Biografiskt Lexikon* (Riksarkivet, 2013).

Drawing upon over fifteen years of data from the EACN and 600 lakes in Scandinavia, Odén claimed that acid rain was contributing to fish extinctions and warned of possible damage to soils and forests.[110] Calling it a "chemical war" waged against his country, Odén sparked a huge national outcry in Sweden and across Scandinavia in the following years.[111]

Odén's article was responsible for stimulating the earliest public attention to acid rain, and provoked considerable debate about the problem among Scandinavian atmospheric scientists and ecologists.[112] His work was likely given such attention not only because of the very public way in which he spoke out about acid rain, but also because many other scientists and government officials in Europe were beginning to fear that sulfur dioxide might be harming the environment in addition to human health. Approximately a year prior to his publication, national representatives to the Council of Europe's Commission for the Conservation of Nature and Natural Resources met to discuss emerging scientific research on the biological dangers caused by fossil fuel pollutants in the air.[113] The first European-wide scientific conference on the environmental effects of air pollution was conceived as a result of their meeting, and brought together more than one hundred scientists from 15 countries as well as official representatives from many influential international organizations, including the OECD, WHO, and Euratom.[114] In the wake of the Swedish uproar, the scientists gathered in April 1968 to synthesize the growing body of ecological and meteorological data on air pollution, with sulfur dioxide a focus of their discussions. Participants noted that the effects of sulfur dioxide had been observed in many places far from industries, and completely non-industrialized areas were collecting rainwater with a pH as low as 2.8, about the same as that of vinegar.[115] Studies reported reduced crop yields in agriculture and forestry throughout affected areas, as well as the complete disappearance of lichens at winter concentrations of sulfur dioxide above 50 micrograms per cubic meter.[116] Based on the research presented, the scientists concluded that it was not possible to delineate limits of sulfur dioxide that would prevent injury to plants, as even minute levels were shown to be harmful.[117] Although the participants acknowledged that large gaps in scientific understanding remained, a consensus emerged that air pollution was causing such serious damages to the environment that "governments must take action."[118]

Despite the scientific evidence for the international dispersion of air pollution, some of Odén's own colleagues, including Eriksson, were not convinced that acid rain was an environmental threat.[119] Only a small number of biological studies had been done specifically on the environmental impact of acid rain, and there was little atmospheric evidence concerning whether and to what extent sulfur dioxide could travel long distances across national borders.[120] Though a few scientists had suggested its residence time in the atmosphere could be as long as five to ten days, others had put the figure at less than twenty-four hours.[121] If the latter was true, then foreign pollution was unlikely to be the source of the problem. Additionally, Eriksson and other Swedish ecologists were not persuaded that the increase in acidity was as alarming and severe as Odén alleged and seemed uncomfortable with his public accusations on "such complex issues."[122] The Swedish government appears to have had a similarly muted reaction. Although Odén presented his work in May 1968 to the Swedish Ministry of Agriculture, which oversaw environmental issues in the country, he was denied further funding from the Swedish government to pursue additional research on acid rain over the next two years.[123] His efforts to bring greater attention to the problem among government officials succeeded only thanks to a small group of Swedish representatives to the UN who sought to focus international attention on environmental problems for entirely different reasons.

Just a few short months after Odén published his findings in 1967, Swedish diplomats proposed holding the first international conference on environmental issues at the UN. The impetus for Sweden's bold proposal, however, was not acid rain. Instead, the diplomats were determined to avoid the organization of another UN conference on the peaceful uses of atomic energy, which Swedish representatives believed promoted the interests of the nuclear industry at the expense of international work toward disarmament.[124] With the environmental movement gaining momentum across Europe and North America, the topic seemed a perfectly timed avenue for steering the UN toward other pressing international issues.[125]

Sweden was no exception to the surge in public and political agitation over the environment. Several pollution problems had generated considerable attention to pollution and toxic chemicals in Sweden over the course of the 1960s, and the government became a pioneer in environmental regulation in response to public pressure over the incidents. First, fish off the southern coast of the country were found to have such high levels of mercury that citizens were discouraged from consuming them more than once a week, and many Swedes in Uppsala were hospitalized after eating pheasants that contained high levels of mercury in their bodies from contaminated feed.[126] Shortly thereafter, Sweden became the first country to ban the use of DDT after Swedish scientists discovered fish, birds, and plants contained rising amounts of the chemical, increasing as one moved up the food chain.[127] Given these domestic developments, it is unsurprising that the pollution problems Swedish representatives first identified as possible topics for international treaties at a possible UN conference on the environment were the discharging of oil and wastes by ocean vessels and pesticides such as DDT.[128] In attempting to generate international support for the conference, Swedish officials seldom even cited air pollution problems like acid rain; initially, the issues of population growth, overuse of natural resources, and water pollution were much more frequently listed as rationales for the gathering.[129] In fact, it was not until 1969 that Swedish government officials first began to complain publicly about acid rain, singling out Britain as the major contributor to sulfuric acid in their country's precipitation.[130]

The inclusion of acid rain as part of Sweden's preparations for the UN conference did not occur until 1970, after the Swedish king formally approved the formation of a national committee to oversee preparations for the Stockholm Conference at a cabinet meeting in December 1969. The national committee was chaired by former Prime Minister Tage Erlander with the minister of agriculture, Ingemund Bengtsson, serving as vice chairman. They subsequently formed three subcommittees to work on separate aspects of the conference: a committee for organizational issues, a committee on press and information, and a committee on research. The latter, officially known as the Committee on Research and Substantial Issues, was created in the spring of 1970 and chaired by Professor Arne Engström.[131]

The primary task of the committee on research was to compile a country report on environmental issues since the UN had asked all participating governments to send background material for the meeting's discussions. After the group's formation, it contacted leading Swedish scientists and researchers requesting suggestions for potential topics.[132] Though the possibility of performing a case study of acid rain was raised in the committee's early meetings among other suggested issues, much of the discussion focused on areas with possible relevance for developing nations, such as natural resource conservation.[133]

The major impetus in choosing acid rain for the Swedish case study came from Odén, who submitted a proposal to make it the cornerstone of the committee's work in May 1970.[134] To bolster his pitch, he argued acid rain was an ideal case study not only because of its regional aspects but also because of the likelihood that developing nations would confront the problem as they industrialized, making it a truly global, international issue.[135] He suggested that Sweden's report discuss current research on the increase in acidic rainfall over Europe since the 1950s and evidence for its impact on soils and biological systems, particularly aquatic life and forests.[136] In addition, Odén thought it would be useful to incorporate cost-benefit analyses on controlling acid rain that looked at the potential long-term effects on the environment and human health to strengthen arguments for controlling sulfur dioxide pollution.[137] In a not so subtle attempt to remedy his lack of funding for acid rain studies, Odén concluded his proposal by adding that the Swedish Ministry of Agriculture might provide financing to get work underway as quickly as possible given the absence of any grants for his research on the problem over the previous two years.

The potential international appeal of Odén's proposition was underscored by a meeting of representatives to the Scandinavian Council for Applied Research a week after the committee received his submission.[138] The Swedish delegate to the gathering was Bert Bolin, a professor of atmospheric science and member of Sweden's Committee on Research and Substantial Issues who had read Odén's proposal just before arriving. Bolin had a longstanding interest in atmospheric chemistry and global climate change, which he developed after a brief but influential period working with Rossby earlier in his career.[139] Upon learning of the Nordic countries' common interest in acid rain pollution at the meeting, Bolin asked whether they would be willing to assist Sweden with a case study of acidification for the 1972 UN Conference, which the other delegates enthusiastically endorsed.[140] Based on the support received from the Scandinavian representatives at this meeting as well as his own personal conversations with Swedish researchers knowledgeable about acid rain, Bolin recommended that Sweden accept Odén's proposal to use acid rain as a case study for the UN Conference provided the report emphasized the regional nature of the problem, not just local ecological effects in Sweden.[141] In this way, acid rain became a beneficiary of the 1972 UN Conference, gaining scientific and political attention as a result of the social and political enthusiasm for tackling environmental problems during this period.

With the endorsement of Bolin, in September 1970 the committee on research approved the selection of acid rain as the topic of Sweden's national report and secured funding for the preparation of a draft.[142] Bolin formed a working group of Swedish scientists shortly thereafter to begin synthesizing the available research concerning acid rain's effects on the environment and its atmospheric transport. Odén was placed in charge of compiling studies on aquatic effects of acid rain and their impact on fish populations, while Swedish biologist Carl Olaf Tamm of the

College of Forestry prepared material on possible decreases in plant life and forest production.[143] Meteorologists at Stockholm University were given responsibility for preparing material on the sulfur cycle and the ways in which sulfur dioxide theoretically could be transported across national borders. This work was largely based on the striking correlation between increasing acidity in Scandinavia and the increase in anthropogenic fossil fuel emissions over the same period, since no atmospheric research had yet been done demonstrating the connection between foreign pollution and acid rain.[144] Other aspects of the problem, including corrosive effects, possible health impacts, and cost-benefit analyses, were also incorporated into the case study.[145] The final report was completed in August 1971 and transmitted to the UN in the hope that the gathering would generate discussion of the scientific findings and greater international attention to the problem.[146]

However, the idea to use the 1972 UN Conference as a platform for any substantive negotiations on the issue of acid rain was looking less and less promising as the UN began its preparations in 1970 and 1971. Encompassing nations across the globe, the UN appeared to be the only intergovernmental body with the capacity to facilitate diplomatic discussions on pollution problems that transcended national boundaries. But from the outset of Sweden's proposal to hold a conference on environmental issues at the UN, many government officials in industrialized countries questioned its ability to lead diplomatic talks on pollution problems.[147] These critiques centered on two major issues: the UN's lack of experience in organizing scientific research and whether political negotiations on environmental problems would be better served by regional organizations. Both issues became insurmountable obstacles to generating substantive discussions of acid rain and air pollution during the 1972 conference.

The importance of involving scientists in international environmental cooperation was highlighted repeatedly throughout the planning of the 1972 meeting, particularly by its director general, Maurice F. Strong. During preparatory meetings for the conference, he frequently stressed "the need for science to provide many of the key answers to today's environmental problems" so policymakers could fully understand the chemicals they were putting into nature and avoid "taking risks that may be irreversible."[148] Strong argued that a partnership between science and politics would be crucial during the conference for effective action, citing Sweden's case study on acid rain as an exemplary model of scientific advising on a global environmental issue.[149]

Given the importance of scientific advising for identifying and regulating dangerous chemical pollutants, many foreign policy officials agreed with Strong that scientists were essential contributors to international cooperation.[150] Yet Strong's idealized image of the use of science in the 1972 conference was deeply at odds with the UN's track record as a forum for scientific and diplomatic work on the environment. In comparison to other intergovernmental groups, the UN had little experience in facilitating scientific research on environmental problems, and critics of the 1972 UN conference believed that this severely handicapped the organization from serving as a leader in international negotiations.[151] Former US diplomat George Kennan even went so far as to recommend governments create an entirely new international organization consisting of ostensibly "independent" scientists to oversee global environmental cooperation given the UN's past ineptitude.[152]

During these debates, the OECD was frequently argued to be a much more efficient and capable organization for handling environmental issues given its prior expertise in organizing scientific studies as compared to the UN.[153] Even Strong conceded that the OECD had a much greater amount of experience and knowledge in carrying out research on environmental issues, though he felt that ultimately the real arena for discussing effective policy measures needed to be a global organization like the UN.[154]

However, the necessity of a global group to oversee environmental negotiations was also heavily debated during the conference preparations because of ongoing Cold War tensions between capitalist and communist countries as well as the stark differences in environmental problems experienced by industrialized nations compared with developing countries.[155] Complaints surfaced among the more industrialized nations that the focus on environmental issues of developing nations took time and resources away from their pollution problems.[156] Additionally, the Cold War appeared to be an insurmountable hurdle to a truly global UN environmental meeting, with the Soviet Union and its satellite states pulling out of the conference when East Germany was barred from fully participating.[157]

Because of these obstacles, industrialized countries on both sides of the iron curtain questioned whether the UN was the appropriate forum to discuss international environmental policies. Among Western governments, the Cold War provided another reason to seek cooperation through other intergovernmental groups in addition to the UN's inexperience with scientific research. Though NATO was proposed as a possible substitute, the OECD's established machinery for scientific cooperation and its membership of Western European nations along-side the US and Canada led many government officials to view it as the ideal setting for environmental negotiations.[158] Foreign policy officials among the OECD's member states also saw it as the most appropriate forum for cooperation because of its use of recommendations rather than mandatory regulations, which they believed intruded on national sovereignty.[159] On the other side of the iron curtain, the Soviet Union harbored similar concerns about the potential for environmental diplomacy to infringe upon national sovereignty, and its representatives to the UN openly criticized the push toward global negotiations on these grounds. As one Soviet diplomat argued during a preparatory meeting for the conference:

> International cooperation on the human environment should be based on equal recognition of and respect for bilateral, sub-regional, regional and global actions taken by the countries participating in the cooperation. An overestimation of global or regional aspects of the problem can fetter actions of countries and affect sovereign rights and feelings of various nations.[160]

Faced with the challenges of Cold War tensions alongside difficulties forging cooperation between developed and developing nations, the 1972 UN Conference came to a close having devoted little time to the problems of chemical pollutants in the environment, including acid rain. As many critics of the conference anticipated, the outcome of the 1972 conference was a set of twenty-six principles on environmental protection, known as the Stockholm Declaration, which largely dealt with issues pertaining to developing nations.[161] The Swedish report on acid rain was not discussed in detail during the conference and resulted in no specific recommendations in the declaration, though "Principle 21" did stipulate that states were generally obligated not to pollute the environment of other nations.[162] More than affirming the need for a global effort on pollution problems, the 1972 Conference thus reinforced beliefs among many countries that the UN was an unwieldy, ineffective organization poorly suited to the technical work that was believed to be crucial for informing environmental policies among Western industrialized governments.

In considering whether to press for further UN involvement in research or political discussions on acid rain, the Nordic governments largely concurred that the hopes for further cooperation were best pursued at the OECD. In meetings of the Nordic Council during final preparations for the 1972 UN conference, Norwegian representatives argued that the OECD was by far the most appropriate intergovernmental organization for leading international efforts on pollution problems because its membership was restricted to industrialized countries and it could most ably oversee scientific and technical work.[163] Though some Swedish representatives believed that the global nature of pollution problems made them more suited to cooperation through the UN, other delegates from Sweden and the remaining Nordic states insisted that the combination of Western, industrialized membership in the OECD alongside its "technical-scientific activities" made it a far better avenue for this work.[164] While Scandinavian representatives hoped that the 1972 Conference guidelines and principles might help direct the work of regional groups like the OECD, it is clear that most of their government officials had few expectations for the UN in fostering either research or diplomacy on pollution.[165]

While helping to redefine acid rain and pollution as global threats, then, the movement to bring environmental issues to the UN during the 1972 Stockholm conference faltered over Cold War divisions and the desire to oversee research and policy together in one institution. The OECD, with its longstanding reputation for facilitating scientific cooperation, appeared to be the obvious choice for such a project on acid rain. In the years to come, however, the acid rain problem would call into question the wisdom of relying upon common economic and political ties to conduct scientific research that would lead to environmental regulations.[166] The decision to prioritize expertise in scientific research presumed that the results of these studies would naturally lead to international treaties to remedy environmental threats like acid rain among allied governments. This ideal of the scientific expert as a solver of political problems proved to be much more difficult to put into practice as the Nordic countries made arrangements to corroborate Odén's claims through further research within the OECD.

2 The Science of Acid Rain

The most serious obstacle to obtaining an understanding of large-scale consequences of atmospheric pollution is sheer lack of information which is in turn the
result of lack of basic data, of facilities and especially of money to support this
type of work . . . It is ironic that almost all available funds go into studies of relatively small regions, that is, in the neighbourhood of cities; important though
these may be, it is a case of gross imbalance so to neglect the general problem
affecting the whole world and therefore the lives of all of us.[1]
James Lovelock, "Air Pollution and Climatic Change," 1971

The hope of the acid rain study, officials said, will be first to establish the facts
in a scientific manner and then to try to devise some satisfactory international
approach on preventative measures.[2]
Clyde H. Farnsworth, "Norse Seek Curb on Acid Rainfall," 1970

When scientists first discovered acid rain was a regional
problem in the late 1960s, how far fossil fuel pollutants
traveled through the atmosphere was still unknown.[3]
Several industrialized countries affected by the problem
soon began to question how they should adapt their research priorities on the environment and energy to this
newly identified environmental hazard. Should governments sponsor collaborative projects on fossil fuel pollution and its ecological impact? How much of this work
should be done in universities, government research centers, or the private sector? Did it make sense to coordinate
these research projects internationally, or should countries

do their own studies and then compare the results? How closely should researchers work alongside environmental officials, and how responsive should this work be to public pressure for political action? Ultimately, what kind of science would be most useful to policymakers and diplomats looking for solutions to global environmental threats?

Norway, one of the countries most at risk from acid rain pollution, and Britain, one of the countries most responsible for the problem, offered two starkly different models for how to approach these questions. In the wake of the Swedish findings, these nations' research communities and environmental ministries became leaders in proposing a model of how to construct an "environmental" science of acid rain and how to use science in formulating environmental and energy policies. In Norway, the government launched the most extensive scientific research effort into the problem following Odén's provocative findings. Their work resulted in two important shifts in scientific research on pollution problems that influenced environmental science across Europe and North America: first, promotion of international cooperation to examine air pollution on a regional scale, and second, encouragement of interdisciplinary collaboration to gain an overarching understanding of acid rain. New projects that emphasized these approaches helped create a "big science" of acid rain and the environment, characterized by large amounts of government funding and new technologies. Researchers were expected to work in teams and to be responsive to the needs of policymakers and the public in trying to address pollution, rather than pursuing what was typically viewed as "basic" research.

Not everyone was in favor of this push toward international, collaborative environmental science. British officials strenuously objected to the Norwegian plan to tackle acid rain and other chemical threats through interdisciplinary projects coordinated among multiple countries. Instead, they offered more of a "bottom-up" approach to studying fossil fuel pollution, with countries engaging in separate research efforts whose results could later be compared with one another.

During the early 1970s, the Organisation for Economic Cooperation and Development (OECD) brokered a hybrid approach between these two visions of environmental science, with implications for its use in environmental and energy policies in the industrialized world. The OECD created the first international atmospheric research project on fossil fuel pollution but declined to include any biological or ecological work in the study. OECD officials, and indeed many other government representatives at this time, assumed documenting and quantifying the pollution transfer between countries would form the foundation of any reduction schemes going forward. Even if differing widely in their responses to acid rain, government officials in Europe and North America shared a strong belief in the ability for new scientific studies to provide clear answers to how societies should respond to the potential threat of acid rain. However, the belief that the OECD project would serve as a guide to enacting public policy on acid rain was sharply questioned following the 1973 oil shocks. The OECD failed to issue recommendations to reduce sulfur dioxide pollution after the ensuing energy crisis, prompting government officials and scientists to rethink how scientific expertise should inform international diplomacy on acid rain and whether the problem could be resolved under the auspices of a Western, economically oriented institution.

Acid Rain and the Development of Environmental Science

The uproar generated by Svante Odén's publication on acid rain coincided with new debates about how scientists should study pollution problems in the wake of the modern environmental movement. While concerns about radioactive contamination had already spurred different approaches to researching environmental contamination within the US during the previous two decades, European nations had only recently begun grappling with the role of science in analyzing chemical contaminants.[4] Scandinavia, where concerns about DDT, mercury, and acid rain originated, became a pioneer in developing innovative scientific approaches to investigating pollution.

The month prior to the publication of Odén's article in *Dagens Nyheter*, the Scandinavian Council for Applied Research, also known as NORDFORSK (Nordisk Samarbeidsorganisasjon for Teknisk-Vitenskapelig Forskning), undertook a review of research and policies concerning environmental pollution in each Nordic country. Their hope was to coordinate work on air and water pollution in light of the political attention to environmental issues across Europe.[5] The Scandinavian Council for Applied Research was founded in the years after the Second World War to assist Scandinavian research institutes and academies of science in exchanging results or conducting joint projects.[6] The organization financed scientific collaborations on a wide variety of topics that had practical applications and were deemed matters of common interest to all countries, ranging from foundry work to air defense.[7] Its creation was part of a broader push to politically and economically unify Norway, Sweden, Denmark, Finland, and Iceland, a movement that was complicated by the deepening Cold War and European integration through the European Economic Community.[8]

Despite setbacks in creating a united foreign policy during the first decades of the Cold War, scientific cooperation among the Nordic states thrived under the Scandinavian Council for Applied Research. As apprehension about the environmental effects of pesticides spread across Europe following the publication of Rachel Carson's *Silent Spring* in 1962, the Scandinavian Council for Applied Research formed its first research committee on pesticides in 1965 and sponsored additional work on pollution problems once Odén's acid rain treatise was released.[9]

Though Swedish scientists had first drawn attention to acid rain and the problem of long-range transport of fossil fuel pollutants, it was Norwegian scientific and environmental groups that lobbied most heavily for the Scandinavian Council for Applied Research to conduct further research. As in its neighbor to the east, Norwegians' enjoyment of outdoor recreation in nature helped shape the country's identity and culture. The importance of protecting Norwegian "friluftsliv," which loosely translates to "open air life," was frequently invoked by environmental groups and concerned citizens in debates about acid rain's potential impact.[10] But in contrast to Sweden, the Norwegian government was much more active in institutionalizing and organizing scientific work on pollution problems. This resulted in a close cooperation between Norwegian scientists and the government that laid the groundwork for Norway's international leadership on acid rain.

In particular, the Royal Norwegian Council for Scientific and Industrial Research (Norges Teknisk-Naturvitenskapelige Forskningsråd, or NTNF) helped move scientists from work on chemical warfare and other military projects into environmental research on pollution problems like acid rain. The NTNF was founded in 1946 to help rebuild Norway's industrial and scientific bases after the Second World War; modeled on the British Department of Scientific and Industrial Research, it was instrumental in developing the Norwegian space program and atomic research in its early years of operation.[11] During the 1960s, the idea to form a scientific institute specifically focused on air pollution research began percolating within the NTNF.[12] Hoping to build up expertise in the area, the NTNF decided to establish the Norwegian Institute for Air Research in 1967 (Norsk institutt for luftforskning, or NILU) for the purpose of conducting atmospheric research on air pollutants.[13]

Brynjulf Ottar, an atmospheric chemist with a long career in military service, was selected as the founding director of NILU. He had earned a degree in chemistry from the University of Oslo in 1941 under the tutelage of Odd Hassel, Norway's first professor of physical chemistry and a Nobel Prize winner.[14] When the Nazis invaded Norway during

FIGURE 2.1 Brynjulf Ottar, director of the Norwegian Institute for Air Research. "NILU gjennom 25 år," Norwegian Institute for Air Research, Archives and Library, Oslo, Norway.

the Second World War, Ottar became active in the country's resistance movement and underground intelligence operations. After the end of the conflict, he joined the Norwegian Defense Research Institute, where he was promoted to director of research in 1952. Under his management, the Defense Research Institute began investigations into the dispersion of toxic warfare gases in valleys in Northern Norway during the 1960s.[15] In the decades thereafter, these experiences served as a foundation for studying air pollution, and many of his colleagues from the Defense Research Institute went on to become his first staff scientists at NILU.[16] There were only seven employees during the institute's first year, growing to 33 in 1970 and nearly 60 by 1972.[17]

Coinciding with NILU's creation in 1967, Odén's acid rain publication sparked Ottar's curiosity about the long-range transport of fossil fuel pollution. He viewed the issue as an important scientific problem that could give his new institute a direction and purpose during its early years, possibly raising its profile both domestically and internationally.[18] To generate support for a study of acid rain, Ottar raised the possibility of organizing a collaborative, international research program on the problem in 1969 with scientific delegates to the OECD from Sweden, Norway, Britain, France, Germany, Finland, and the Netherlands.[19] The scientists agreed to consider a proposal on the matter, and Ottar subsequently enlisted the help of the Scandinavian Council for Applied Research to convene a meeting in Copenhagen the following February in order to discuss plans for the project.[20] Upon receiving a positive response from other Scandinavian scientists and government officials, Ottar worked with the Secretariat of the OECD to develop an outline of the research on behalf of the Nordic countries. In the spring of 1970, the OECD formally accepted Ottar's proposal and named him director of the study on fossil fuel emissions, known as the Long-range Transboundary Air Pollution Project.[21] The ambitious endeavor soon became front page news in Norway, with Ottar pledging that acid rain would be NILU's top research priority in the years ahead.[22]

An international study on fossil fuel pollution was unprecedented, and the project had the potential to greatly enhance scientific knowledge about its atmospheric transport.[23] Yet its organizers also hoped that the study could serve as a catalyst to policy recommendations on fossil fuel emissions among the OECD's members. Officials at the intergovernmental group planned to use the study's results to determine if national reductions in emissions would be enough to diminish or eliminate potential problems from acid rain or if an international approach would be necessary.[24]

In many respects, the project was an attempt to unite environmental science with diplomacy through the OECD, which was then widely viewed as the only intergovernmental institution with the experience to do so. After the OECD's member countries began creating new ministries of the environment following a wave of environmental activism in the 1960s, in 1970 the organization transformed its "Committee for Scientific Research" into an "Environment Committee," anticipating the increasing role environmental science was expected to play in foreign policy.[25] The newly formed Environment Committee was initially led by an American, Hilliard Roderick, a former nuclear physicist at the Atomic Energy Commission who had served as deputy director of the United Nations Natural Sciences Department during the late 1950s and 1960s.[26] His vice chairman was Norwegian Erik Lykke, a former Foreign

FIGURE 2.2 Erik Lykke, on the right, served as director general of Norway's Ministry of the Environment from 1972 through the late 1980s, as well as chairman of the OECD Environment Committee and UN Economic Commission for Europe Environment Committee. He is pictured with Gro Harlem Brundtland, Norway's minister of the environment from 1974 to 1979 and later prime minister of Norway during the 1980s. "NILU gjennom 25 år," Norwegian Institute for Air Research, Archives and Library, Oslo, Norway.

Service officer and representative to the North Atlantic Treaty Organization (NATO) prior to his appointment as representative to the OECD Environment Committee.[27] A key contributor to the 1972 United Nations Stockholm Declaration, Lykke became an outspoken advocate for the need to study acid rain and regulate the pollutants causing the problem. Both men assumed their positions at the OECD hoping to create an international regulatory regime for environmental problems based on scientific expertise, and believed the results of the acid rain study would pave the way for diplomatic negotiations on fossil fuel emissions.[28]

To accomplish these policy goals, the project set out to determine the importance of local versus distant sources of sulfur dioxide, the main pollutant causing acid rain, to the air pollution over a region.[29] There were three main aspects to the program: emission surveys, aircraft sampling, and ground measurement stations. All three areas of data collection were needed to track pollutants from their industrial sources through the atmosphere to their final absorption in rainwater. Ottar and his colleagues at NILU were tasked with coordinating the data collection among the participating countries and developing mathematical models of pollution transport to estimate the relative contribution of different industrial sources to acid rain. If foreign sources were implicated in causing acid rain over Scandinavia and other areas in Europe, the OECD could then seek to negotiate recommendations on reducing the offending pollutants.[30]

Delegates from Britain's Department of the Environment, however, were highly critical of this plan. They argued that the study was far too ambitious since few countries had even undertaken domestic research on pollutants like sulfur dioxide outside of urban air quality monitoring.[31] Shortly after Ottar submitted plans for the OECD's acid rain project, they prepared a counterproposal to the Scandinavians' submission recommending that countries first develop their own national programs to study the atmospheric chemistry and fate of sulfur dioxide domestically, which could form the foundation for examining the overall regional picture. However, the other OECD member states believed the British plan did not sufficiently address the problem of acid rain given its suspected international nature, and felt it was entirely possible to construct national measurement stations in tandem with international collaboration and analysis of acid rain.[32]

The split revealed a fundamental disagreement about how environmental research should be constructed and its role in the diplomatic process, one that would continue to play out over the coming decades in debates about acid rain. For British officials, faced with accusations of causing a serious environmental problem for its neighbors, studying the complex chemistry involved in fossil fuel emissions was necessary before any larger claims could be made about its culpability in acid rain. Little was known about how sulfur dioxide reacted with other chemical compounds in fossil fuels emissions as they were transported through the atmosphere, and to what extent these processes could influence the formation of acid rain. For Scandinavian scientists, while these questions were certainly important, the bigger issue was whether pollution from foreign power plants was causing acid rain and if so, what proportion of the pollution could be attributed to various countries. This was a scientific question heavily dependent on creating mathematical models of atmospheric circulation in cooperation with other countries. For OECD officials and delegates to the Environment Committee, the latter approach seemed the best and most effective way of responding to public agitation over acid rain and the broader threat of fossil fuel emissions. By the summer of 1971, the steering committee held its final planning meeting and prepared to launch the first measurement phase based on the Scandinavian proposal.[33]

Crossing Boundaries: Constructing a Science of Acid Rain

A multitude of earth processes are involved in the production of acidic precipitation and its environmental effects, ranging from atmospheric transport to leaching of nutrients in soil. Attempts to study the phenomenon in its entirety resulted in several developments in scientific research that marked a departure in the way scientists approached problems of fossil fuel pollution. First, acid rain prompted new cross-disciplinary collaborations among Norwegian scientists in order to analyze air pollution from its source through deposition onto aquatic and terrestrial ecosystems. Ottar and other Scandinavian scientists involved in the OECD project realized early on that ecological research needed to be an integral component of scientific studies on acid rain. Yet uniting atmospheric and ecological disciplines would not be nearly enough to unpack the complexity of a transboundary pollution problem. Tracing its formation at the source to its environmental impact would require a range of specialists from atmospheric physicists to soil chemists to forestry researchers. To foster such interdisciplinary projects, the Norwegian government and its applied research institutes invested considerable amounts of scientific manpower and financial resources to construct a national environmental research effort of unprecedented scale. There were no comparable studies launched in any other European country at the time, and this approach to environmental research on air pollution eventually served as a model for governments across Europe and North America, especially the United States.

Second, the OECD project entailed international cooperation on atmospheric research using new techniques and monitoring stations that were quite different from previous practices of meteorological data collection.[34] Few countries outside of Scandinavia had experience monitoring pollution beyond urban areas or tracking it through the atmosphere, and heavy polluters were reluctant to assist with the OECD research, notably Britain, France, and West Germany. This prompted NILU to work closely with other Scandinavian scientists through the Scandinavian Council for Applied Research in order to build up a Nordic network of stations and prove that research techniques were feasible prior to introducing them into the larger project.[35] NILU's ability to organize a large observational network in Scandinavia as well as test out techniques proved crucial in effectively carrying out the study despite resistance from the heavily polluting OECD member states.

Though the OECD project was first conceived as an atmospheric program, Ottar and other meteorologists involved in the study all concurred that research into acid rain's environmental effects should be incorporated as well.[36] Scandinavian scientific representatives petitioned the organization to include ecological studies with the atmospheric program, but the OECD decided to omit environmental impacts from the project because the organization did not consider ecological studies to be transnational, a key criterion for work taken up by the intergovernmental group.[37] Instead, it encouraged member states to share the results of national research over the course of the project. The exclusion of ecological studies elicited heavy criticism from many scientists, especially Swedish researchers who had prepared their country's case study on acid rain for the 1972 UN Stockholm conference. Upon learning that the project would not examine the environmental effects of acid rain, Bert Bolin and several of his fellow scientists at the University of Stockholm wrote a highly critical letter to the organization stating that studies of the ecological aspects of the issue deserved just as much attention as the atmospheric processes involved.[38]

Ottar shared many of these concerns about the OECD's decision, and argued that interdisciplinary collaboration was essential both for understanding the phenomenon of acid rain pollution and for presenting a persuasive case in future policy negotiations.[39] From the beginning of the OECD project's preparations, Ottar publicly pledged that NILU would coordinate domestic ecological investigations while serving as the coordinating center for the OECD project.[40] True to his word, Ottar soon hired a biologist specializing in the measurement of tree growth, Nils Brandt, to undertake such work for the institute.[41] In fact, Ottar originally hoped NILU might oversee all research aspects of acid rain from atmospheric to biological studies.[42] Yet as Ottar attempted to build up a management team with expertise in physics, meteorology, and biology, officials in the Royal Norwegian Council for Scientific and Industrial Research started to suspect that organizing a national research effort on acid rain would require far more resources than NILU alone could provide.[43]

Fortunately for Ottar and his colleagues, scientific interest in organizing national work on acid rain's environmental effects was already growing in Norway. Odén's publication had sparked considerable discussion among Norwegian biologists about the possible effects of acid rain on the environment, leading to a series of meetings between scientists at Norway's applied research institutes to follow up on his findings.[44] Members of the Norwegian Forestry Research Institute were particularly concerned about the possible impact of acid rain on forest growth and agricultural productivity, and the head of its ecological department, Kristian Bjor, initiated discussions with Ottar and Nils Brandt about the impact of acid rain on forest ecology during preparations for the OECD project.[45] Shortly thereafter, Brandt set up a meeting between researchers from NILU, the Norwegian Forestry Research Institute, the Norwegian Institute for Water Research, and the Norwegian Meteorological Research Institute to discuss initiating parallel work on the ecological impact of acid rain.[46] Together, they formed a research consortium to study acid rain's environmental effects through a project called "Sur nedbørs virkning på skog og fisk" (Acid Rain's Effects on Forests and Fish), eventually known as the SNSF project.

Although the inclusion of government research institutes was likely a product of high levels of interest among their members, Ottar seems to have deliberately limited involvement to organizations specializing in applied research and excluded university-based scientists. He did this out of concern for the need of researchers to work collaboratively with colleagues from multiple fields on shared projects, believing that university scientists lacked the ability to work as part of a team and integrate their research into comprehensive evaluations.[47] Drawing on his experiences of "big science" from Norway's Defense Research Institute, Ottar encouraged the Royal Norwegian Council for Scientific and Industrial Research to model acid rain studies after scientific projects on chemical warfare and atomic energy, and the project's initial research designs reflect the importance of such interdisciplinary cooperation on the making of a "big science" of the environment. For example, several forestry researchers initially suggested that studies should be conducted into the ways acid rain impacted the transport of nutrients in the soil, particularly the leaching of important ions like calcium for flora and fauna.[48] Subsequently, soil microbiologists collaborated with foresters in releasing radioisotopes within areas of heavily acidified soils to track nutrient uptake by spruce and pine trees as well as the effects of acid rain on the nitrogen cycle in forest catchments.[49] In addition, NILU atmospheric chemists and physicists helped foresters and limnologists pinpoint areas in the Southern region of the country likely to receive large amounts of acidic precipitation. NILU scientists then set up monitoring stations in these locations to measure the acidity of precipitation, while forestry researchers tracked changes in the forest stand and limnologists took measurements of water runoff to detect possible washout of nutrients from the soil.[50]

Beginning modestly with about a dozen full-time researchers, the Royal Norwegian Council for Scientific and Industrial Research rapidly scaled up the SNSF project after the initial pilot phase was completed. It hoped to have the results ready in time for the OECD project's conclusion so findings from both research efforts might be used to pressure heavy polluters for pollution reductions.[51] The SNSF project soon involved over 150 scientists working on interdisciplinary investigations with an annual budget of $2 million, or about $10 million today.[52] As the studies progressed, the Norwegians involved in the research cited the interdisciplinary nature of the environmental program as its most valuable aspect, since it allowed them to work outside "ordinary" scientific boundaries between different specialties and institutions.[53] Indeed, the interdisciplinary, collaborative nature of the project earned the SNSF program international attention and respect from other scientists working on pollution problems. It became an important tool in promoting the project's results among scientists in other countries and stimulating parallel work, notably with the US Environmental Protection Agency (EPA) and American ecologists researching acid rain, many of whom visited Norway to learn directly from the Norwegians' experience.[54]

Fossil fuel pollution, by virtue of its ubiquitous, invisible nature, thus seemed poised to reshape the way the scientific community understood the environment and the earth's processes. Similarly to the way atomic research transformed physics and even biology into government-sponsored, high-technology fields, acid rain studies led to the emergence of a self-conscious "environmental" science that could tackle the coming pollution problems of industrialized society. But in growing to encompass disparate fields and stretching to accommodate researchers across national divides, the OECD and SNSF projects would force scientists and environmental officials to grapple with the boundaries of science and policy as well as how to police national interests in government-funded research.

The End of the "Heroic" Era

The first challenges to the OECD's international environment research program came from polluting nations, who were not eager to assist with work that might implicate their industries in causing pollution problems. Although it appears that all member governments were originally supportive of the acid rain project, Britain, France, and West Germany soon began to express reservations, which Ottar suspected was because of the "far-reaching consequences" for themselves.[55] This was especially evident with Britain. Toward the end of the planning process, the British delegate abruptly informed the other members of the OECD project steering committee that his government refused to participate in the study unless West Germany and France agreed to the project first. Several delegations voiced their disappointment at Britain's sudden, unexpected resistance and found it downright "strange."[56] Internal British government documents reveal that their representatives feared France and West Germany might pull out of the study, and they did not want to invest in the work if Britain would be the only major polluter participating.[57] Eventually all three governments agreed to join, but further conflict occurred over funding just a few months before the study's pilot phase was supposed to launch in 1972. The French delegate told the committee that his country could not contribute financially until the next year because the deadline had passed to submit a proposal to Parliament, and Britain's representative reiterated its position that it would not participate until both West Germany and France came onboard.[58] These issues forced the Scandinavian governments to provide nearly all the initial funding for the study.[59]

The fact that the project had been in the planning stages for two years raises the question of whether such delay tactics were deliberate. But whatever the case, NILU was forced to absorb so much of the costs of the OECD study that it went considerably over its budget by the end of the project, at one point almost having the power turned off at its headquarters because it could not find the money to pay the electricity bill.[60] Despite its "international" veneer, the OECD project's financing revealed the limits of government cooperation on pollution problems that had previously flourished in earlier years. It was one thing to cooperate on urban smog and other local pollution issues by sharing results and measurement techniques, but quite another to ask major polluting countries to fund a study that might lead to additional expenses for their industries.

Ambivalence about the study was also evident in the limited data submitted by France, West Germany, and Britain. During the first months of the study, France did not submit the data NILU requested to the coordinating center, forcing Ottar to "spend some time in chasing the various national authorities" to receive the information.[61] As NILU began preparations for the full year-long measurement phase, it reported that more observatories were needed to account for areas of high rainfall in France, which had six online, but that France would not provide them with the additional stations.[62] West Germany's actions were even more discouraging. Scientists from West Germany submitted almost no data during the first months of the study, and the minimal data they did transmit was considerably delayed and not in accordance with the agreed upon measurement techniques. When NILU informed the West German delegate that ten more observatories were required to account for topographic factors in its region, the country did supply them, but equipped the stations with only the absolute bare minimum in measurement technology.[63]

Britain, in turn, furnished only two stations for the trial period and offered just one more for the full measurement period.[64] To their credit, the Department of the Environment did provide five additional sites during episodes of heavy pollution lasting several days.[65] More importantly for the research, scientists at the government's Harwell laboratory and Met Office did schedule flights over the North Sea to sample pollutants at several heights in the atmosphere's mixing layer, which became useful data points as the study progressed.[66] Britain's Minister of Technology had instituted the atmospheric pollution program at the Harwell laboratory in 1967 under the Atomic Energy Act to use the lab's expertise on radioactive fallout as a basis for studying other regional and global air pollution problems.[67] It constructed continuous sampling devices to detect sulfate in the atmosphere, and also undertook laboratory experiments using radioactive tracers to study the formation of chemical pollutants and the rates of reaction.[68]

Despite the British investments in flight sampling, the Scandinavians were nevertheless frustrated by the country's limited monitoring stations since Britain had hundreds of sampling stations for urban air pollution. More problematic, though, was the British government's insistence that if the first year of the project did not produce sufficient evidence of pollution transport it would try to "kill the programme completely."[69] Britain's environmental officials were strongly pressured by the government-run coal industry to resist letting the project "snowball" since they believed it was extremely unlikely the country was responsible for acid rain in Scandinavia and wanted the government to invest as little as possible in the research.[70]

Faced with these obstacles from the major suspected polluters, NILU worked closely with the Scandinavian Council for Applied Research to build up an extensive Nordic network during the project's early years. Ottar hoped these investments would show the project could generate evidence of the long-range transport of air pollutants, which would make it much more difficult for Britain, France, and West Germany to refuse to participate.[71] Through this cooperation, Finland, Denmark, Sweden, and Norway developed much larger networks for taking observations than other participants, especially when compared to their landmass and population size. They ranged from eight in Finland to almost thirty in Norway. These countries also equipped several stations with additional technology to account for the influence of other pollutants on the acidity of precipitation, such as nitrogen oxides, and employed round the clock sampling devices. In contrast, just one station in France was equipped with technology to examine other pollutants while Britain and West Germany had none. Furthermore, virtually all stations in Britain, France, and West Germany took measurements only every six hours.[72] The aircraft sampling during the full measurement phase also reveals the disparity among the participating countries; whereas Sweden and Norway flew thirty-five and thirty-seven flights, respectively, Britain participated in twenty-two, Germany in eight, and France in five.

To the surprise of British environmental officials, by the end of the first measurement phase in 1973, the study had successfully generated sound estimates of each country's contribution to the pollution levels of its neighbors, in no small part as a result of the intensive efforts by NILU and their Scandinavian partners.[73] Five of the eleven participating nations were clearly subjected to greater amounts of air pollution from foreign countries than from their own industries, with Scandinavia as a whole receiving two-thirds of its pollution from outside sources.[74] The study also made notable advances in determining the varying influences of other pollutants, such as particulate matter and nitrogen oxides, in causing acid rain. In light of these results, British officials determined that it would no longer be "acceptable politically" to withdraw from the program of research, and NILU drafted an initial report on the study's findings.[75] With these results in hand and the SNSF project in progress, Norway's Ministry of the Environment began preparations to approach the OECD about including recommendations on transboundary air pollution in a meeting of environmental ministers the following year.[76]

The holding of a "ministerial" meeting on the environment promised to bring scientific expertise to bear on concrete policy recommendations for the first time. Other ministerial meetings had frequently been held at the OECD on matters such as trade and economic development, but no recommendations had ever been prepared on the environment. The OECD set in motion the necessary work for this kind of high-level agreement back in 1970 when it formed its Environment Committee, which began discussing the potential policy implications of acid rain and transboundary air pollution as the scientific project got underway. Unlike the organization's previous work on air pollution in urban environments, transboundary air pollution raised difficult questions about countries' legal liability for damaging the environment of their neighbors and how to allocate the considerable costs in reducing air pollution in market economies.

As a result of the substantial interest shown by member governments in these economic and policy aspects of international fossil fuel pollution, the Environment Committee decided to commission economic reports and legal advice from its member countries shortly after its inception as well as preliminary analyses by a subcommittee of economic experts.[77] These specialists, who were drawn both from the OECD staff of economists and lawyers as well as from academic institutions throughout member countries, took the lead in promoting what was known as the "polluter-pays principle." This important concept in environmental law means that the polluting industry, rather than the government, must pay for the costs associated with controlling emissions.[78] Legal scholars and economists developed the polluter-pays principle during this period in order to avoid distortions in international trade and unfair competitive advantage for the industry of one country over that of another.[79] Based on the potential fruitfulness of these investigations for addressing problems like acid rain, in September 1973 the Environment Committee further enlarged its program of activities on transboundary pollution to include the administrative, legal, and institutional aspects of the problem.[80]

With progress in the acid rain study and on the economic, legal, and policy aspects of pollution, the OECD Secretariat believed it would be the ideal time to build upon the momentum of the 1972 United Nations (UN) Conference on the Human Environment and hold its own high-level meeting of each country's Minister of the Environment in November 1974.[81] The OECD officials on the Environment Committee hoped the meeting would move beyond the broad pledges made at the 1972 UN Conference and establish concrete directives on environmental policies for the next decade.[82] Where the UN had faced criticism for bureaucratic inefficiency, bending to the needs of developing countries and falling prey to Cold War politics, the OECD envisioned its environmental meeting as applying scientific expertise to clear directives for its member states.

However, as plans got underway in the fall of 1973, the world economy was rocked by economic recession because of the Organization of Petroleum Exporting Countries (OPEC) oil embargo. In October of that year, OPEC began cutting oil exports to industrialized nations that were supporting Israel in the Yom Kippur War, including the US, the Netherlands, and Denmark, who were eventually subjected to a total embargo in December as payback for their military assistance during the conflict. Although not subjected to a complete cutoff of oil, Western European countries were also seriously affected by the crisis. Shortfalls in oil supplies and high prices seemed to threaten countries like Britain and West Germany with economic devastation not seen since the Second World War.[83]

Despite the ongoing oil crisis, the national representatives to the OECD's Environment Committee pressed ahead with plans for the environmental meeting and spent much of 1974 drafting recommendations on ten topics for the ministers to review and finalize in November. In addition to acid rain and transboundary air pollution, these included policy recommendations on the environmental effects of chemicals, noise pollution, water pollution, energy, and the polluter-pays principle.[84] Among all these areas, transboundary air pollution was the most contentious by far. The proposed recommendations were almost completely discarded before the meeting because of controversies among the delegates to the Environment Committee over language on the existence of the problem and how to control fossil fuel emissions given the ongoing energy crisis.[85]

In preparation for the meeting, the Environment Committee delegates had finished a draft of the transboundary air pollution recommendations along with the other nine environmental declarations by June 1974, five months before the meeting was scheduled. But throughout preparatory sessions in the summer and into the fall, opposition from the British delegation derailed finalization of the transboundary pollution guidelines.[86] Britain repeatedly tried to alter the wording of the recommendations so that it would not include any references to acid rain or the need to reduce fossil fuel emissions, and it was referred back and forth between the Environment Committee and Executive Committee of the OECD several times.[87] During these deliberations, the British delegation voiced its displeasure over what it perceived as an absence of financial considerations, and informed the rest of the committee that it would reserve its agreement to the transboundary pollution recommendations.[88] It continually demanded the deletion of all references to pollution "beyond national frontiers," calling them "superfluous."[89] The American chair of the Environment Committee, Hilliard Roderick, openly chastised the British delegate for these remarks, noting that the OECD was specifically designed to deal with problems that were international in character.[90]

British criticisms centered on the principle of "equal right of access," the fourth title of the transboundary pollution declaration. It would have given foreign parties affected by pollution the same rights to remediation for environmental damage as citizens possessed within the courts of the polluting nation.[91] This principle was proposed in conjunction with the principle of "non-discrimination," which meant a country would agree to regulate emitters who contributed to transboundary air pollution no less stringently than domestic polluters. In doing so, governments would be agreeing to avoid actions such as placing power plants along a border, sparing their own citizens pollution but subjecting a neighboring state to the toxic fumes. The recommendations would also have opened the door for countries such as Norway, Sweden, and Canada to petition major polluters like Britain, West Germany, and the US for greater emissions cuts. Environmental officials could argue that even if these countries had sufficient controls to satisfy local air quality, further measures were necessary to protect their ecosystems from foreign fossil fuel emissions that contributed to acid rain.

Many delegations hoped that the codification of these two "principles" in the legal system of each member country would be a first step in improving upon the UN Stockholm Declaration of 1972.[92] That agreement recognized environmental harm as a human rights issue, stipulating that the protection of the environment was essential to "the enjoyment of basic human rights."[93] However, there were no binding mechanisms to ensure compliance with the principles laid out in the Stockholm Declaration.[94] The OECD Secretariat believed that the adoption of the transboundary air pollution recommendations by its member countries would turn protection of the environment into "a sort of new human right" with legal enforceability, potentially liberating environmental policies from "territoriality" and state sovereignty.[95] During the preparatory discussions for the Ministerial meeting, the Norwegian delegates in particular underscored the need for the same legal rights to protection and compensation for foreigners as citizens were given within the polluting country.[96]

With the exact wording still in dispute as the November gathering approached, Norwegian delegates made repeated diplomatic overtures to the British representatives to see whether a compromise could be reached on the transboundary air pollution document and to better understand their objections to it. Enlisting the support of Gérard Eldin, the deputy secretary general of the OECD, they sought to break the deadlock and change the British attitude toward the recommendations.[97] These conversations revealed that British opposition was rooted in the ongoing economic crisis caused by the 1973 oil shocks, which had sent energy prices soaring in Britain.[98] British officials admitted that they had no genuine objections to the text itself, but rather feared that the members of OPEC would exploit Britain's agreement to any environmental recommendations in order to drive up oil prices even further. They believed these oil-producing governments would argue that if Britain was willing to accept a financial burden for environmental protection, it could also afford to pay higher prices for oil, potentially crippling the government's attempts to deal with the energy crisis.[99]

Eldin and Norwegian officials found these claims to be far-fetched and were frustrated by the British attitude, believing that environmental protection measures should not be set aside simply because of additional financial responsibilities.[100] Although the Netherlands suggested trying to "gang up" on Britain to pressure them into signing on to the recommendations, Norwegian officials felt that there was little hope of getting more out of Britain "in the OECD context."[101] In part, this was likely because the OECD recommendations were unbinding; what good would it do to compel Britain to sign on to recommendations it could simply discard in the name of financial exigency? But it was also because of a growing consensus within the Norwegian government that pressure on Britain, and indeed all the major polluters of Western Europe, would require more than a bilateral partnership between themselves and the Netherlands. For now, the officials felt it would be wise to accept the British position without further antagonism while waiting for an opportunity to tackle these problems with countries in both West and East Europe, especially since they expected to have the results of the first phase of the SNSF project in a year's time.[102]

However, the British delegates had succeeded in weakening the recommendations to such a great extent that the plan to make any declaration at all was almost thwarted by the Canadian representatives, who felt that the document would turn "the clock back 65 years for the North American members of the OECD."[103] Although acknowledging that less progress had been made in Europe than between its country and the US, it insisted the transboundary pollution recommendations should be more than a sanctioning of existing practices, which many delegates felt were inadequate. Lamenting the close deadline of the Ministerial meeting and the serious deficiencies of the proposal, a Canadian representative argued that the circumstances required completely jettisoning the recommendations, "whose inadequacies we may later come to regret."[104] Other delegates, notably the Americans, agreed with the Canadian assessment of the document.[105] However, West Germany, the Netherlands, and Norway felt that it was the "cornerstone" of the submissions to the Ministers, and that removing the document entirely would deprive the meeting of much of its usefulness.[106] Canada relented in the face of opposition from those who wished to include at least some declaration related to transboundary pollution, however flawed, but the delegation attached a scathing dissent as an appendix to the document. Characterizing the statement as a "retreat from the Stockholm declaration which we all accepted two and a half years ago," the Canadian representatives asserted that the text of the document should have been preserved in its original form, particularly in reference to the section on equal right of access.[107] The controversy became so heated in the days leading up to the meeting that the document had to be transmitted to the Ministers with a special note attached explaining that major points of disagreement were still not resolved among the delegates to the Environment Committee.[108]

In comparing the UN Stockholm Declaration to the OECD recommendations, the Canadian delegates may have been slightly overstating their case in saying that the latter did not move beyond the former in any way. The OECD principles on transboundary pollution contained more details regarding appropriate notification and consultation between countries when polluting facilities were constructed along borders, the creation of warning systems for sudden increases in emissions, exchange of monitoring measurements, and the application of the polluter-pays principle to all facilities regardless of whether their emissions affected domestic or foreign peoples. It even retained a section about the issue of an equal right of access, despite the acrimonious deliberations.[109]

Yet the Canadians' frustrations were most certainly tied to the extremely watered-down language when comparing the two drafts of the OECD recommendations. The original emphasized the importance of harmonizing environmental policies between member states; the final text contained no statements to that effect. While the initial document stipulated that neighboring countries should have the right to request information on new pollution sources if it was not given freely, the final version eliminated those provisions.[110] Wording about the "rights" of victims was replaced with that of "no less favorable treatment." References to an "equal right" to be heard and appear in court were replaced with language encouraging countries to afford victims the option to petition for "standing in judicial or administrative proceedings."[111] This alteration left courts the option to deny hearing such cases using the principle of territoriality of laws, which some countries had interpreted to mean that foreign persons had no rights in their legal systems on the grounds that domestic legislation applied only to actions within the territory of a state and all persons or interests therein.[112] The famous Trail Smelter case, in which American farmers petitioned the US government to take action against Canada over fumes emitted from a smelting plant, is the most prominent example; Canadian courts refused to invoke jurisdiction over damage to land abroad and the case had to be settled through arbitration.[113] The OECD principles would have allowed similar future incidents to be settled in the courts of the offending nation. Regarding the principle of non-discrimination, while the initial document called for countries to "adopt" the principle, the final version asked countries to "initially base their actions" on this principle.[114] But perhaps the most glaring change was the removal of all references to the need to reduce or eliminate transboundary air pollution and fossil fuel emissions.[115]

At the high-level ministerial meeting in November 1974, OECD officials were clearly disheartened by the outcome, which had fallen far short of expectations. In his remarks to the attendees, the OECD Secretary General pointed out that the energy crisis had led many diplomats and environmental officials to question how to move forward on global environmental cooperation. The meeting's deliberations, he said, reflected the onset of a new, different stage in environmental policies quite different from the earlier "enthusiasm" which characterized the 1972 UN conference. The first "heroic phase," which had seen the establishment of environmental authorities in different countries during the late 1960s and early 1970s, had ended. Difficult policy choices were now necessary, with the potential for conflicts to emerge among industries, economic growth, scientific research, and environmental impacts.[116] For OECD officials, these conflicts seemed to require a reexamination of economic and legal principles on the environment and transfrontier pollution in particular. Only by further probing the legal aspects of environmental rights and the economic costs and benefits of regulation could countries move forward together to face the crisis posed by fossil fuels.[117]

But for the elected chairman of the meeting, Norwegian Environmental Minister Gro Harlem Brundtland, the lessons of the stalemate were somewhat different. Brundtland, who was a physician by training, assumed her position as Norway's Minister of the Environment after serving as a medical consultant to the government on the issue of abortion from 1968 to 1974. Her experience as a young feminist wading through contentious expert debates on abortion practices had shaped her perceptions of scientific expertise in the political process. Brundtland saw it as imperative for policymakers to use current scientific information as "preventative medicine" to avoid horrible outcomes for society.[118] She thus approached negotiations on acid rain with an eye for seeking concrete solutions based on the best advice available at the time.

Brundtland opened the conference by arguing that the scientific evidence available from the OECD's own acid rain research made it clear countries should agree to reduce their fossil fuel emissions. Although acknowledging the costs of undertaking pollution reductions, Brundtland asserted that they would not be prohibitive given the development of new and cheaper technologies. Without action, she stressed, the assimilative capacity of Norwegian soil would soon near complete exhaustion.[119] Undeterred by the resistance from major polluters like Britain, France, and West Germany, Brundtland was determined to lead industrial countries toward energy policies based on the burgeoning science of the environment.

After her return home, the Norwegian media lauded Brundtland and Erik Lykke for "standing strong" internationally despite the unsuccessful attempt to implement emissions reductions through the OECD. In cooperation with Brynjulf Ottar, they were beginning to envision the construction of a partnership between environmental science and diplomacy that would move beyond the limitations of both the UN and the OECD. As they began to regroup with their Scandinavian allies, however, new scientific and political alliances were forming between the American and British energy industries. They would call into question how science could be used in environmental policymaking, what it meant to do "independent" research on problems like pollution, and how to weigh environmental risks against the economic costs in regulation.

3

Energy Industry Research and the Politics of Doubt

We're not just shooting from the hip. I can't tell you that the emissions from your automobile fell on Hubbard Brook. Then again I can't say they didn't. But it's like the relationship between cigarette smoking and lung cancer. I can't prove it but it's common knowledge and I believe it's true.[1]

Gene Likens, American ecologist, 1975

Essentially with the acid rain issue, the start of acid species reduction came through the Clean Air Act, which really was aimed initially at cities, population exposure, and the local effects of polluters. After the mid '70s people began to realize that there were large scale pollution effects as well. But the industry at that time was ambivalent about that. They didn't think they were causing problems beyond local exposure, and they said, "Well, we've done all these plant improvements and controls, and installed tall stacks to get rid of the local ground level concentrations. So what more can be done for low exposure questions?" So they really supported EPRI's [Electric Power Research Institute] acid rain research program because they were very nervous about the academic and government science that was pushing from the other side. They recognized the value of a science program to challenge the objectivity of other investigators.[2]

George Hidy, prior director of EPRI's Environmental Program, 2013

The American White Mountains, located in the northeast part of the country, are the premier range for nature preservation and recreation in the region. Stretching across two states, they have long served as a popular attraction for hiking, camping, and scenic escapes from the urban centers of Boston and New York City. But in the summer of 1974, research from two American ecologists published in *Science* suggested that the White Mountains, and likely much of the US eastern seaboard, was receiving large amounts of pollution from acid rain. Working at the Hubbard Brook experimental station in the New Hampshire section of the White Mountains, Gene Likens and F. Herbert Bormann discovered that precipitation there had much higher levels of acidity than other areas in the US.[3] Their article was immediately picked up by major news outlets, including the *New York Times*, and catapulted acid rain onto the American political scene.[4] Arriving in the midst of tense gridlock at the Organisation for Economic Cooperation and Development (OECD) over possible recommendations on reducing fossil fuel emissions, the findings put even more pressure on industry and policymakers to address the problem.

While Likens and Bormann's paper generated the first significant political attention to acid rain in America, the idea that the phenomenon was also occurring across the Atlantic had been percolating for some time. Likens began monitoring acidity in rainfall at Hubbard Brook in 1963 and, like Sweden's Svante Odén, had noticed lower and lower levels of pH by the end of the decade. Fortuitously, Likens received a North Atlantic Treaty Organization (NATO) senior scientific fellowship to visit universities in Britain and in Sweden in 1969, which included a sojourn in Uppsala, Sweden, with one of Odén's colleagues. While there, he met Odén for the first time, and mentioned his interest in precipitation chemistry as well as his ongoing work at Hubbard Brook. Upon learning of Likens' research, Odén invited the young scientist to accompany him on an overnight train to Oslo, where he was scheduled to give a lecture on acid rain. The two men stayed up talking until dawn about Odén's proposition that foreign power plant emissions were causing acid rain in Scandinavia. The conversation had a profound impact on Likens, who subsequently installed additional monitoring stations in New York after his return home to see whether acid rain was also a regional problem in the US.[5] This work, combined with his data from Hubbard Brook, provided a powerful piece of evidence on the need for industrialized countries to address acid rain cooperatively.

Likens' research did not escape the notice of the energy industry in the US or Britain. With data demonstrating fossil fuel pollutants could cross international borders from the OECD's acid rain project and a growing body of ecological studies on acid rain's possible environmental effects in Scandinavia and the US, leaders of electric utilities in both countries began to grapple with the implications of acid rain for energy production and their bottom line. Would it be possible to reduce the pollutants causing acid rain? How expensive would it be? Could they even

FIGURE 3.1 Gene Likens installing a bulk precipitation collector at Cornell University in 1970. Photograph from Gene Likens.

be certain that they were the culprits in this problem? How could they ensure their interests were represented in any political debates over what to do about acid rain?

The creation of "doubt" over scientific findings concerning threats to public health or the environment occurred throughout the twentieth century. From cigarettes to lead paint to climate change, private companies and government-owned industries have worked to discredit research that implicates their products or pollutants in causing harm.[6] Acid rain was no different in many respects. Beginning in the early 1970s, as Likens, Bormann, and other scientists pursued environmental research establishing the potential dangers of fossil fuel pollutants, the coal industry questioned whether pollutants could truly travel such long distances in the atmosphere, whether some "natural," non-manmade occurrence could be causing the observed damages, and whether the environmental harms were severe enough to justify the costs of regulating emissions.

A key difference in the acid rain case, however, was *how* the energy industries in Britain and the US sought to engage in these debates. Rather than simply lobbying public officials or attacking research results, the coal industries in the US and Britain decided to develop their own body of experts in the natural sciences to serve as an industry "defensive" research establishment. While the desire to invest in environmental research on acid rain emerged separately in each country, the British and American coal industries quickly began cooperating in these efforts by exchanging researchers, sponsoring joint projects, and strategizing on how to avoid costly regulations on acid rain. It went far beyond a handful of scientists or industry officials campaigning against regulations or raising issues with the current scientific consensus.[7]

The irony is that these investments in environmental science did not go hand in hand with investments in technologies to reduce pollution. At the same time as the British coal industry began developing its environmental research apparatus, it declined to provide new funding for projects on key control technologies.[8] In the US, the research arm of the electricity industry did continue to finance some projects on controlling emissions, but the funding and resources available for those contracts paled in comparison to the money spent on environmental research. As just one example, two years after Likens and Bormann's *Science* publication on acid rain, American utilities were spending over $6 million on research to examine the atmospheric transport and ecological effects of acidic deposition while spending just over half a million on research into controlling acidic pollutants.[9] In other words, spending on environmental projects was more than ten times that of control technologies during this period.

By the end of the decade, the British and American coal industries had built up the largest monitoring programs on acid rain in their respective countries, with vastly more resources than university scientists. In the US, the coal industry was nearly outspending even the Environmental Protection Agency (EPA) on pollution research during the 1970s, while in Britain they far surpassed spending by the government's Department of the Environment.[10] Because of their enormous investments in building up a cadre of environmental experts and monitoring networks, both the British and American industries earned prominent seats at the table for national and international negotiations over fossil fuel regulations in the coming decades. It is therefore worth examining how they came to invest in these programs, the results of their scientific studies, and the role they began to play in environmental politics during the 1970s.

The question of industry objectivity in acid rain research is a complex, difficult one. In interviews, many scientists who worked on projects funded by the British and American coal industries insist that the research was unbiased, divorced from advocating particular policy positions, and ultimately improved our understanding of acid rain.[11] To a certain extent, there is merit in these claims. Critiques by industry scientists did push ecologists in Scandinavia and the US to refine their understandings of precisely how acid rain damages freshwater and forest ecosystems. For instance, it was only after industry scientists critiqued Norway's ecological studies on acid rain that researchers there began to examine the role of soil in mediating acid flushes into rivers and lakes. As a result, they uncovered the importance of acid rain in releasing aluminum from soils into freshwater systems, which proved to be one of the most damaging impacts of acid rain on biological life.

However, it's also the case that there are many more examples of industry scientists misrepresenting research results to policymakers and the public as well as suppressing results in Britain that implicated the coal industry in causing acid rain. By stressing uncertainties, claiming supposed "alternative" factors could be responsible, and characterizing the environmental benefits of reductions as minimal, these scientists cannot be properly described as unbiased scientific investigators. Some have since argued that environmental scientists have similar if opposite prejudices, deemphasizing costs of control or failing to publicize results that did not support limiting fossil fuel emissions. There is no evidence for such charges in the acid rain case. To the contrary, environmental scientists in universities or government laboratories released findings that acid rain did not harm the environment as much as suspected in certain respects, such as with forest damage. The same cannot be said for industry researchers.

Divesting from Pollution Control Technology

The beginning of industry involvement in acid rain science, and the decision to prioritize this work over developing control technology, took shape in Britain at the outset of the first international research program on acid rain at the OECD. The OECD's approval of a Scandinavian research proposal to study the atmospheric transport of acidic pollutants in 1971 triggered an intense internal debate among British government officials about the implications the results might have for British energy policy. Norway and Sweden had singled out Britain for the amount of sulfur dioxide pollution it emitted, which at the time was nearly equal to the quantity produced by France, West Germany, Belgium, the Netherlands, and Luxembourg combined.[12] Additionally, Britain relied much more than other nations on dispersing pollution through high chimneys rather than trying to reduce its emissions at the source. This strategy was based on the assumption that pollution from fossil fuels would be less harmful if spread over larger areas. Though high stacks dispersion mostly prevented further smog disasters and reduced nearby concentrations, some officials in the British government were concerned that the Scandinavian claims about acid rain would force them to rethink this approach and implement pollution abatement technologies, notably flue gas desulfurization.[13]

Flue gas desulfurization is a broad term used to describe the process of removing pollution from power plant emissions as they pass through chimneys, known as flue gas. The earliest theoretical proposals to remove sulfur pollution from gases originated in England during the nineteenth century, and British engineers pioneered flue gas desulfurization technology in the early twentieth century.[14] Several processes of removing sulfur dioxide were tested at a number of power plants around London, but were deemed unworkable because they lowered the buoyancy of the gases, resulting in less dispersion of the plume into the atmosphere and higher ground-level concentrations.[15] Subsequently, researchers in Britain as well as in the US began investigating dry processes of flue gas desulfurization in the decades after the Second World War, while deploying high chimney stacks in the meantime to prevent future air pollution disasters like the 1952 London smog.[16] At the time the OECD acid rain project was organized, the US government was financing several of these projects in Britain because of its superior, longstanding expertise with flue gas desulfurization. However, these projects were about to be discontinued since the US government wanted to further develop the technology domestically.[17] There was no other British work on desulfurization in progress, so the government needed to decide whether they should make their own investments in desulfurization projects in light of the acid rain issue.[18]

Britain's Central Electricity Generating Board (CEGB), which managed the country's nationalized electricity industry, exerted the greatest pressure against continued research into desulfurization technologies. The CEGB was formed in 1957 to oversee Britain's electricity industry, which was nationalized a decade earlier along with several other industries after the British Labour Party came to power following the Second World War. Prior to acid rain, it had little involvement in environmental science and politics. The organization's main research arm, the Central Electricity Research Laboratories (CERL), consisted of a small group of scientists and engineers working on problems related to electricity distribution and "keeping the lights on no matter what the cost."[19] The problem of acid rain would transform the involvement of the CEGB, and by extension the CERL, in environmental science.

The implications of acid rain for British energy policy and coal use were immediately evident to CEGB officials. Shortly after the publication of Odén's work in 1967, CEGB leaders began claiming publicly that British emissions posed little threat to the air quality of other countries.[20] It also commissioned a joint report on acid rain alongside Britain's National Coal Board, which managed the country's mining operations, the British Steel Corporation, the Institute of Petroleum, the Department of Trade and Industry, and the Warren Spring Laboratory, a government research institute. Published in 1971, the report was adamant that scientific evidence linking British emissions to acid rain in Scandinavia was "wholly non-existent" and that "simple physical arguments seem to preclude the possibility that such major effects could be experienced so far from the source when no similar effects have been seen to occur close to the source where emissions are much more concentrated."[21] It asserted that fossil fuel pollutants like sulfur dioxide differed from pesticides and other chemicals because they were "natural" components of the atmosphere and plant nutrition with a "cleansing cycle."[22] By this logic, even if studies confirmed the transport of fossil fuel pollutants to other countries, the authors claimed fossil fuel pollutants like sulfur dioxide would actually be beneficial to agriculture and the environment.[23] The idea that fossil fuel emissions were somehow harmless because they were "natural" would become a recurring argument in the early 1970s by the coal industry.

Skepticism that "natural" pollutants from fossil fuel burning could truly cause a phenomenon like acid rain was also present in Britain's Department of the Environment, which largely supported the CEGB's desire not to invest in pollution control. Many members of the Department of the Environment were confident the OECD study on acid rain would show British pollution did not travel far enough through the atmosphere to be responsible for an increase in acidic precipitation. Because of this, they did not believe Britain should finance research into desulfurization unless scientific studies had "proven" acid rain was a serious environmental problem that required reductions in pollution emissions.[24] As one environmental official put it, "research had better be concentrated on improving our grasp of the truth rather than on abatement technology" because the Scandinavian claims could very well turn out to be false.[25] Given their belief that the scientific accusations might be spurious, members of the Department of the Environment argued that funds would be better spent controlling other pressing environmental pollution. They felt that since high stacks on power plants sufficiently maintained Britain's air quality, research funds should be allocated to other environmental problems that were more serious threats to public health, such as water pollution.[26]

These sentiments were supported by officials at the Department of Trade and Industry, which was technically responsible for overseeing Britain's broader energy policies in this period, though the CEGB exerted enormous influence as well. Members of the Department of Trade also believed that pollution abatement was too expensive to consider until research had confirmed it was a wise course.[27] In addition, they expected that the use of coal would decrease gradually over the next decade and be replaced by less polluting natural gas and oil, making control technologies unnecessary. These officials felt it was extremely unlikely that other countries would implement pollution reductions given the costs of removing sulfur from flue gases, making it difficult to turn a profit from exporting the technology and putting British industries at a competitive disadvantage.[28] Furthermore, should Britain implement such restrictions, they suggested that it would be much more economical to rely on Britain's supply of low sulfur fuel in the North Sea than try to lower sulfur dioxide emissions from its coal resources, though the feasibility of this option was not seriously tested.[29] The first major discovery of oil in the North Sea had occurred only the year prior and it was not yet known whether it would be commercially viable.[30]

With doubts over the existence of the problem and concern about investing British resources in other environmental areas, the CEGB had little trouble securing the support of other government departments in allowing the flue desulfurization projects to expire. But these views were not shared by all British government officials, a number of whom cautioned the CEGB that Scandinavian research had already made a strong case for acid rain's deleterious environmental effects.[31] In fact, several officials within the Department of the Environment believed that it was likely the OECD project *would* implicate Britain in causing acid rain pollution, thus putting pressure on the country to make emissions reductions instead of relying on dispersion through high stacks.[32] One person who advocated for investing in pollution control technology because of this risk was the director of Britain's Central Unit on Environmental Pollution, Dr. Martin Holdgate. A field biologist trained at Cambridge University, Holdgate's engagement in the acid rain debate would deepen over the coming years and increasingly put him at odds with the CEGB, particularly after he was made chief scientist in the Department of the Environment in 1976.[33]

In 1971, the same year the CEGB and its allies released a report arguing there was no credible scientific evidence to support the existence of acid rain, Holdgate was given a "very tentative" first set of figures from Britain's Warren Spring Laboratory concerning the amount of sulfur dioxide the country exported. The laboratory had begun conducting flights over the North Sea to sample sulfur dioxide pollution as it left the British Isles, and the results indicated that Britain exported a "substantial part" of the country's total sulfur dioxide emissions, likely from one-quarter to one-third.[34] Armed with this "straw in the wind," Holdgate made repeated attempts to convince fellow officials in the department that transboundary air pollution was occurring and that it would be foolish to abandon research into flue gas desulfurization.[35] He found support among a number of his colleagues, including Britain's representative to the OECD Environment Committee, Leslie Reed, who believed flue gas desulfurization would be an important "insurance policy" should regulations go into effect.[36] Reed had originally trained as a mechanical engineer before gaining experience on acid rain at the Warren Spring Laboratory.[37] He then began working as a "scientist-diplomat" at Britain's Central Unit on Environmental Pollution, simultaneously assisting with the OECD's acid rain research project and serving as Britain's representative to the OECD Environment Committee.[38]

Private industry seemed unlikely to invest in these technologies unless the government made it clear that they intended to restrict emissions, though even if the government exerted such pressure, many in the Department of the Environment were not optimistic that private companies would want to spend money on technologies that laid the foundation for pollution reductions.[39] There were also concerns that if Britain chose to enter the European Communities, other member states who wanted to reduce their sulfur dioxide emissions might compel Britain to do the same.[40] Though Britain's National Coal Board mostly supported the CEGB in winding down pollution control projects, some officials there argued that Britain should invest in flue gas desulfurization because the European Communities might no longer allow reliance on high stacks to disperse pollution in the years ahead—a prescient prediction, as we will see in later chapters.[41]

But although Holdgate was successful in convincing some members of the Department of the Environment that investment in abatement technologies should be done, he was unable to mobilize the department against the CEGB and other government officials who disagreed with this approach.[42] Ultimately, the view that the country should not invest in abatement technology unless scientific research showed British emissions were causing acid rain prevailed within the government, and Holdgate resigned himself to keeping a "watching brief" on the issue as the OECD study got underway.[43]

Britain appeared to be taking a strange gamble; they had a history of expertise in desulfurization technologies and therefore a lot to lose by abandoning work in this area should the OECD study demonstrate that British emissions resulted in acid rain over Scandinavia. And with considerable research showing that radioactive fallout could travel extremely long distances in the atmosphere, it seems baffling that many officials were confident sulfur dioxide pollution could not be similarly transported.

However, there is a key difference between sulfur dioxide and fallout that explains many officials' self-assurance. While manmade chemicals and radiation had been shown to be dangerous to humans and the environment, sulfur dioxide was produced through natural processes like volcanoes and had a global cycle.[44] The idea that power plant emissions of a naturally occurring chemical compound could harm the environment seemed illogical to many British officials responsible for the country's energy policy, and so they let the US-backed projects on flue gas desulfurization expire, believing that the OECD project would vindicate their country of the Norwegian and Swedish accusations.

Simultaneously with these British debates in the early 1970s, private American power companies began a parallel move against investments in flue gas desulfurization even as the US federal government sought to encourage research into the technology. The dynamics across the Atlantic were slightly different given the corporate nature of the American industry and its separation from government regulators, as well as the fact that acid rain had not yet been identified as an environmental threat to North America. Instead, industry mobilized against the technology because of a proposed revision to the 1970 Clean Air Act that appeared likely to require use of flue gas desulfurization for compliance with air quality standards. In response, US coal companies began a $63 million advertising campaign against installing the technology to meet the proposed pollution levels and successfully petitioned Congress and the federal government to instead utilize high stacks and smaller plant improvements in burning coal to maintain local air quality, as the British had done following the London smog disasters.[45] Eventually, these strategies came under criticism from environmental researchers and activist groups for contributing to the long-range transport of pollutants and acid rain.[46] Following Likens and Bormann's publication, the National Academy of Sciences and the Environmental Protection Agency both warned Congress of the long-term dangers acid rain could pose to the environment in separate assessments on sulfur dioxide emissions and the need for further emission controls.[47]

In 1974, after the OECD study's initial measurements showed that long-range transport of fossil fuel emissions was in fact occurring on a large scale, the British CEGB could have reversed course. Yet there does not appear to have been any proposal to reinvigorate a flue gas desulfurization program in the British government. Similarly, despite Likens and Bormann's work indicating that acid rain was also occurring in the US, American coal companies continued their campaign against installing flue gas desulfurization technologies on power plants.[48] Instead of investments in pollution control, both the British and American coal industries would launch new programs to combat scientific research on acid rain with their own experts in atmospheric physics, chemistry, and biology.

The Energy Industry Enters the Environmental Science Field

When Gene Likens and Herbert Bormann published their *Science* paper on acid rain in 1974, American ecological research on the dangers of fossil fuels was extremely limited. It was only in the years following the Second World War that "ecology" as a field was organized around the ecosystem concept, driven by concerns about nuclear fallout and waste at US national laboratories. By the late 1960s, ecologists had made great strides in studying the movement and accumulation of radioactive pollutants and toxic chemicals in the environment.[49] Yet there were no large-scale programs devoted to monitoring environmental effects from power plant emissions. EPA's first regional pollution monitoring network, the Community Health and Environmental Surveillance System, or CHESS, focused on collecting epidemiological health data rather than ecological measurements of pollution impacts. After beginning operations in 1971, it soon came under fire for poor measurement techniques and analysis methods, including from the National Academy of Sciences.[50] As the agency struggled to assert its authority over research into pollution and public health, it took no steps to begin investigations into acid rain in the years following Likens and Bormann's crucial findings. It faced criticism for this lack of attention to ecological research from Congress and university researchers, who called on the EPA to take a more proactive stance in addressing the problem.[51]

Britain was even less equipped to tackle a problem like acid rain because of the state of government and academic environmental research.[52] While the human health impacts and the corrosive effects on buildings from air pollutants had been studied extensively since the London smog of 1952, potential damages to the environment had been scarcely examined prior to the 1970s. The most extensive studies were carried out on the response of lichens to sulfur dioxide pollution, in the hopes of using such data as an indicator of pollution levels in both urban and rural areas.[53] After the OECD project began, however, British officials thought at least some investigations into the environmental effects of acid rain might be warranted.[54] At a coordination meeting for the project with officials from the Department of the Environment, CEGB, Meteorological Office, Warren Spring Laboratory, and Harwell laboratory, Leslie Reed suggested additional rainwater analyses in Scotland could be useful to establish the acidity of rainfall within the country, which appears to have been the first time British officials considered the potential need for national studies of acid rain.[55] Alan Holden, a chemist at Scotland's Freshwater Fisheries Laboratory, offered to undertake such an examination of pH in rainfall at one location in Northern Britain.[56] In addition, the Warren Spring Laboratory assisted in analyzing some of these measurements as part of the OECD project and unearthed some disconcerting results, including the lowest pH reading on record in Europe at a station in Scotland in April 1974.[57] But besides the limited work of these laboratories related to international collaborative research, no other government departments undertook investigations into acid rain's environmental effects.

The absence of federal research in the US and the limited involvement by British government offices left an opening for private American industry and the British CEGB to step in. The notion that industry should fund its own environmental experts first arose in Britain during the early 1970s at the CEGB's main research laboratory, the CERL. It had conducted only a few biological studies since the 1950s, mostly focused on how high temperature effluents harmed aquatic life. That changed in 1973 with the appointment of Peter Chester as the new laboratory director of the CERL.[58] Described as a bright, domineering figure often determined to get his way, Chester had trained as a physicist and worked at the CEGB since 1960 in the solid state physics section.[59] While lacking training in environmental science, Chester astutely anticipated the important role environmental experts would come to play in acid rain policies, and believed the British CEGB needed to launch its own research to investigate whether power plant emissions were causing ecological harm.

In 1973 and 1974, Chester heavily recruited scientists to study acid rain and expanded the environmental research apparatus of the CERL, including the formation of new atmospheric physics and biology divisions.[60] Additionally, Chester helped found a new department called "environmental chemistry" shortly after the OECD project commenced in order to begin investigating the ecological effects of acid rain.[61] This department was run by Anthony (Tony) Kallend, who had previously worked in the chemistry division on problems of corrosion at nuclear power plants, and it conducted a large portion of CERL research into the freshwater and forest impacts of acid rain. The biology section, led by Gwyneth Howells, also assisted with work on the environmental impacts of acid rain through examining its effects on organisms.[62] Many projects were run jointly between the divisions, with a certain amount of interdepartmental rivalry that seems to have been exacerbated by the appointment of Howells, a woman, as one of the section heads.[63]

As Chester began building the CEGB's environmental research arm, a newly formed American industry organization undertook a similar effort to create its own environmental experts: the Electric Power Research Institute (EPRI). EPRI, a nonprofit funded by American coal companies, was first established in 1973 shortly before the publication of Likens and Bormann's acid rain findings in *Science*. The initial impetus for EPRI's founding was to coordinate new research into technologies that could help improve distribution of electricity and generate more energy from fuel.[64] As the result of an infamous power outage on the east coast in 1965, the industry faced pressure from Congress to invest more in long-term research in the late 1960s or else face a tax to fund a federal program.[65]

After a prolonged search, EPRI chose Chauncey Starr, a former Manhattan project physicist, as its new president and head of research operations. Starr had previously served as Dean of UCLA's School of Engineering, where he was an outspoken advocate of greater government attention to issues of energy and public policy. According to EPRI officials, Starr caught the attention of the president of the Edison Electric Institute thanks to an article he published in *Scientific American* in 1971. Entitled "Energy and Power," it framed US economic success as dependent on high energy consumption. Starr feared there would be increased demand for energy resources because of world population growth, and believed that smart investments as well as careful balancing of environmental pollution against the costs in emissions control would be necessary to maintain US standards of living.[66]

Upon assuming his post at EPRI, Starr pushed top officials at the Edison Electric Institute and other companies in the energy industry to include an environmental division in the new cooperative research effort.[67] Despite vigorous opposition, he convinced the CEOs involved in creating EPRI that scientific expertise on environmental problems would be essential in debates on future energy regulations. As Starr later put it in an interview, "the resulting support of a broad intellectual community could be an important element in the influence of EPRI nationally."[68] To recruit top scientists and promote EPRI's authority among policymakers, Starr proposed founding a central office in Palo Alto, then booming with talent from emerging high technology industries, as well as a Washington liaison office.[69] He would become a key figure in casting doubt on acid rain and climate change science, eventually joining the board of the George Marshall Institute, a Washington think-tank that expressed skepticism over environmental threats.[70]

With current US pollution controls for human health focused on particulate matter rather than sulfur dioxide, acid rain appeared to be the only reason the federal government might implement stricter limits on these emissions.[71] In light of this, EPRI's research on sulfur dioxides primarily dealt with acid rain during the early years of its existence. It set up its own regional network of sulfur dioxide monitoring stations along the US eastern seaboard, and hoped to investigate whether there were serious ecological effects from current pollution levels.[72] While it ran some projects itself, such as the monitoring network, and supported a small group of scientists at its headquarters, EPRI generally distributed grants to university scientists and other contracted researchers rather than build up its own research laboratories like the CERL. EPRI began its grant program during 1976, and by 1980 its budget reached over $5 million with an expected doubling to $10-$12 million in the next few years, much of which would go toward studies on acid rain.[73]

A partnership between the British and American coal industries in environmental research on acid rain formed in the months following Likens and Bormann's *Science* paper. In December 1974, the chief scientist in Britain's Department of Energy, Walter Marshall, wrote Starr a letter to ask whether EPRI might be interested in partnering with Britain's CEGB, the Department of Energy, and other relevant government agencies in research and development projects concerning the environment.[74] Marshall, the former head of theoretical physics at Britain's Atomic Research laboratories, was a formidable intellectual and political figure in Britain in the early 1970s. The youngest member ever elected to Britain's Royal Society at the time, he assumed his position in the Department of Energy in 1974 with the aim of bolstering the countries' coal and nuclear industries. Marshall was eager to forge cooperation with EPRI in order to combat possible regulations of fossil fuel emissions in both countries as well at the international level, and he saw investments in environmental science as a key way to protect British interests. In service of these aims, Marshall forged a close working relationship with Peter Chester at the CERL, who ultimately became the main point person for facilitating collaboration between the energy industries of the two countries. Starr, for his part, thought the CERL would be an excellent partner in EPRI's work on acid rain, particularly given their shared concerns about future regulations domestically and internationally.[75]

Thanks to Marshall's overtures, during the ensuing decade EPRI, the CEGB, and Britain's Department of Energy collaborated on research into acid rain's dispersion and environmental impacts, as well as the prospects for regulation to address acid rain in each of their respective countries.[76] These included holding joint workshops and strategizing on research approaches, exchanging researchers, and engaging in cooperative projects on the atmospheric and ecological aspects of acid rain. For example, Robert Goldstein, a chemical engineer who came to EPRI in 1975 after working in the radiation ecology division of the US Oak Ridge National Laboratory, spearheaded the earliest cooperative ecological programs with the CERL's Gwyneth Howells during 1976.[77] In addition, he hosted a CERL researcher, biologist David Brown, for two years at EPRI headquarters to facilitate collaboration on research between the two organizations; the work of Brown, Howells, and Goldstein would prove extremely powerful in the debates on acid rain's ecological effects in the years ahead.

Significantly, EPRI also awarded the CERL several enormous research grants for studies of the problem.[78] The most elaborate and expensive of these projects was known as the "flying chemistry" program, which sought to understand the transformation of sulfur dioxide and nitrogen oxides into sulfate and nitrate over the North Sea. The research was billed as a way to determine whether reducing sulfur dioxide would result in a proportional reduction in the acidic form of the compound, sulfate, and received extensive coverage in the British and American press.[79]

The idea of questioning whether reductions in sulfur dioxide would result in proportional reductions in acidic pollutants had to do with the importance of fossil fuel pollutants known as "oxidants" in producing acid rain. Oxidants, such as ground level ozone, are the main pollutants responsible for the presence of photochemical smog. When exposed to sulfur dioxide emissions in the atmosphere, they act as catalysts for turning sulfur dioxide into sulfate and nitrogen oxides into nitrates, which eventually return to earth as "acid rain." In meetings between Starr and Chester during the early years of their cooperation, they came up with the idea to investigate whether these oxidants could be limiting factors in creating acid rain. If they were, then large reductions might result in only very marginal benefits for affected regions. This argument came to be known as the "linearity" issue, and by the late 1970s EPRI and the

CERL had successfully placed it on the scientific agendas of both countries. But although it came to have an impact on scientific and political debates, more so in America than in Europe, it was primarily in the realm of ecological effects that the EPRI and CERL partnership would provide the most dividends.

A "Silent Spring" for Acid Rain?

In her singular work, *Silent Spring*, Rachel Carson used the powerful image of a world without songbirds to argue society should stop spraying chemical pesticides. As acid rain became a widespread concern among European and North American governments during the 1970s, ecologists started to equate the lifeless lakes from acid rain to the soundless landscape described in *Silent Spring* as way of galvanizing support for reducing fossil fuel pollution. Norwegian researchers were the first to draw parallels between the two issues, notably the director of the SNSF project (Acid Rain's Effects on Forests and Fish, Sur nedbørs virkning på skog og fisk), Lars Overrein, who made the comparison explicitly in 1976. Citing the complete elimination of freshwater fish and declines in numerous species across inland lakes and rivers, Overrein argued that the documentation of severe ecological impacts should prompt concerns about long-term environmental damages from acid rain. Comparing the SNSF project to Carson's writings on pesticides, Overrein wrote:

> Where does this end? We may be approaching "the silent spring" which may be an appropriate term to refer to the outlook for many lakes and rivers in Scandinavia as the linking of these effects along trajectories from major emission sources in Europe becomes more and more apparent.[80]

Yet there was an important difference between acid rain and pesticides. Unlike the latter chemicals, acid rain was never shown to have any direct health impacts and could not draw upon immediate self-interest to motivate political action.[81] Like climate change, the impact on people would come indirectly from a change in their environment, which placed ecological work at the forefront of scientific and policy debates over the extent of environmental damages.

The potential importance of ecology to elucidating the costs and benefits of regulation led EPRI and the CERL to begin projects on the environmental effects of acid rain during the mid-1970s. These research efforts, even more than work on the atmospheric processes involved in

acid rain, became an important tool in creating doubt about whether society should reduce fossil fuel pollution. They were designed to discredit
Norwegian ecological research completed during the SNSF project as
well as the work of American ecologists in the White Mountains of New
Hampshire and the Adirondack Mountains of New York. Two projects
begun in 1975 and 1976—a CERL study of the Tovdal River in Southern Norway and an EPRI study of lakes in the Adirondack region of
America—would assume enormous importance as environmental scientists began pressing for national and international regulations during the
mid- to late 1970s. The research programs also reveal some of the problems with industry objectivity in environmental studies and the ways in
which the CERL and EPRI were able to exert an outsize influence on
scientific and political discourse about acid rain.

Chester initiated plans to conduct the first CERL field research in
Southern Norway, the area most affected by acid rain, as the Norwegian
SNSF project's first phase of research was coming to a close in 1975 and
1976.[82] He had expressed an interest in Norway's ecological studies of
acid rain soon after assuming the CERL directorship, and some Norwegian government officials were initially optimistic about the possibility
of educating Chester about their work.[83] In fact, Chester and Howells
had originally asked to collaborate with Norwegian researchers on ecological studies of acid rain in the summer of 1975.[84] That year, Howells
and another biologist at the CERL, Alan Webb, traveled to the region of
the Tovdal River to examine ongoing studies and begin preparations for
their own projects.[85]

Despite the initial interest in working alongside one another, tension
and animosity between CERL and SNSF scientists developed shortly after the publication of an SNSF report in *Nature* in 1976. The report
was based on research carried out primarily along the Tovdal River in
Norway to examine the impact of snowmelt on fish deaths. The SNSF
biologists' main argument in the article was that the fish died after rapid
snow-melting lowered the pH in the river, causing an impairment of the
active transport mechanism for sodium and chloride in the gills, which
are crucial for nerve conduction and enzyme processes.[86] Data on the
content of sodium and chloride in fish blood plasma was taken from
samples of fish at five field sites with various amounts of dead fish, ranging from none at the first site, to some at the third and fourth sites, to
extensive numbers of dead fish at the fifth site.[87] It demonstrated a clear
relationship between loss of sodium and chloride nutrients in fish blood
plasma and the number of dead specimens. Subsequent laboratory analyses showed that fish from the worst affected location, site five, that were

transferred to tanks with water of the same quality but lower acidity regained normal sodium chloride levels. The research provided crucial evidence not only connecting acidified precipitation to environmental harm, but showing a possible mechanism by which it could kill fish.

These findings received substantial attention from the international scientific community as well as government officials in both Europe and the US, and they caused considerable unease among CERL researchers.[88] Upon publication of the results in *Nature*, Webb and Chester asked the SNSF scientists if they could see the original data set.[89] After closely parsing the SNSF report, CERL scientists identified what they viewed as a serious deception on the part of the Norwegian biologists. While the *Nature* publication implied that each of the five field sites measuring fish deaths were located along the Tovdal River, the original report specified that the third site was stationed along a neighboring river, the Gjøv.[90] It's not clear that this was an intentional omission, and the removal of this site's data from the analysis would not have changed the overall trend of increasing fish deaths with loss of blood sodium and calcium.[91] However, CERL biologists believed that the fact that one of the measurements of fish death had been taken in a different river from where the acidity measurements of snowmelt were made cast doubt on the validity of the study; several of these scientists would later claim that this discovery led Chester to distrust all the SNSF ecological research in Norway thereafter.[92]

To what degree Chester seized on this discrepancy because of the CEGB's interests in disproving the seriousness of acid rain's environmental effects is unclear, but there's evidence he had been searching for any alternative explanations for the ecological damages observed in the region. At that time, Chester was conducting an extensive literature review on acid rain studies to determine what gaps remained in current scientific knowledge, which brought the work of a prominent Norwegian critic of the SNSF project to his attention: Ivan Rosenqvist. In the summer of 1976, just before the *Nature* article appeared, Rosenqvist had published a scathing critique of the SNSF research and the connection between acid rain and fish death in Southern Norway. He subsequently translated and forwarded this work to prominent magazines and journals like the *New Scientist* and *Chemical and Engineering News*.[93] In his report, which had been prepared at the request of the Norwegian Research Council for Science and the Humanities (NAVF), Rosenqvist argued that changes in land use were the true cause of increased acidity in freshwater ecosystems, not acid rain.[94] These included forestry and logging practices, which Rosenqvist believed acidified soils, and a reduction in traditional dairy farming, which he argued led to less grazing of vegetation in large pastures and an increasing biomass of trees and vegetation that also acidified soils.[95]

The underlying tension between the SNSF researchers and Rosenqvist, as well as professors on the NAVF who financed his work, appears to have been partly caused by the exclusion of university researchers from the SNSF project. As explained in chapter two, university scientists had originally been left out of the first phase of the research because Brynjulf Ottar, the director of the OECD project, and other leading acid rain scientists believed they lacked the ability to work as part of a team and integrate their work into comprehensive evaluations.[96] In fact, Rosenqvist's infamous report originated out of a meeting where university scientists were invited to comment on the findings from the initial phase of the SNSF project in an effort to obtain their views prior to publication of the work. At the gathering and in the months that followed, Rosenqvist was highly critical of the SNSF researchers' expertise, stating that he found it alarming that university scientists had not been included in the project and "researchers are embarking on fields where their skills are weak without seeking academic assistance where it may exist."[97] He also wrote to members of the Norwegian Parliament urging them not to throw "good money after bad" with the second phase of the research and instead seek the guidance of academics.[98] His perspective was shared by other NAVF scientists, who echoed these criticisms of excluding university professors in their forward to Rosenqvist's report.[99] The Ministry of the Environment was happy to involve academic scientists as consultants and cooperate with them. But officials such as Erik Lykke believed that having them participate directly would be problematic since the program's purpose was to "provide documentation for further international negotiation" under time pressure, a task which they were deemed ill-suited to complete.[100] The Secretary of the Norwegian Council for Scientific and Industrial Research, Hans Christensen, believed such differences were what truly motivated Rosenqvist's attacks on the SNSF project. He argued that university scientists engaged in basic research did not understand that the goal of such a large, applied, interdisciplinary project was not to "explain all the details" but rather "clarify the causal relationships within a limited period of time."[101]

The vitriol of the debates fractured the formerly close relationship of Ottar and Rosenqvist, who had been longtime friends, and led to an irreparable isolation of Ottar from the academic community of research scientists that lasted until the end of his life.[102] From the outset of this conflict, Ottar along with other members of the Norwegian Institute for

Air Research, the Ministry of Environment, and SNSF steering committee were troubled by the possibility that Rosenqvist's critiques could
be used by industry scientists to undermine arguments for the need to
reduce acid rain.[103] Their fears were not misplaced, as Chester invited
Rosenqvist to brief CERL scientists on his work shortly after it began
receiving widespread media attention.[104] Rosenqvist's arguments about
the possibility that land use changes could be responsible for acidifying
freshwater ecosystems in Norway soon became one of the primary research areas of the CERL and EPRI.

The following year, Chester sent Webb and two other CERL scientists, David Glover and David Brown, to the Tovdal River to undertake
their own investigations into whether acids in the soil produced by land
use changes were acidifying the river.[105] These were known as "weak
acids" as opposed to "strong acids" from atmospheric pollution, and
could be differentiated through chemical analysis. Despite Rosenqvist's
confidence in his claims, prior to this fieldwork there had been no previous data collected on weak acids in the Tovdal River area.[106]

The CERL researchers traveled to Norway during the spring snow
melt in 1977 to take samples over a period of several weeks in order to
test whether weak or strong acids were predominately contributing to
acidity in the region.[107] In contrast to Rosenqvist's hypothesis, Webb and
his colleagues found that during periods of snow melt, strong acids from
the atmosphere were indeed primarily responsible for the acid pulses
killing fish.[108] Upon returning to CERL headquarters, Webb prepared a
detailed report on the research that stated that the contribution of weak
acids to the pH of surface waters appeared to "decrease markedly as the
strong acid content increases."[109] This supported the Norwegian claim
in *Nature* that acid rain, not changes in forestry or land use, was the
dominant source of the environmental damages observed.

After submitting his report to his superiors to obtain permission for
publication of this research, Webb was called into a meeting with Chester and the manager of the Chemistry and Biology divisions, Anthony
Hart.[110] They asked Webb why he had included statements that the
strong acids from atmospheric pollution were predominantly responsible for the increase in acidity in freshwater ecosystems. When Webb explained that the CERL work at the Tovdal River supported this finding,
he was told not to "hand that over on a silver platter" to the Norwegians.[111] He subsequently sat on the report for several months, unwilling
to make Chester's requested changes, until Chester pressed him to submit it for publication in 1978 with the parts supporting the Norwegian
claims removed, which Webb eventually consented to do.[112]

Though this seems to have been one of the more serious examples of CERL scientists altering a scientific publication to support the CEGB's political position, the difficulty CERL researchers faced in releasing data that corroborated acid rain's damage to the environment was not unique to Webb. Simply informing Chester and CEGB leadership about studies that did not conform with the CEGB's stance on the science appears to have been a common issue.[113] These problems point to a culture where many scientists could not conduct independent research and in some instances had their work suppressed when it conflicted with CEGB economic and political interests.

While there are no clear incidents of suppression of research results within EPRI's ecological studies during this same period, its oversight of publications calls into question the purported objectivity of its sponsored work. Although university scientists who received research contracts were encouraged to publish the results of their findings, all publications had to be reviewed and approved by EPRI before they were submitted for peer review. In the event of a "fundamental disagreement" over the contents of a publication, the researcher could move forward with the paper only if no reference was made to EPRI's sponsorship of the work.[114] Such a procedure would, in theory, have allowed results to go forward that were not in the coal industry's interest. However, the practice of pre-approving publications was unusual among granting agencies in Western, developed countries, the vast majority of whom did not prescreen scientists' papers before publication. These include American institutions like the National Science Foundation and National Academy of Sciences as well as Scandinavian research councils such as the Scandinavian Council of Applied Research and Norway's Council for Scientific and Industrial Research.

In addition, EPRI's own scientists misrepresented the data and conclusions of their projects to the US government. Soon after Chester began pursuing Rosenqvist's idea that other environmental changes could be causing the damages attributed to acid rain, EPRI began a similar, $4 million project on the Adirondack lakes to demonstrate that the Norwegian conceptualization of acid rain's ecological effects was flawed.[115] Led by Robert Goldstein and a private environmental consulting firm, Tetrach Tech, the study examined three neighboring lakes that received similar amounts of acid rain but had strikingly different levels of pH in water samples.[116] The Sagamore, Woods, and Panther lakes were all within 30 kilometers of one another, and were chosen to understand how differences in plant communities and soil impacted the susceptibility of particular lakes to acid rain.

Since the discovery of acid rain in the late 1960s, ecologists had suspected that some ecosystems were more sensitive than others to acid rain. For example, soils in the eastern US and Scandinavia were largely composed of rock formations such as granite that have little "buffering capacity," meaning that they cannot counteract acidity in rainfall as it moves through soils and into a freshwater ecosystem. In contrast, soils with other types of rock, such as limestone, have a greater ability to buffer acidic precipitation; as a result, the pH of surrounding lakes is not as affected. In EPRI's study of the three lakes, the hope was to try to model exactly how certain properties of the soil and catchment area impacted pH in the freshwater ecosystem. While the model's general approach was viewed as scientifically sound by environmental researchers in the US and Scandinavia, there were concerns about its ability to evaluate long-term damage from acid rain and identify the "major processes" involved.[117] Like the line of research pursued by Chester based on Rosenqvist's critique of the SNSF project, the EPRI study hoped to prove that "weak acids" already present in the soil could be at least partly responsible for the damage instead of the "strong acids" deposited through rainfall containing fossil fuel pollutants.[118] Data collection began in 1977 with the participation of scientists from ten different institutions and concluded in 1981, but the study's results were already getting attention in Washington, DC, before the completion of the work.[119]

In 1980 and 1981, the US Congress held a series of hearings on acid rain to assess the current state of scientific research and what regulations were needed to protect the environment. During these hearings, EPRI scientists used the Adirondacks project to support very different conclusions than those of other researchers conducting the fieldwork. The discrepancy turned on a simple point: did the results of EPRI's ecology study show that certain geochemical factors made some ecosystems more susceptible to damage than others, or did they demonstrate that there were factors in the soil *other than acid rain* that were causing the damage? The results in fact demonstrated that the former was likely the case and explained the differences in pH readings between the lakes, but it was the latter argument that EPRI officials made again and again to policymakers.

For example, in testimony before the House of Representatives in 1980, EPRI scientist Ralph Perhac claimed that because the three lakes received the same amount of acid rain but had a wide range in pH "obviously some factor other than precipitation is responsible for the acidity."[120] That same year, the head of EPRI's ecological studies program, Robert Brocksen, offered nearly identical remarks to the Senate under questioning over whether the US needed to act now to prevent serious damage. Brocksen asserted:

> Certainly acid precipitation can change the character of lake and stream waters, but so can other factors. We know of areas where the acidity of rain is the same across the area, yet lakes within the region have markedly different pH values. In EPRI's lake acidification study, we have found three lakes in the Adirondack Mountains of New York State which have very different acidities . . . obviously some factor other than precipitation is responsible for the acidity.[121]

That factor, EPRI officials believed, might be residing in the soil, echoing Rosenqvist and the British CERL's recent arguments in Europe. Yet at this same hearing, one of the outside scientists contracted to work on the Adirondack project told congressional officials that while local factors were important in making ecosystems sensitive to acid rain, "there is very strong evidence that the regional cause of acidification indicated on these maps is the acid precipitation problem," a direct contradiction to Brocksen's testimony.[122] In his prepared written remarks, Brocksen also argued that there was no evidence that could support an overall increase in acidity in the past several years, claiming that Likens' findings were flawed and not "statistically significant." He also stated that there was contradictory evidence on whether Britain was causing acid rain in Scandinavia using data from the CEGB, which seemed to cast doubt on the existence of the problem throughout the developed world.

In an unusual move, Britain's CEGB submitted comments to Congress as well through the US National Coal Association, which were not balanced with any statements from Scandinavian researchers or other scientists from elsewhere on the continent. It emphasized that the only ecological damages shown to be caused by acid rain were in limited, "sensitive" ecosystems, going on to describe "major areas of uncertainty and difficulty in the interpretation of available data." They denied that any research had shown an increase in acid rain over the previous several decades, including Likens and Bormann's measurements from Hubbard Brook, Odén's measurements from the European Air Chemistry Network, and the more recent data from the OECD project. These arguments were bolstered by a submission from the Edison Electric Institute, one of EPRI's key funders, which pointed explicitly to the debate between Rosenqvist and Norwegian ecologists as illustrating "the very limited extent of our knowledge about lake acidification processes, and that the research currently being undertaken needs to be greatly expanded."[123]

By the end of these hearings, congressional officials were left feeling that the uncertainty surrounding acid rain's ecological threat required a more robust national research project on acid rain in the coming years to better determine what controls, if any, were necessary. The outcome had vast benefits for the energy industry's bottom line. As EPRI scientist Ralph Perhac explained in a later interview, "some very severe restrictions were being proposed, and just the fact that legislation, I mean that research was going on, tempered that. It held back Congress a little bit. So, the research played a role in that sense."[124] By suggesting that the ecological harms from acid rain were unclear through work such as the Adirondack lakes project, EPRI was able to mislead Congress about the actual state of the evidence that acid precipitation threatened freshwater environments.

In the coming years, both EPRI and the CERL would continue challenging the environmental research on acid rain's effects. Crucially, because EPRI and CERL researchers were the top funders of acid rain research in both countries, they were able to obtain key positions in future government studies and negotiations on the problem. In the US, EPRI scientists were invited to partner with the EPA in a new national program to investigate acid rain, known as the National Acid Precipitation Assessment Program (NAPAP). During this work, industry representatives were allowed to review draft documents on the state of the science concerning acid rain and add additional information to the reports. Robert Goldstein, who ran EPRI's project on the Adirondack lakes, served as the EPRI representative to the EPA group.[125]

These developments and others at the outset of the 1980s, including President Reagan's replacement of EPA personnel responsible for acid rain research when he took office, would lead ecologists such as Ellis Cowling to fear that the EPA was disregarding university research in favor of industry science.[126] Scientists under contract with EPRI, such as George Hidy from a private environmental consulting firm, also eventually helped write a US National Academy of Sciences report on acid rain without disclosing their relationship to the organization.[127] And President Reagan would eventually appoint Ralph Perhac, EPRI's director of environmental research, as one of three members of the administration's Acid Precipitation Task Force in 1982.[128] In Britain, where acid rain was not yet a domestic concern, CERL researchers went on to serve as scientific representatives to intergovernmental groups working to develop an international agreement on acid rain. They would exert influence on two major reports about acid rain at the OECD and UN, as well as take the lead in environmental research on acid rain at the European Communities.

Although industry expended considerable resources and effort to cement their authority in the acid rain debates, the responsibility for their eventual dominance over environmental science and policy on pollution must also rest with the British and American environmental agencies. Industry scientists were able to step into highly prominent and influential roles in research and advising in part because of the absence of government investments in acid rain studies during the 1970s in both these countries. This created a vacuum of expertise that EPRI and the CERL were only too happy to fill. As the coal industry marshaled their resources to strengthen a defensive research establishment on acid rain and began to build partnerships with the British and American governments, environmental scientists in Scandinavia were thus forced to grapple with how they could bolster their own work to combat the claims of EPRI and CERL experts. These efforts would lead Norwegian scientists to form their own unusual partnerships, as they began to consider reaching across the iron curtain to develop European-wide acid rain studies and generate international support for reducing fossil fuel emissions.

nuclear weapons development, but the environmental sciences were not immune from the Cold War's reach. Though there were some cultural exchanges and notable achievements in international earth science cooperation between the capitalist and communist blocs, such as the International Geophysical Year, many of these efforts were enmeshed with military objectives.[5]

On the surface, these Cold War divisions may seem like they should have doomed attempts to solve the problem of acid rain. Environmental researchers were already in conflict with the energy industries in Britain and the US by the mid- to late 1970s over whether acid rain was a serious enough ecological threat to justify cuts in fossil fuel emissions. Adding the need for communist cooperation would appear to create an additional hurdle for scientists and policymakers. But as the decade drew to a close, it was paradoxically assistance from the communist bloc that helped propel scientific and political momentum on acid rain, particularly in Europe. Joint scientific projects also helped decrease tensions between the two Cold War factions and served as the first concrete collaboration between the Eastern and Western blocs during the period of détente. These research efforts eventually paved the way for the first international diplomatic treaty on fossil fuel pollution, the 1979 United Nations (UN) Convention on Long-range Transboundary Air Pollution, which counted 29 European nations, the US, Canada, and the European Communities as signatories. Hailed as "an outstanding example of cooperation among countries with different economic and social systems," it served as a bulwark against attempts by the coal industries of Britain and the US to squash research on acid rain and efforts to regulate pollution emissions.[6]

The road to scientific cooperation across the iron curtain and the signing of the 1979 Convention was far from smooth for the pioneers of this work. The initial impetus came from Norwegian environmental scientists and diplomats, who had become more and more discouraged about the prospect for achieving emissions reductions from major Western emitters given the breakdown in OECD negotiations. Their hope was that improving knowledge of the atmospheric transport of air pollutants would lay the groundwork for eventually reducing acid rain on a European-wide basis. However, cooperation with Eastern Europe posed a host of unexpected obstacles for Norwegian scientists because of a large disparity in technological capabilities between the capitalist and communist countries as well as Cold War tensions over sharing data with the enemy. These challenges would eventually come to reshape the construction of environmental monitoring and the importance of relying upon atmospheric models in policy discussions of emissions reductions. Cold War politics arguably saved international cooperation in research and policy on acid rain, but it also revealed the limits of technology in solving the problem and called into question how scientists should function as experts in the diplomatic process.

Overtures to Eastern Europe

The idea to cooperate with Eastern European countries on acid rain was first proposed in 1973 by Brynjulf Ottar, who was wrapping up his first year as director of the Organisation for Economic Co-operation and Development (OECD)'s project on acid rain while serving as head of the Norwegian Institute for Air Research (NILU). The project was intended to establish whether and to what extent foreign nations were causing acid rain in Scandinavia, and had begun its first measurement phase in November 1972 with the participation of Austria, Belgium, Britain, Denmark, Finland, France, the Netherlands, Norway, Sweden, Switzerland, and West Germany. The results of the first phase confirmed Ottar's suspicion that East Germany, Czechoslovakia, and Poland could be the source of a considerable portion of the acid rain over Scandinavia in addition to the contributions from Western states.[7] Theoretically, it might have been possible to estimate their contribution by examining public data on fossil fuel consumption and emissions in Eastern Europe and then calculating the amount of sulfur dioxide transported from these countries to Scandinavia under various atmospheric conditions. However, Ottar believed that published estimates of emissions data from Eastern Europe were much too low, making it difficult to gauge the total amount of sulfur dioxide pollution reaching Scandinavia from these countries.[8]

Inaccurate or incomplete data from Eastern Europe threatened to pose problems in addressing acid rain on both scientific and political fronts. It would be far more difficult to develop an accurate sense of how much each country was contributing to the acid rain of its neighbors without better data from the communist bloc, which Ottar feared could further embolden the coal industry to object to any emission reductions. Over the course of his short tenure as leader of the OECD's project, Ottar had grown keenly aware of the resistance to reducing air pollution on the part of major emitters.[9] He was particularly frustrated by statements from Britain's Central Electricity Generating Board (CEGB) that sulfur dioxide pollution was in fact beneficial and posed no danger to the environment.[10] If Norway and other Scandinavian countries were to seek pollution reductions from Britain and other polluting countries,

Ottar believed they would have to be able to fill in any important gaps in knowledge likely to be exploited by major emitters. Accounting for the communist bloc's role in the problem would therefore be crucial.

To remedy the absence of good data on Eastern Europe's emissions, the following February he asked Erik Lykke, director general in the Norwegian Ministry of Environment, for assistance in obtaining this information through Norwegian diplomats in the Soviet Union, Poland, Czechoslovakia, and East Germany.[11] Lykke contacted the Ministry of Foreign Affairs to reach out to these four governments, but it was unable to establish the necessary contacts with government bodies or research institutes that might have such information.[12] The lack of existing institutional frameworks for cooperation amid ongoing Cold War tensions seems to have been largely responsible for the failure of Norway's embassies to assist Ottar and Lykke with their plan. Given these roadblocks, in June representatives from the Ministry of the Environment and Parliament met with Ottar and the secretary of the Royal Norwegian Council for Scientific and Industrial Research (NTNF), Hans Christensen, to decide how to proceed.[13]

With few diplomatic channels available to facilitate talks with the communist bloc, it was not clear who should make the next overtures to Eastern European governments. The intergovernmental organization that was perhaps best positioned logistically to take up the challenge was the UN's Economic Commission for Europe, though its involvement in environmental diplomacy was quite recent. The UN had begun to assume a much higher international profile on environmental affairs since the 1972 UN Conference on the Human Environment, but it was still largely run by a handful of sinecures with few resources and the Norwegians were unsure it was the appropriate forum to oversee their démarches to Eastern Europe. During these discussions the question of obtaining emissions data quickly broadened into conversations about what the larger significance might be in including Eastern European countries in the scientific research. All agreed that in the long run, it would be necessary to address air pollution problems on a European-wide basis, and contacts with Eastern Europe could serve as a crucial first step in achieving these aims.[14]

As these talks continued throughout the coming months, however, Ottar began to prepare a far more ambitious plan than just exchanging data. Instead, he became determined to formalize a scientific project on acid rain that would actively involve communist countries in studying the problem. For Ottar, there were several key reasons to include Eastern European countries in a program of research, but most important was combating the doubt-mongering of the coal industry in countries responsible for acid rain. Representatives of major emitters were likely to raise two issues in future disputes and negotiations on the problem: whether enough had been learned scientifically about acid rain to take action and whether the costs in reducing emissions outweighed the benefits to the environment.[15] To address the first issue, Ottar suggested that it would be crucial to try to guide both meteorological and ecological research to be able to fill in any important gaps in knowledge likely to be exploited by groups such as Britain's CEGB.

The second issue, he felt, was more challenging. Opponents of emissions limits would easily win in a game of one-upmanship on the costs versus benefits of reducing air pollution, since it would be far easier to quantify greater energy costs than the benefits of better health or less damage to fish. Ottar believed that addressing this problem would require not only studies of the ecological impact of acid rain, but also detailed information on which sources in Europe posed the most danger to Scandinavia. If these key pollution contributors could be identified throughout Europe then reductions could be more efficiently distributed, perhaps making Western emitters receptive to curbing their emissions.[16] For these reasons, Ottar proposed that the Ministry of Environment should seek not just to obtain adequate data on emissions from Eastern Europe, but also to invite these countries to participate in the OECD project. This could both reduce scientific uncertainties in atmospheric transport of acidic pollutants and address the cost-benefit argument. What started as a need to obtain emissions information thus evolved into a strategy for including Eastern European countries in scientific cooperation to precisely pinpoint the power plants most responsible for causing acid rain and lay the groundwork for emissions reductions.

Ottar's arguments proved persuasive within the Norwegian government, and in the summer of 1973 they commenced preparations to establish a formal cooperative project with the communist bloc. Since the OECD project had only just begun, Norway's Ministry of the Environment and NTNF decided it would be wise to at least approach the OECD's Environment Committee about the possibility of including the Eastern European countries in its ongoing research. At a meeting of the Environment Committee in June 1973, Lykke raised the question of whether the OECD might be willing to invite communist countries to participate in the project.[17] The Environment Committee had previously bandied about the question of collaborating with Eastern Europe in February 1972 during the planning phase of the project, but ultimately decided it would not be a suitable forum for direct cooperation.[18] However, the

Norwegian government evidently suspected that this would be the probable outcome of their overtures to the OECD and had already set a plan
in motion to use acid rain as a "test case" for East-West cooperation at
an upcoming preparatory meeting for the Helsinki Conference on Security and Cooperation in Europe (CSCE).[19]

The CSCE grew out of the Soviet Union's periodic offers to organize
talks on European security problems, which it had made since the 1950s;
Western governments, however, had been resistant to the idea of holding
such discussions. Ongoing political and military confrontations, coupled
with the Soviets' sometimes overt hope of using the conference to hin-
der Western European integration, contributed to reservations among
Western governments.[20] A new proposal from the Warsaw Pact governments in 1969 known as the "Bucharest Declaration" eventually paved
the way for preparatory talks on organizing the CSCE as détente eased
relations between the US and the Soviet Union.[21] The discussions were
divided into three general areas, known as "baskets": first, questions related to security in Europe; second, cooperation in the fields of science,
economics, and environment; and third, humanitarian issues.

Discussions on environmental matters took place within a Subcommittee on Environment, which held a series of ten meetings from October 22 through November 6, 1973, to prepare a text for inclusion
in the final conference declaration. Hans Christensen served as one of
the Norwegian delegates to these meetings and was instructed to try to
"sell" the long-range transboundary air pollution project through informal contacts with Eastern European countries.[22] After presenting a proposal for scientific cooperation on air pollution research, Christensen
met informally with representatives from the Soviet Union, East Germany, Czechoslovakia, and Poland. The Soviet delegate responded enthusiastically to the idea, telling Christensen that cooperation on acid
rain research was well-suited to the conference and could eventually
lead to investigations into the transport of all "hazardous substances"
throughout Europe. He also seemed interested in the Norwegians' reference to limiting emissions of sulfur dioxide, and later informed Christensen that he had contacted Moscow to further conversations about the
project in the coming months. Talks with East Germany were similarly
positive, although those with delegates from Poland and Czechoslovakia went less smoothly; the former due to a language barrier, and the
latter because Czech officials felt such a specific project did not belong
in the discussions. Overall, ten European countries lent their support to
Christensen's pitch by the end of the environment meetings: the Soviet
Union, Denmark, Malta, France, Britain, West Germany, Sweden, East

Germany, the US, and Hungary. West Germany spoke openly about the importance of having a concrete project to show that countries were taking the Conference seriously, which the US representative also emphasized.[23] The Norwegian proposal for a European-wide network to study acid rain was the only specific plan presented at these meetings.

Despite receiving widespread praise at the CSCE, privately, Norway's overtures to Eastern Europe alarmed many officials in the OECD and Western governments. At a meeting of the OECD Environment Committee one week after the CSCE discussions ended, there was "considerable turmoil and confusion" among the delegates about what exactly Norway's "intentions" were in reaching out to the Eastern European nations.[24] OECD officials warned Christensen that trying to arrange the direct participation of communist countries in the ongoing OECD acid rain project would meet strident opposition. Ottar, who also attended the meeting, subsequently inquired about whether he could at least receive authorization from the attendees to give emissions data from OECD governments to Eastern European countries in return for their information.[25] The OECD officials and delegates from member states agreed to this but reiterated that cooperation should remain limited to an exchange of data only.[26]

The tense conversations with the OECD member states made it clear that Norway's bid to put acid rain at the forefront of East-West talks would need to be handled delicately.[27] Nevertheless, Ottar, Lykke, and other members of the Ministry of the Environment were buoyed by Christensen's successes at the CSCE and Eastern European interest in becoming involved in a collaborative project. They decided to proceed on two fronts. First, at Lykke's suggestion, the Norwegian government agreed to sponsor a meeting of scientific experts in Norway to explore the technical aspects of the program. Simultaneously, the Norwegian government also began reaching out to their Nordic allies about the proposed cooperative project with the communists.[28] During a consultation in December with representatives from Finland, Sweden, and Denmark, Lykke and Ottar attempted to explain the potential research in greater detail and gain their support for cooperation with the Eastern European countries, beginning with the proposed scientific meeting in Norway of relevant experts.[29] A consensus was reached that such a meeting could be held the following year, but should be limited to a "review and exchange of experience" in atmospheric research on air pollution for all interested countries.[30]

With the support of their Nordic allies, the Norwegian Ministry of Environment subsequently composed a formal letter to the OECD's

Steering Committee for the acid rain project conveying their intention
to submit a proposal to the CSCE on the creation of a European-wide
air pollution monitoring program.[31] Their success in capitalizing on the
Helsinki talks to make contact with Eastern Europe highlights the importance of détente, and the CSCE in particular, for opening the door
for scientific and technological cooperation on acid rain. But it also reveals the importance of Norwegian scientists and diplomats in the process of easing relations between Eastern and Western governments. The
wary reactions of their fellow OECD members as well as their Scandinavian neighbors are evidence of the considerable mistrust Western
governments had toward the communist bloc even as they participated
in the roundtables of Helsinki. This mistrust and skepticism would become even more evident as Norway sought to develop a technical plan
for environmental monitoring of acid rain with the Soviet Union as well
as solicit firm commitments to join the scientific research program from
countries across the continent.

Environmental Monitoring and the Limits of Détente

The Norwegian government's venture to establish acid rain research on
a European-wide basis sharply tested the easement of relations between
the capitalist and communist blocs as well as the fledging international
apparatus for environmental diplomacy during the mid-1970s. As the
Helsinki talks proceeded over the course of 1974, Ottar, Lykke, and
other scientists and diplomats sought to establish a formal partnership
with the Soviet Union on an environmental monitoring program through
the UN's Economic Commission for Europe.[32] Its environmental directorate was led by Amasa Bishop, an American nuclear physicist who
took over the newly formed division after several stints brokering international cooperation in atomic research. Upon learning of the proposed
project, Bishop agreed to pass along Ottar's credentials to the division's
Soviet Union delegate in the hopes of beginning a dialogue between scientists from both countries.[33] Shortly thereafter, the Soviet government
agreed to begin talks on forming an acid rain research project and nominated a geophysical scientist to collaborate with Ottar on developing the
program.[34] Their first meeting took place in November that year, during
which the pair discussed not only the possibility of exchanging information between their respective countries but also embarking on a collaborative research program. It would represent the first intergovernmental,
permanent form of scientific cooperation between the Soviet Union and
a Western government.[35]

These early scientific and diplomatic feelers were not disclosed to the public, making the Soviet Union's later embrace of environmental cooperation on acid rain seem odd to observers at the time given the lack of scientific or public interest in the issue among communist states. In the context of the détente talks, however, the Soviet Union's support for the research program becomes much more understandable. Norway's acid rain proposal gave the country an opportunity to prove their commitment to lessening tensions with Western governments, one that was nowhere near as contentious as the other Helsinki basket issues of nuclear armaments and human rights.

The communist bloc's support for cooperative acid rain research was made clear when Ottar convened the first expert meeting on the possible UN project later that year.[36] He intended to use the meeting both to assess interest in creating a European monitoring network among scientists in Eastern and Western Europe as well as to discuss the current state of the field in order to prepare a program of research.[37] Ottar's colleagues in the Ministry of Environment and the NTNF hoped that wide attendance from scientists and environmental officials across the continent would build upon the momentum of the Helsinki talks and generate enough political support among Western governments for the UN to formally approve the creation of a monitoring network.[38]

To the amazement of the Norwegians, more than fifty participants representing 16 European countries came to the December meeting, including delegates from Poland and Hungary in Eastern Europe, the US and several international organizations. While the Soviet Union opted not to send representatives at this early stage, the Comecon headquarters in Moscow did prepare a statement emphasizing the importance of tackling these problems on a European-wide basis.[39] Though acknowledging that acid rain had only recently been discussed among the communist countries, the statement noted that "the general reduction of tensions creates real ground for a practical implementation of broad and long-term cooperation with the developed capitalist countries" on environmental problems.[40]

Over the course of the gathering, the scientific delegates each reported on current atmospheric studies within their respective nations and discussed the aspects of air pollution transfer most in need of further investigation. Although the OECD's project on acid rain had made a substantial contribution to understanding the long-range transport of air pollutants, there were still many remaining questions about the physical and chemical mechanisms of their circulation. For much of its history, meteorological research had concentrated on large-scale mechanisms of atmospheric processes and short-term weather forecasting, the latter of which was aided by the advent of computers.[41] Atmospheric circulation that took place in the middle range of these processes, known as the "mesoscale," was the most important in understanding pollution transport but had received the least amount of scientific attention.[42] Mesoscale atmospheric processes are influenced by terrain and weather patterns, and it was not clear to what degree pollution dispersal depended on factors such as topography, vertical height of the emissions stack, wind fields, and atmospheric turbulence. In addition to these physical mechanisms, the chemical reactions involved in oxidizing sulfur dioxide to sulfuric acid or ammonium sulfate were not well understood, particularly the importance of catalytic or photochemical processes.[43]

Ottar's initial proposal for a "European Monitoring System" on acid rain circulated at the December meeting built off the research undertaken during the OECD project and laid out a plan for addressing some of the remaining questions about mesoscale atmospheric processes on a European-wide basis.[44] It involved the use of emissions inventories, monitoring stations, and atmospheric models to track pollutants across the continent from their source to eventual deposition. Countries would provide data on their fossil fuel emissions in mapped grid squares of about 100 square kilometers. Measurement stations with new sampling technologies would then be constructed in all participating countries to examine rainfall for pH and sulfate; specialized, limited duration programs would measure additional pollutants, such as ammonia, nitrate, magnesium, and sodium, which would assist in improving understanding of the multiplying role of various pollutants in acidity.[45] Based on this data, atmospheric physicists could develop and test models of pollution dispersion to refine their precision. Ottar's goal was to have the network not only approximate the general dispersion of pollution but also gain enough specificity to calculate the contribution of particular source areas to acid rain during a given set of weather conditions.[46]

The plan was positively received by the majority of scientific representatives at the meeting, and Norwegian officials subsequently petitioned the UN's Economic Commission for Europe to begin the technical work to develop the network.[47] In light of the constructive exchanges at the Oslo gathering and Norway's offer to facilitate the development of the project, representatives to the organization's environment division approved a preliminary work program in February 1975.[48] In addition to Norway and the Soviet Union, the countries of Austria, Canada, Denmark, Finland, Hungary, the Netherlands, Poland, Portugal, Sweden, and West Germany were the first to commit to joining the project.[49]

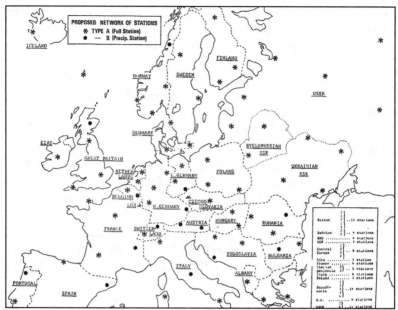

FIGURE 4.1 The initial proposed network of measurement stations, July 1975. "Letter from Brynjulf Ottar to Erik Lykke," July 23, 1975, RA-S-2532-2-Dca-L0292, Norwegian National Archives.

However, despite the success of these first forays in scientific and diplomatic cooperation, the Norwegians soon faced huge obstacles in getting the European-wide research effort off the ground. These challenges came from all sides: skeptical Western governments, the inexperienced UN Economic Commission for Europe, and Soviet officials mistrustful of sharing data with "the enemy." Some of the first pushback came from Western European governments who refused to formally commit to the UN's acid rain project. These included France, Belgium, and Switzerland, which eventually came onboard in 1976, as well as Britain, which did not join until 1977 and was the last major Western power to agree to participate.

Though these countries frequently sent representatives to the preparatory technical meetings, their attendance masked their deep cynicism about the monitoring program. In addition to expressing doubts about the role of foreign fossil fuel pollution in causing acid rain, representatives from these four countries and other members of the European Communities questioned the Soviet Union's seriousness in participating in the monitoring program as well as the quality of information that could be expected from Eastern Europe.[50] Their concerns were not entirely misplaced. Ottar's attempts to receive information from Eastern European countries in order to finalize plans for the network's construction frequently went unanswered.[51] Many Eastern European governments were not in possession of a substantial portion of the information he sought, such as what stations might be able to supply data to the network, what pollutants were routinely measured, and the type of measurement techniques used.[52] Of greatest concern was the fact that the communist bloc governments did not have emissions data readily available in a grid system, which was required to track emissions from their sources; they sought to produce the necessary charts over the course of 1975.[53]

The Norwegian efforts to establish a European-wide acid rain project were also made more difficult by the outsized ambitions of the executive secretary of the UN's Economic Commission for Europe, Janez Stanovnik. Buoyed by the new authority and international prominence bestowed on the agency through the détente talks and eventual 1975 Helsinki Accords, Stanovnik had begun to press Ottar to include the US and Canada in the monitoring program. Part of his motivation in making this request was to raise the profile of the intergovernmental organization. However, Stanovnik was also receiving pressure from the US State Department to include the American government in any kind of collaboration with the Soviets. State Department officials reached out to Lykke directly to lobby for their inclusion in the acid rain program because of the recent discovery of acid rain in North America.

Lykke and Ottar believed this was a terrible idea. They were wary of scaring off their Soviet partners and concerned about the UN assuming such a high-profile negotiating role in the acid rain program. Ottar especially thought including the Americans was far too risky given the difficulties with simply getting both Eastern and Western Europe onboard and the UN Economic Commission for Europe's inexperience with facilitating cooperation on environmental issues. As he explained in a letter to Lykke:

> I generally do not mind, in and of itself, if people have big ideas, but it worries me that Janez does not seem to be interested in my cautious objections. Based on what has happened in recent days, it is obvious that the UN ECE has not even consulted with the USSR regarding the proposed meeting. . . . It thus appears that their enthusiasm is greater than their ability to organize.[54]

Lykke informed the US State Department that Norway was not interested in the US's involvement at this stage, though their scientific experts were welcome to attend forthcoming technical preparatory meetings.[55] He explained that the Norwegian government wanted to establish the program on a European-wide basis before potentially expanding it to other countries, as Ottar recommended.[56]

The prospect of doing so through the UN's Economic Commission for Europe was made further difficult because of its dire financial situation at the start of the planning process. With an extremely small budget and staff, the UN's Economic Commission for Europe could not contribute to the startup costs of the monitoring program. Consequently, the Scandinavian Council for Applied Research agreed to provide a grant to NILU for 300,000 kroner in 1976 to assist with preparations for the program.[57] The non-Scandinavian Western governments, while eventually paying for monitoring stations within their countries, did not assist with any of the preparatory costs. The bulk of the budget came from NILU itself, which earmarked 1 million kroner of its own funds during 1976 and 1977 to defray the startup expenses.[58]

Norwegian scientists and environmental officials thus shouldered a considerable burden in organizing and financing the network because of the mistrust of Western governments and the UN's ineptitude. Despite these obstacles, Ottar persevered in his attempts to solicit information from the communist bloc and finalize preparations for the network. He believed it would be very difficult for governments to turn down a concrete proposal on the heels of the 1975 Helsinki Accords and was determined to use the détente talks to his advantage in order to get the acid rain program off the ground, even if it needed a powerful scientific and financial push from NILU in order to ensure its establishment.[59] Once it was clear the monitoring network was feasible, he hoped, reluctant Western European nations would come on board.[60] Based on the information he was able to gather at the Oslo expert meeting and from his queries to Eastern Europe, Ottar presented a formal program of research to the UN's Economic Commission for Europe at an October meeting in Geneva in January 1976. Norway subsequently received approval from the UN to put together a task force of scientists to begin setting up the monitoring program.

But although Norwegian environmental scientists and government officials may have been able to use Cold War negotiations to create an avenue for starting the European-wide program, the ongoing conflict soon caused a number of problems as Ottar sought to get the program running. After the discussions in Helsinki, it proved incredibly difficult to simply communicate with Soviet officials across the iron curtain, either through correspondence or telephone.[61] The Norwegians also came to believe that scientists and environmental officials in the Soviet Union lacked the requisite expertise in atmospheric science.[62] In fact, as Ottar was drafting the monitoring program proposal in 1975 for presentation to the UN's Economic Commission for Europe, he decided to cancel meetings in Moscow between himself and Soviet experts after conferring with Stanovnik and the Norwegian Ministry of the Environment about the futility of talking with the Soviets given the disparity in research experience between the two countries.[63] Ultimately, Norway was granted complete leadership of the project as a result of these issues.[64]

But perhaps the most significant obstacle to smooth Soviet and Norwegian collaboration was lingering Cold War suspicion. As Norwegian scientists discovered in the course of finalizing the research plans for the monitoring network, the Soviets were wary of releasing data on their power plants. This information would be necessary to have any hope of tracking emissions thought to be responsible for acid rain. Anton Eliassen, a Norwegian atmospheric physicist recruited to the project in 1975 to guide the modeling aspects of the program, learned about the Soviet trepidations firsthand. Eliassen had previously worked with Ottar on the OECD project shortly after receiving a degree in meteorology from the University of Oslo.[65] He was the son of Arnt Eliassen, a renowned atmospheric physicist who had assisted John von Neumann in developing the first computer-generated weather forecasts after the Second World War.[66]

Described as the "godfather" of the UN's monitoring program and a brilliant mathematician, Eliassen had already achieved recognition for his modeling expertise through his work on the OECD project.[67] He traveled to Moscow shortly after he began assisting Ottar with the program's preparations to discuss the possibility of receiving emissions data from the Soviets. In a room with a dozen Soviet statesmen and atmospheric scientists, Eliassen was told in no uncertain terms that they would not be handing over a map of their emissions sources. Informally, after some vodka had relaxed the mood slightly, the Soviets let slip that releasing the locations of their country's emission sources would be akin to giving a map to the Americans for places to bomb the country.[68] Though Eliassen remarked that it was likely the Americans already knew where to strike, the Soviet representatives did not waver in their decision. These issues with Soviet security concerns, combined with the litany of challenges in working with Western and Eastern governments, resulted in a marked shift away from an emphasis on building a measurement network. The investments in technologies and release of data required a confidence in cooperation that was not present despite the opening in communication provided by détente.

Pollution Modeling without Target Maps

As Ottar and Eliassen began to finalize plans for the environmental monitoring network in 1976 with the UN Economic Commission for Europe's scientific task force, these constraints on data availability and the communist bloc's relative inexperience in air pollution research led to a reconsideration of their approach to the project. Rather than seeking to create a similarly robust set of monitoring stations in Eastern Europe, Ottar instead decided to emphasize the use of newly developed atmospheric models. The shift was not an obvious one. Though atmospheric physicists in Norway had pioneered the creation of atmospheric models throughout the twentieth century, even helping American meteorologists use them during combat operations in the Second World War, the original plans for the UN's environmental monitoring program involved a far larger set of data from power plants and monitoring stations.[69] This would have aided scientists in seeking to identify the precise locations of emissions contributing to acid rain. Models could help fill in some of these gaps in data, but they presented their own challenges, especially when used in the policymaking process.

Ottar began to adjust plans for the UN's environmental monitoring program in light of the constraints of working with Eastern Europe in the spring of 1976, outlining the rationale for emphasizing models over expanding measurement stations because of these countries' limited resources and the Soviet's security concerns.[70] In discussions with his Norwegian colleagues, he argued that the use of models did not depend on national measurement programs being equally strong in all countries as long as a select number of countries built up enough stations. With a robust Nordic network already in place from the OECD project, Ottar believed this was the best way to proceed and developed customized research plans based on the participating countries' interests and capabilities. The collected data could be combined with current trajectory estimates and emissions surveys available from the OECD project and serve as the point of departure for applying dispersion models until the number of monitoring stations could be increased.[71]

In concert with greater utilization of models in the UN's acid rain monitoring program, Norwegian researchers also struck a deal with their Soviet counterparts that addressed their insistence on not revealing

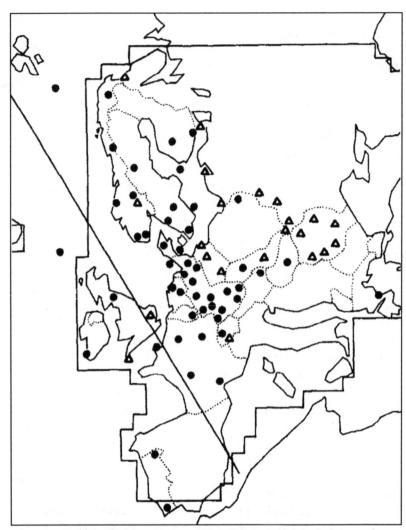

FIGURE 4.2 Map of the UN's European environmental monitoring network during its first years of operation. Note the Soviet Union's monitoring stations along their border, rather than the country's interior. The black dots represent stations whose data were used in a particular modeling analysis, while data from the triangles had not been used in this specific instance. Anton Eliassen and Jørgen Saltbones, "Modelling of Long-range Transport of Sulphur over Europe: A Two-year Model Run and Some Model Experiments," *Atmospheric Environment* 17, no. 8 (1983): 1457–1473.

the locations of their power plants or releasing data on the country's air quality. Rather than providing a map of emissions sources, the Soviet Union would be allowed to construct monitoring stations along the perimeter of the country and simply report emissions "fluxes" over its western border.[72] This compromise was later contested by the US, but it failed to persuasively argue that the Soviet Union needed to divulge more information since the overall purpose of the program was to track transnational pollution trajectories.[73]

Ottar's revised plan was accepted without modification at the program's first task force meeting of national scientific experts in Geneva during May 1976.[74] The task force recommended that Norway should convene a second, final gathering of atmospheric scientists to assist countries who wished to construct monitoring stations and determine the most appropriate atmospheric model for use in the program.[75] This second meeting was scheduled for October before the expected launch of the first measurement phase in January 1977.[76] Formally named the "Cooperative Programme for Monitoring and Evaluation of the Long Range Transmission of Air Pollutants in Europe," the network was scheduled to operate for at least the next few years.[77] Representatives to the UN's Economic Commission for Europe formed a steering body of scientific representatives from member countries to coordinate the research effort, and most countries planned to have their first measurement stations in operation by September 1, 1977.[78] NILU was tasked with analyzing and calibrating measurement data while two meteorological centers, one at the Hydro-Meteorological Institute in Moscow and the other at the Norwegian Meteorological Institute, were chosen to coordinate the atmospheric modeling work.[79]

But while the Norwegians may have solved their delicate Cold War dance with the Soviet Union to secure its involvement, the compromises they made to the network proved troublesome as it got underway. In the early years of the network's operation, meteorologists in Moscow would send their calculations of transboundary pollution flows by telex to meteorologists in Oslo every Tuesday and Thursday. The aim was to integrate the results from the Soviet border into the monitoring program's atmospheric model for the entire continent. In practice, however, this turned out to be nearly impossible. Eliassen, who was put in charge of modeling at the Norwegian Meteorological Institute, discarded the flux model results because they were so difficult to incorporate mathematically into his atmospheric model. He found greater success simply by guessing the extent and location of the Soviet Union's emissions and checking these numbers against measurements of pollutants produced by the monitoring stations.[80]

Because of the challenges in reconciling the Soviet's problematic flux data with the network's model as well as the unreliable data from elsewhere in Eastern Europe, atmospheric physicists at the Norwegian Meteorological Center were unable to complete the anticipated improvements in modeling pollution transport. After the UN program began operations, Eliassen and his colleagues made several revisions to the model based on data from the network, including changes in the velocity of dry deposition, the mixing height of pollutants in the atmosphere, and the rate of pollution "washout" through deposition. However, these enhancements were counterbalanced by errors and missing data from the communist bloc.[81] A number of errors were introduced from using data obtained through different sampling techniques, which Ottar and other atmospheric chemists at NILU progressively corrected through proper calibration of monitoring technologies at the new stations in Eastern Europe. While this work was carried out, atmospheric physicists at the Norwegian Meteorological Institute struggled to refine the accuracy of their models and check their predictions.[82] As a result, the monitoring program's initial findings did not come close to identifying the key sources of acid rain pollution as Ottar had originally envisioned.

These shortcomings notwithstanding, the Norwegians were buoyed by their success in capitalizing on détente to advance cooperation on acid rain research. The next logical step to Lykke, Ottar, and others was to see whether political negotiation might follow on the heels of scientific diplomacy. Perhaps the Soviet Union, they wondered, could prove useful not only in pressuring countries like Britain and France to continue sponsoring scientific work on the problem but also in reaching some form of agreement to reduce fossil fuel emissions. The idea to expand their alliance with the communist bloc into the political realm took root as the network became operational in 1977, and seemed an attractive option in light of the continued resistance to reducing acid rain among Western governments.

5   Environmental Diplomacy in the Cold War

A decade ago, the idea that a group of nations—with clear differences of interest and ideology—could unite to protect the environment would have seemed
doubtful at best. Indeed, many would have said it was impossible. Yet we have
made a significant start. And that should be a source of pride for us all. Even so,
the question remains: are we acting rapidly enough? . . . We must recognize that
the earth does not march to a human clock: the ozone now disappearing from the
stratosphere will not be replaced for a century or more; the lakes whose life has
been destroyed by acid rain could remain in that condition for decades, even if
we eliminate immediately all the emissions that produce it. . . . We must broaden
the scope and hasten the tempo of our cooperation—for human activities are
shrinking the size of our global commons at a geometric, not an arithmetic,
pace. The humankind stands at twelve o'clock. What lies beyond, we are not sure.
    Douglas M. Costle, administrator, US Environmental Protection Agency, 1979.[1]

As the United Nations' (UN) environmental monitoring network for acid rain prepared to begin operations
in the summer of 1976, officials in Britain's Foreign and
Commonwealth Office issued a worrisome report about
the Scandinavian alliance with the Soviet Union on environmental matters. Noting the Soviet Union's longstanding desire to disrupt the political integration of Western
Europe through the European Communities, experts on
Eastern European affairs argued that the Soviets' cooperation on transnational pollution might be driven by larger
geopolitical aims. The acid rain problem, they wrote,
was "a source of some friction within the Western camp,

notably between the UK and the Scandinavians, and Soviet motives here could be to aggravate dissentions."[2] It was crucial, the report argued, to keep a close watch on further efforts to move environmental problems into East-West discussions.

They appeared unlikely to get much support from the US government in blocking the increasing attention to international pollution. Since the Environmental Protection Agency (EPA) was only just beginning investments into research on acid rain, the State Department in the Carter administration had worked closely with university scientists to try to learn from the substantial European experience in atmospheric and ecological investigations.[3] The Department's Bureau of Oceans and International Scientific and Environmental Affairs had also expressed a strong interest in promoting environmental cooperation on pollution within intergovernmental organizations, and saw acid rain as a natural component in any discussion of environmental problems with the communist bloc. As one of its top officials, Lindsey Grant, wrote to leaders at the UN, European Communities, and Organisation for Economic Cooperation and Development (OECD) in 1976, the US and Canada were eager to integrate their bilateral project on acid rain into the ongoing work of these institutions. More importantly, diplomatic progress on acid rain would give immediate content to the environmental tasks assigned to the UN by the recent Helsinki Conference and "offer some real hope of bringing the USSR and East Europe into a cooperative effort."[4] For the US, acid rain could consequently facilitate the détente process at little domestic cost in light of their own government's efforts alongside the Canadians.

The US and Scandinavian interest in promoting acid rain cooperation with Eastern Europe appeared at an inopportune time for the British government and its nationalized power industry, overseen by the Central Electricity Generating Board (CEGB). The European Communities was then in the process of drafting regulations for sulfur dioxide pollution, putting pressure on Britain both from its Western allies and the communist bloc. The European Communities had first begun to consider standardizing environmental policies among its member states after the 1972 UN Stockholm Conference. Though the oil crisis of 1973 paused attempts to address these issues, in 1974 the European Communities renewed its efforts. One major proposal became known as the "sulfur fuel oil" directive.[5] Originally drafted in 1973, it called for members of the Communities to limit the content of sulfur in fuel oils with the aim of decreasing sulfur emissions. The decision to implement air pollution policies at the Communities' level stemmed from concerns that member states who had already passed domestic legislation limiting sulfur contents would be putting their industries at a competitive disadvantage with other governments who had no such regulations.[6]

Britain was by then the only government in the European Communities without national regulations on sulfur content in fuels or other technological means of reducing air pollution.[7] Instead, the CEGB continued to rely on high stacks to disperse pollution and lower ground-level concentrations. Seemingly besieged on all sides, CEGB officials and scientists developed a plan for combating pressure to reduce their emissions within these intergovernmental organizations. Their tactics would have a profound impact on the course of scientific and diplomatic progress on acid rain as Western governments sought to leverage détente politics to move toward an international agreement on the problem.

Economic or Environmental Catastrophe

The idea to build on scientific cooperation with the communist bloc to obtain an international treaty on acid rain originated during consultations between Norway and Sweden in November 1975.[8] With the Helsinki Accords signed a few months prior, the two governments decided to set up a joint Norwegian-Swedish working group to devise a long-term strategy for securing an international agreement on transboundary air pollution.[9] As part of this strategy, environmental officials from both countries sought to exert political pressure on Britain through the European Communities' sulfur fuel oil directive negotiations. Though neither were members of the Communities, Norwegian officials believed they could argue that the sulfur fuel oil directive should include references to transboundary air pollution to facilitate future negotiations with their countries.[10] Swedish representatives agreed it would be worthwhile to try this approach, and both governments reached out to the European Communities to discuss the matter. The two governments were especially troubled that the British government, notably representatives from the CEGB, had pushed strongly to receive an exception to the directive if high stacks were used to disperse pollution—a key cause of acid rain.

Their diplomatic overtures were well received by European Commission officials, who worked in the executive arm of the European Communities and were responsible for finalizing the sulfur fuel directive. The officials subsequently agreed to insert provisions on transboundary air pollution given the Norwegian and Swedish concerns about acid

FIGURE 5.1 Members of the European Communities from 1973 to 1981, highlighted in black. "Enlargement of the European Union," Wikimedia commons, May 28, 2009.

rain.[11] It was submitted in December 1975 to the Council of the European Communities, which negotiates and adopts laws for member states based on proposals from the Commission. As Britain had requested, the European Communities did include an exception to requirements to use low-sulfur fuels for installations that utilized high stacks. However, the directive went on to require that member states using the high stacks exemption should "take all necessary measures" to ensure these installations did not cause transboundary air pollution.[12] It also required that such installations maintain a reserve quantity of low-sulfur fuel in case ground level concentrations exceeded levels that could harm human health or the environment.[13]

As soon as members of the CEGB received a copy of the submitted directive, they reached out to Britain's Department of Energy to express alarm about its provisions regarding high stacks in relation to transboundary air pollution. They also feared that a legal requirement to limit sulfur in fuel oil could open the door for similar restrictions on the sulfur content of coal.[14] Britain relied heavily on coal for power production, and the CEGB believed that the cost of installing pollution control technology in its power plants would greatly increase electricity prices and potentially cripple the British economy.[15]

The organization was so concerned that the CEGB chairman, Arthur Hawkins, contacted the vice-president of the European Commission in early 1976 to express his displeasure with the directive and fears about its extension to coal.[16] Hawkins argued that if the directive were applied to coal then half of member states' current reserves would be unusable without desulfurization, resulting in a Community-wide economic catastrophe "as great as it is possible to imagine."[17] The European Communities took the CEGB's objectives very seriously, and sent representatives to meet with the CEGB in Britain and discuss their objections to the directive.[18]

The CEGB guide appointed to manage this visit was Peter Chester, director of the CEGB's Central Electricity Research Laboratories (CERL), along with staff scientists Doug Lucas and Tony Clarke. This decision was deliberate. By 1976, Chester had already earned a reputation for utilizing CEGB research to cast doubt on environmental studies documenting the role of fossil fuel emissions in causing acid rain and its ecological effects. To argue for the imprudence of the directive, Chester arranged for European Commission officials to tour a power plant with high stacks and visit the CEGB's research laboratories for meetings and discussions on meteorological dispersion from such plants.[19] In the course of these conversations, CERL scientists claimed that high stacks were an effective form of pollution control not only for local air quality but also for transboundary air pollution and acid rain.[20] In fact, Chester, Lucas, and Clarke told the European Commission representatives that their studies proved high stacks were equally successful in reducing pollution that crossed national boundaries. They claimed that high stacks dispersed pollution so well that just one percent of their power plant emissions fell on foreign soils. How the CERL arrived at the number was not explained, nor was the fact that a more relevant figure would be the proportion of pollution in Norway and other countries originating from Britain.

The European Commission officials, while not versed in the details of the scientific debate, appear to have been largely unconvinced by the technical arguments of Chester and the CERL scientists. After their visit, Commission representatives concluded that the CERL's statements on

Britain's role in causing acid rain were not credible, and they also had doubts about the CEGB's cost estimates of controlling sulfur dioxide in emissions.[21] However, when the European Council finally met to review the directive, the provision for transboundary air pollution proved very contentious. Danish and Dutch officials tried to persuade their colleagues that removing the stipulation would encourage governments to disperse their pollution to other countries.[22] But despite their support of the Norwegian and Swedish efforts, disagreements about the sulfur fuel directive persisted for years among the European Communities members and it was eventually withdrawn from consideration.[23]

Although the sulfur fuel oil directive turned out to be an ineffective avenue for negotiations, Scandinavian environmental officials were soon presented with another opportunity to compel international action on acid rain. In March 1976, Soviet leader Leonid Brezhnev sent a proposal for holding a series of "pan-European" conferences on energy, environment, and transportation to diplomats from across the continent. Many governments saw Brezhnev's proposal as a means of diverting attention away from security and humanitarian issues at follow-up negotiations in 1977.[24] However, Norway's director general for the environment, Erik Lykke, thought the Soviet Union's suggestion for an environmental conference provided a fortuitous opening to start negotiations on reducing acid rain with both Eastern and Western governments. Lykke discussed the option of approaching the UN about this conference with representatives from Sweden's Agriculture Department, who agreed to make a joint pitch for holding Brezhnev's "environmental" conference on the topic of transboundary air pollution during the forthcoming meeting of the UN's Senior Advisers on Environmental Problems in February 1977.[25]

There was one hurdle, however, that threatened to derail international negotiations over acid rain. The largest scientific study on the atmospheric transport of fossil fuel pollutants, overseen by the OECD in partnership with the Norwegian Institute for Air Research (NILU), was then in its final year of measurements. Early findings from the project had clearly demonstrated that fossil fuel pollution from one country could be transported to others and fall as acid rain. But precisely how much nations contributed to their neighbors' pollution needed to be agreed upon by scientific representatives from all countries participating in this work. These numbers could provide the necessary evidence for an agreement or undermine it altogether, which set the stage for a charged, high-stakes finalization of the OECD study's results among British and Norwegian researchers.

Scientists as Diplomats

The CEGB, still reeling from its dispute with the European Communities on the sulfur fuel oil directive, was watching the finalization of the OECD report closely alongside Britain's Central Unit on Environmental Protection. The latter had primary responsibility for coordinating international pollution policies for Britain, and its director Leslie Reed was receptive to the CEGB's concerns about the report's impact.[26] Following a series of initial meetings in 1976 among OECD scientific delegates to discuss a draft report on the project's findings, Reed and members of the CEGB conferred about the accuracy of the report and its potential to implicate Britain in Scandinavia's acid rain problem. CERL scientists had run many parts of the project within Britain, and under Chester they had become leaders in environmental research on the problem. As a result, when several scientists at the CERL raised questions about calculations concerning the amount of sulfur dioxide Britain exported to Norway, British representatives to the OECD refused to sign off on the current report from the study.

Their rejection of the draft report led Brynjulf Ottar, director of NILU, to organize an informal meeting between meteorologists from Britain, Sweden, and Norway at NILU to review the study's calculations in May 1976.[27] The British delegation was led by Ron Scriven, a meteorologist at the CERL who had already earned a reputation for trying to poke holes in Norwegian atmospheric studies.[28] Although the meeting began with the intention to agree on Britain's contribution to acid rain in Scandinavia, the CERL scientists' critiques of the study's findings resulted in a complete impasse with the Scandinavians. Ottar became increasingly frustrated by the British delegates' insistence on separating "fact from theory" and on emphasizing the need for more research in the report's conclusions.[29] The CERL scientists argued that a number of factors had not been properly accounted for in the analysis, principally the background contribution of sulfate and acidity from winds off the Atlantic Ocean, uncertainties about the rate of conversion from sulfur dioxide to sulfate in precipitation, and the effects of local sources in Norway itself.[30]

If not for a couple independent-minded British scientists, the CEGB might have succeeded in preventing a consensus from being reached on the OECD's final report. However, in late January 1977, two new representatives from Britain arrived in Oslo in a last ditch effort to review the OECD results with Norwegian meteorologists.[31] Bernard Fisher, an atmospheric physicist at the CERL, and Ronald Barnes, an atmospheric scientist recently hired by the Central Unit on Environmental Pollution, arrived for two weeks of meetings at NILU during a harsh winter storm and below freezing temperatures.[32] The cold temperatures outside mirrored the chilly atmosphere between the British and Norwegian scientists, who regarded one another with a fair amount of suspicion. Barnes had been warned by those at the CERL that NILU scientists were an "evil empire" bent on closing British power plants and were deliberately biasing the results, while the Norwegians were unsure whether their new British guests would adopt attitudes similar to those of Lucas and Scriven.[33]

Ottar tried to extend an olive branch to the new team upon their arrival, inviting Barnes and Fisher to dinner with him and his wife. The next day the two men were brought to NILU to begin discussions of the CERL's objections to the report with the atmospheric physicist Anton Eliassen, the meteorologist Harald Dovland, and the chemist Arne Semb. Fisher and Eliassen, who were the foremost modelers of the group, spent the first few days talking one on one about the first and second criticisms of the CERL, which concerned the background contribution from northwest winds over the ocean and the sulfate transformation rate.

Fisher, an extremely well-respected atmospheric modeler, came to the negotiating table with a dramatically different approach than his colleagues at the CERL. Unlike Lucas, Scriven, and others at the research laboratories, Fisher did not dispute that a sizeable portion of British emissions reached Norway and caused acid rain.[34] However, he had done an analysis of wind trajectories over the North Sea and believed that a portion of the emissions carried on these winds could not be directly attributed to Britain given their direction, and thus should be included in an "undetermined" category of deposition from ocean evaporation, long-range transport from North America, and recirculated air from Europe.[35] Though Fisher's suggestions resulted in the model lowering the British contribution to acid rain in Norway by about one-fifth, Eliassen was persuaded that this was necessary given Fisher's analysis of the wind trajectories. British emissions were subsequently estimated to be responsible for about one-quarter of the total acid rain in Norway. This was still six times the amount Norway received from the next largest foreign contributors, which were West and East Germany, and a major acknowledgment of Britain's role in the problem.[36]

Eliassen and Fisher's success in resolving Britain's contribution to acid rain in Norway soon began to lessen the apprehension among the participants and eventually led to a warm, lasting friendship between the two men despite the opposing political positions of their respective countries on acid rain.[37] The group's final deliberations dealt with the CERL's question about the potential effects of local emissions in Norway on acid deposition, and were the only time the conversations seemed to grow a bit acrimonious. The Norwegians vigorously disputed the CERL's claim that NILU had underestimated the importance of local sources in Southern Norway, while the British scientists kept insisting that Norway's total did not sufficiently account for their own industrial emissions. Finally, during a pause in the debate, the problem became clear when Eliassen remarked offhandedly that when one's in an airport, it says "international" departures and "domestic" departures.[38] Suddenly, Barnes realized that the problem was not one of math, but of language: the Norwegians had used "domestic" to refer to both industrial and residential sources. He explained to the NILU scientists that in British emission surveys, the word "domestic" refers only to emissions from households, not those of the entire country.

The discrepancy had been the result of miscommunication and language barriers, not deliberate misrepresentations. Despite their excellent command of the English language, the Norwegians were unaware of this alternate use of the word "domestic" when talking about pollution. Their tally had in fact included industrial sources under this heading and thus accounted for all local emissions in Norway. After a lot of laughter, Barnes suggested using the word "indigenous" instead, putting the last area of dispute to rest. Thereafter, when the scientists saw one another at international conferences, they met for a drink and toasted to the word "indigenous."[39]

In their two weeks of talks, these scientists embodied some of the highest hopes for using science in diplomacy. The ideal of the objective scientist, able to work across political divides, had been a powerful force in foreign affairs since the Second World War but often difficult to put into practice.[40] The success of their meetings, however, did not prefigure a smooth reception of the results by either government, and the scientists' superiors did not react positively to the consensus produced from these discussions. When Anton Eliassen explained to Brynjulf Ottar and members of the Ministry of the Environment that total British contributions to Norway's acid rain would need to be lowered based on Fisher's findings, environmental officials called him a traitor to his country. Ottar allegedly did not stand up for Eliassen at the time, but soon came around and supported making the emissions adjustment.[41] The Norwegian government, on the advice of Ottar and Eliassen, lent its support to its publication shortly thereafter.

The reaction of the CEGB was far harsher. Upon returning to work at the CERL, Bernard Fisher was unceremoniously reassigned from his research lab to more administrative work at the CEGB's headquarters. There was little doubt in his mind that this was a reaction to the fact that he had signed off on a report officially documenting Britain's responsibility for acid rain in Scandinavia. Although Fisher's successful collaboration with the Norwegians earned him a reputation for objectivity and integrity in the British scientific community, his opportunities for advancement within the CERL suffered considerably after his participation in the OECD discussions.[42]

Despite vociferous opposition from the CEGB, the British government gave its consent to the final report. Loath to overrule its own scientists, it formally agreed to issue the report in 1977 on the advice of Ronald Barnes, Leslie Reed, and others at the Central Unit on Environmental Pollution. The successful publication of the OECD report was no small matter. It represented the first official, public admission of Britain's contribution to pollution in Norway and elsewhere, and it also implicated French and West German emissions in causing acid rain as all participating countries were required to consent to its findings. Though the exact amount of pollution transported from one country to another at a particular time was not yet determinable, the project showed that any attempt to substantially lower acid rainfall would require international cooperation.[43]

But while the British government admitted that some of its country's fossil fuel pollution could travel across international borders, it did not concede that this should require power plants to reduce their emissions. In fact, the CEGB anticipated it was likely to lose its battle over the OECD report and was already preparing a plan to avoid an international agreement regardless of the outcome. Details of the CEGB's newly hatched strategy are available only because an anonymous British source turned over confidential internal documents detailing the coal industry's plans to the Norwegian embassy shortly before the OECD report was finalized.[44] The source was kept a secret to protect him or her as well as to ensure that as few people as possible learned that Norwegian environmental officials had obtained a copy of the document. The report survived, however, in Norway's government archives. It reveals that the British coal industry was determined to discredit scientific research on acid rain by any means possible to prevent restrictions on power plant emissions.

The internal memorandum's main argument was that the CEGB needed to alter its strategy of attacking evidence on the atmospheric transport of pollutants and instead focus on ecological studies documenting acid rain's environmental impact. It was written by Doug Lucas and Tony Clarke, two CERL scientists who had previously met with European Communities officials about the sulfur fuel oil directive, as well as another CERL researcher.[45] Their report stated the OECD study would likely force Britain to admit that some of its pollution did reach Norway, so that future arguments "will centre on the effect of airborne pollutants on the Norwegian environment."[46] The authors then reviewed the existing literature on acid rain's environmental impacts to assess how they could undercut the most robust support for the notion that Britain's fossil fuels were the cause of the observed damages. As one example, the CERL scientists proposed that since ecologists had substantial studies on how acid rain caused the death of freshwater fish, the CEGB should argue that other factors besides acid rain could have led to these population losses. Possible alternatives they mentioned included pollution from sewage or overfishing, even though there was no published research suggesting these could be contributing factors.[47]

There was no attempt to assess the current state of scientific knowledge to answer the question of whether ecologists in Scandinavia and the US were perhaps correct in their assessments of the problem. The authors focused exclusively on "gaps" in the literature in the hopes of arguing that not enough was known about acid rain's environmental impacts.[48] For these CERL scientists, the Norwegian pursuit of emission reductions was a "crusade" that placed the value of fish above that of Britain's economy and its people.[49] They felt it was their duty to do all they could to prevent their national interests and the coal industry in particular from being harmed by minor environmental damages of neighboring states.

Why didn't the CEGB just argue as much, instead of attacking environmental studies on acid rain? Likely because they recognized there was a fundamental breach of values between polluters and recipients of acid rain emissions. It wasn't just about "fish" for Norwegians or the White Mountain forests for Americans in New England. It was about the potential for environmental damages like fish declines to have ripple effects through an ecosystem. It was about protecting a way of life that involved experiencing nature and sustaining it for future generations. Rather than engaging in difficult negotiations over the responsibility of the coal industry to incur costs on behalf of a foreign citizenry, it was comparatively easier to simply attack the scientific evidence indicating there was a problem. For these reasons, the strategic shift to arguments over environmental damages resonated with officials throughout the British government and other major polluting nations as pressure increased for an international treaty on acid rain.

Thwarting a Convention with Teeth

After the finalization of the OECD report, in February 1977 Norway
and Sweden made a formal request at the UN to hold an international
conference on acid rain. Diplomats from both countries suggested the issue would make an ideal first topic for the "pan-European" conferences
Soviet premier Leonid Brezhnev proposed as follow-up to the Helsinki
Accords.[50] Receiving a positive response from the Soviet Union, the US,
and their fellow Scandinavians to this idea, the Norwegian Ministry of
Environment was tentatively optimistic that the conference could bring
about a framework agreement to harmonize European air pollution policies.[51] The Norwegian Ministry of Environment quickly began drafting
a declaration on behalf of the Scandinavian governments to eventually
reduce sulfur dioxide pollution, the major cause of acid rain, with the
understanding that the convention's framework might later be extended
to a range of air pollutants.[52] It was modeled after the recently concluded
UN marine conventions, with generally formulated goals and principles
in the primary convention text and more precise commitments stipulated
in annexes.[53]

As a first step, the draft proposed that signatories agree not to permit
any increase in the existing level of sulfur dioxide emissions, after which
they would reduce the emissions by a certain percentage within a set period of time.[54] Since it would be difficult, if not impossible, to differentiate between emissions leading to acid rain and those which had mainly
local effects, the draft convention stipulated that governments should reduce total national emissions levels. This could be accomplished through
several available methods to control sulfur emissions, including setting
maximum limits to the sulfur contents of fuel oils, flue gas desulfurization, coal cleaning, and increasing efficiencies in energy production and
consumption.[55]

The prospect of a legally binding convention on air pollution, let
alone one that included a timetable for mandatory pollution reductions, deeply unsettled the British representative to the UN meeting, Leslie Reed. Reed directed Britain's Central Unit on Environmental Pollution, which helped execute the OECD acid rain project, and also served
as Britain's representative to environmental committees of the OECD,
UN, and other intergovernmental groups.[56] Upon learning of Norway's
plans, Reed sounded the alarm among Britain's diplomats as well as environmental and energy officials. "The Norwegians are aiming for a convention with teeth," he wrote to one of his colleagues, and convened half
a dozen members of Britain's Department of Energy and Foreign Af-

fairs offices to figure out how to respond.[57] Reed had walked a fine line between supporting international research cooperation on the problem and an agreement to reduce emissions that might not be in Britain's interest. With the Scandinavian governments now pressing for action, his position on the issue seems to have hardened and he came out strongly against the proposed international accord.

His compatriots all shared Reed's anxious reaction to the Norwegian proposal. Like officials at the CEGB, they believed implementing the technologies to reduce air pollution would place an excessive economic burden on their nation, and would be "a heavy price to pay for measures which go way beyond our own environmental needs."[58] To stop Norway's proposal from becoming international law, Reed suggested that the British government should negotiate secret alliances with the French and West German governments.[59] With their assistance, he argued, it might be possible to unite members of the European Communities against the proposal.

In order to be able to forge an alliance, however, British environmental officials felt their government needed to dramatically modify its current approach to the European Communities sulfur fuel oil directive and compromise with the other member states on the regulations.[60] They believed that the French and West Germans were likely prepared to accept some kind of obligations on pollution emissions, and the cost of the European Communities measures would be far less in comparison to freezing or reducing sulfur emissions.[61] Conceding to the terms of the sulfur fuel oil directive was thus seen as a means of fending off the much more drastic measures likely to emerge from a Scandinavian proposal.[62] Reed also hoped that a strong alliance could serve to undermine Soviet support for the Scandinavian proposal. The Soviets did not appear to have a genuine interest in acid rain, and they seemed unlikely to allow the Scandinavians to force Western polluting countries into a corner if it might jeopardize the conference itself and their broader foreign policies under détente.[63] However, how the British government could argue against a UN convention remained in question.[64]

Initially, Reed proposed that they might draft an alternative agreement with the French and West Germans that would replace the proposed treaty with a set of non-binding directives. It would attempt to placate the Norwegians by agreeing to increase efficiencies in energy production in addition to cooperating with other nations on the development of alternative energy sources and less expensive methods of lowering the sulfur content of fuels.[65] These closely paralleled the sulfur fuel oil directive provisions, and Reed hoped that such concessions would stave off the more economically burdensome emissions reductions.[66]

Yet if this strategy failed, Britain would need a second line of defense against pressures to reduce their emissions. Reed and his colleagues decided that their "plan B" would be to adopt the CEGB's strategy of attacking research on acid rain and claiming that the science was "uncertain."[67] In the summer and fall of 1977, British environmental and energy officials carefully crafted how to undermine the current state of knowledge about acid rain and the ways this might help them in the UN negotiations. Like members of the CEGB, they believed that the OECD project had made it extremely difficult not to concede that some British emissions crossed national boundaries, but they thought they might be able to successfully argue that there was insufficient proof of environmental harm to warrant mandatory reductions in air pollutants.[68] Their internal briefs on the issue conceded that damages to freshwater ecosystems had been extensively documented throughout Southern Scandinavia from acid rain. However, they believed that these environmental effects did not justify the costs their country would incur by reducing sulfur dioxide emissions.[69]

In meetings that fall, British environmental officials traveled to France and West Germany to try to convince their counterparts in these governments to forge an alliance against the proposed convention. If the three major polluters united against the treaty, it seemed likely that the rest of the members of the European Communities would fall in line. During these conversations, British representatives floated both attack strategies: either creating a non-binding agreement similar to the sulfur fuel oil directive already under consideration in the Communities or attacking the state of scientific knowledge about acid rain to avoid a treaty altogether.

The French response to the British proposals played a decisive role in pushing Britain toward its "plan B" strategy of attacking the state of scientific knowledge concerning acid rain in the UN negotiations. When Reed presented his idea to put forward a counterproposal mirroring the sulfur fuel oil directive requirements, the French balked at discussing energy policies at the UN and expressed a "Gaullist dislike" of any sort of international negotiations to that end.[70] Faced with this reluctance, Reed suggested that perhaps their governments should instead argue there were still scientific uncertainties surrounding acid rain. He informed the French delegates that the CEGB's studies of acid rain had cast doubt on whether its environmental effects were as bad as the Norwegians claimed. According to Reed's notes, the French representatives agreed with this alternative strategy.[71]

The British received a much more tepid response from officials in West Germany's Ministry of the Interior during a subsequent meeting in Bonn to discuss a possible alliance against the treaty. Like the French, the Germans were also extremely hesitant to conduct discussions on energy policies at the UN, though for different reasons. West German delegates were presently engaged in talks on energy issues with the Soviet Union and wanted to keep these negotiations private. Since their government had already proposed air quality standards for sulfur dioxide pollution, they told the British delegates it was far more desirable for West Germany to accept a convention on acid rain and emissions restrictions.[72] There was hope in West Germany that environmental issues would result in increased opportunities to build upon Chancellor Willy Brandt's *ostpolitik* and expand contact with Eastern European countries, a desire they had privately expressed to Norwegian environmental officials as well.[73] The rapprochement would ideally lead to greater economic growth as well as cooperation in the areas of energy, trade, and technology.[74]

Since West German representatives were far more amenable to the Nordic proposal than the French, the British delegation tried to persuade them that CEGB research on acid rain showed there were still considerable uncertainties on its role in causing ecological harms, making international action premature. In contrast to the French, the Germans minced no words in telling the British delegates they thought this was the wrong way to approach negotiations. They believed the OECD study had left little doubt that their countries were largely responsible for acid rain pollution in Scandinavia, and they expected that additional research would merely, as they put it, "rub this fact in."[75] The West German officials further cautioned the British that they regarded the environmental effects of acid rain as well established by scientific research. Rather than sparing their governments from pollution reductions, further studies on the ecological problems of acid rain might result in even greater pressure to implement policies to reduce acid rain pollution.[76]

The British left these meetings slightly chastened about the prospect of building a diplomatic alliance through the European Communities, but they need not have been. In spite of all these reservations, the West Germans ultimately agreed to form a pact with the British and French to do all they could to avoid an international agreement on acid rain. In certain West German government agencies, there seems to have been a fear that the country might need to rely more on its coal reserves in the future and increase their emissions. Although the opportunity to open dialogue with Eastern Europe was clearly appealing, maintaining flexibility in energy usage took priority over these larger foreign policy aims and West German officials gave their consent to Britain's plans.[77]

Once in agreement about the necessary approach, the three governments did not disclose their behind the scenes talks to any other countries in the European Communities, opting to keep it a secret lest they alienate the remaining member states in their bid to unite the group against the Nordic countries.[78]

Over the coming months, Reed drafted a paper outlining how the European Communities should respond to the Nordic proposal, which he cleared with the French and Germans.[79] In March 1978 in Brussels, the three allies made their case to the other member states that a convention on acid rain was ill-advised given insufficient scientific proof, and successfully convinced all member states except Denmark to take a cautious approach to convening international negotiations on the problem.[80] They did not have a difficult task convincing the group to oppose an international accord. Many bureaucrats at the European Communities and member state representatives believed the Norwegian proposal was too ambitious when few countries had enacted national regulations limiting emissions levels.[81] Some delegates felt it would be premature for the group to commit to international agreements on sulfur dioxide reductions when they still could not agree among themselves on the proposed sulfur fuel directive.[82] A number of member states also expressed skepticism about the trustworthiness of Eastern European governments in following through with recommendations for emissions reductions, considering the lack of data and measurement stations as well as the generally poorer air quality in the region.[83]

At ensuing UN meetings in the spring of 1978, the European Communities members jointly expressed the group's opposition to even holding talks on acid rain. Their united front seriously dampened prospects for an accord. As British officials had hoped, the group's intransigence rapidly eroded Soviet support for the Nordic draft agreement, leading the Norwegians to conclude that the Soviets were "willing to pay any price" for a convention.[84] Faced with a united European Communities wary of committing to negotiations on acid rain, the Nordic countries were only able to obtain consent for the UN to form a "special group" to undertake preparatory work for a high-level meeting without any firm decisions on whether such a meeting would be held.[85]

The UN special group first convened in July to discuss what appropriate steps should be taken in advance of a possible high-level meeting on acid rain, notably preparing any legal or scientific materials that would be needed should negotiations move forward.[86] Norwegian diplomats suggested that it might be useful to have a report on the effects of acid rain, while the European Communities countries insisted participants also have access to information on the costs of abatement measures.[87] Reed then proposed that the UN should hire a special consultant— someone independent from the process—to compile a "synthesis report" on the current status of scientific knowledge in service of the negotiations.[88] It was recommended that Ron Barnes, who had worked for Reed at Britain's Central Unit on Environmental Pollution but was now employed by the Esso oil company, could compile the report as an independent consultant to the UN.[89] While his British loyalties might have given some pause, Barnes' successful cooperation with Scandinavian scientists on the OECD report seems to have limited any objections to his selection. To facilitate his work, the UN delegations agreed to provide information on their country's air pollution policies, abatement measures, costs, and scientific research on transboundary air pollution and its environmental effects.[90] These would be distributed to all other governments as well as to Barnes, who would present his analysis at one of the forthcoming meetings.

In addition to Barnes' independent report for the meeting, countries would be allowed to submit their own documents for consideration in possible negotiations. Reed had already begun preparing for this eventuality, drafting a document outlining the remaining "scientific uncertainties" in acid rain research as part of Britain's plan B strategy for avoiding emissions reductions. He and his colleagues agreed it would be essential to also argue that acid rain caused no serious economic losses and appeared to affect only "leisure" activities, as well as noting the "peculiar geographical, topographical and meteorological factors" which made Norway vulnerable to acid rain.[91] These statements were presumably intended to paint the Norwegians as placing recreation and aesthetics ahead of the welfare of the British people, as well as to minimize the danger of acid rain to other areas.

Shortly after the 1978 July meeting at the UN, the Norwegians received Britain's submission, dubbed "Doc. A." It was clearly written to undermine the notion that scientific research on acid rain supported making emissions reductions through a convention. The report raised remaining "uncertainties" on the atmospheric transport of pollutants, many of which had ostensibly been settled during discussions on the OECD report, as well as on acid rain's alleged environmental effects. In regard to the atmospheric work that had been conducted through the OECD, the British paper argued that the reliance on models, while necessary, involved considerable assumptions that introduced uncertainties in the quantitative estimates of the pollution transfer.[92] Though not stated explicitly, the implication of these remarks was undoubtedly that the models could not be relied upon to make policy decisions. Doc. A went on to critique the amount of pollution of "unknown origin" in the OECD report as well as uncertainties about the role of nitrogen oxides in acid rain and the conversion of sulfur dioxide to sulfate in the atmosphere. These gaps in knowledge, they stressed, could mean that big reductions in emissions could lead to little environmental benefit, though there was no direct evidence to support this contention.[93]

Turning to acid rain's ecological dangers, Doc. A claimed it was "beginning to appear that factors other than acid rain" were the cause of changes in water pH in affected regions in Southern Scandinavia.[94] They cited no scientific studies for this statement.[95] The report concluded by arguing that the risks to the environment needed to be weighed against the risks of economic damage resulting from the costs of reducing emissions.[96] The "considerable uncertainty" surrounding acid rain's effects meant that there was "no prospect of securing international agreements to introduce changes until all the issues had been fully evaluated."[97] The conclusion was clear: further scientific proof was necessary before any accord could be properly evaluated at the UN.

The British submission dealt a serious blow to the UN talks and left Scandinavian scientists and environmental officials feeling outraged. At the next UN meeting in September, Norwegian diplomats pulled Barnes aside to tell him they found it "deliberately provocative and inflammatory," not to mention highly misleading on the state of the scientific research since it did not take into account the latest ecological studies conducted internationally or even within Britain.[98] In comments they circulated to all delegates, Norway argued that Doc. A only presented uncertainties and gaps in knowledge instead of available scientific research.[99] They asserted that there was strong scientific evidence for the cause-and-effect relationship between acid rain and ecological damage in freshwater ecosystems, and that none of the alternative causes Britain proposed could explain why areas suffering fish decline coincided with areas that had been subjected to acidic precipitation and subsequent drops in the pH of surface waters.[100] Though they agreed that there were a number of unanswered questions that should be investigated, they felt the work of the special group needed to focus on what scientists did know, not just "gaps in knowledge."[101] They argued that if scientific data related to environmental problems had to remove all possible doubt, the effects of such problems "may by then have become irreversible, or, at any rate, far more difficult and expensive to put right."[102] When the Norwegians confronted Reed about the British report, he claimed that it had been sent out without his knowledge or consent and that he would not press for its discussion at the next UN meeting.[103] Given Reed's internal involvement in drafting the document these statements seem misleading at best, though it's unclear whether Nordic officials ever suspected Britain's environmental agencies were coordinating with the CEGB to derail the talks.

The Scandinavians appear to have felt there was little more they could do other than try to rebut Britain's arguments with evidence for acid rain's harms and the role of foreign power plants in causing environmental damage. To facilitate this, they invited Barnes to visit Norway, Sweden, and Finland to become acquainted with the most recent ecological research on acid rain. The following month, he attended three days of briefings in each of the countries, meeting with top ecologists conducting research on acid rain. However, he was accompanied by another scientist: David Brown, a biologist who worked for the CEGB.[104] Despite the ominous conflicts of interest this posed to the independent assessment, the Scandinavians made a concerted effort to cover every aspect of the issue for Barnes and Brown. Norwegian ecologists reviewed the environmental effects of acid rain, Swedish experts discussed emission control policies and their costs, and Finnish scientists detailed the corrosive effects of acid rain on buildings.[105]

As these meetings took place, however, the European Communities' member states submitted an alternative draft declaration to the UN. It promised little more than further scientific cooperation on atmospheric and ecological questions in service of future consultations on pollution policies.[106] Representatives of the European Communities argued that additional research could give a more "accurate and balanced assessment" of the causal relationships between fossil fuel emissions and environmental damages.[107] These remarks left the Scandinavian delegations incensed, with Norwegian diplomat Oddmund Graham giving an impassioned speech that criticized how "certain delegations" had stressed the need for additional research.[108] He delineated the numerous studies on acid rain undertaken in recent years, particularly the OECD project. Graham made a point of listing each Western European country that participated in the OECD research and quoted from the agreed-upon conclusions in the final report that sulfur compounds do travel long distances in the atmosphere and measurably affect the air quality in other countries.[109]

On the question of environmental effects from acid rain, Graham described the consensus reached at the recent International Conference on the Effects of Acid Precipitation in Telemark, Norway, in 1976, which was attended by scientists and delegates from 20 countries and 6 intergovernmental organizations. The conference delegates concluded that acid precipitation was responsible for a decrease in pH in aquatic ecosystems with especially deleterious effects on fish populations in Southern Scandinavia and Eastern North America, as well as posing a potential threat to forest ecosystems. Graham added that studies undertaken since 1976 had contributed additional details to the international and national understanding of regional acidification but had not altered the main conclusions from the Telemark Conference. Given all this research, he said, his government could not understand the logical basis for the contention that there was insufficient knowledge to take international action.[110]

As tensions ran high at the UN following the European Communities' submission, Barnes delivered his final report on acid rain. While containing some details about the ecological damages that could occur from acid rain, the report suggested that the harms were not very serious and that there was little scientific consensus on the role of acid rain in causing them. His evidence for these claims came primarily from CERL scientists and Ivan Rosenqvist, a controversial ecologist in Norway who disputed evidence that acid rain damaged freshwater ecosystems.[111] The report concluded by stating that although sulfur dioxide and nitrogen oxides did travel over long distances, the pollutants did not "have any significant adverse environmental impact."[112] Barnes' submission caused considerable uproar among the Nordic, Canadian, and Soviet delegations.[113] The Norwegian delegation, for example, argued that the conclusions were subjective and oversimplified, and did not reflect the research cited in the report.[114] After reading Barnes' final report, the Soviet Union, Canada, and Nordic governments asked that it not be used as a further basis for discussion or negotiation, and if it were used, that they be given time to prepare a detailed rebuttal to it.[115] Though the remaining UN delegations agreed to set aside the report, it had a chilling effect on support for the Nordic draft proposal.[116] Although the Norwegians tried to do damage control over the following months with members of the European Communities to break the deadlock, notably with private consultations among French delegates and European Communities bureaucrats, they were unsuccessful in making any constructive headway toward an agreement.[117]

If not for the intervention of the US State Department, Britain, France, and West Germany would almost certainly have succeeded in pushing through their draft agreement committing to scientific cooperation and little else, or perhaps even sabotaged the talks altogether. But at the eleventh hour, US State Department officials put enormous pressure on Britain, France, and West Germany to compromise with the Scandinavians. The Americans had largely stayed on the sidelines during the contentious discussions between the European Communities and the Nordic nations over the previous year. However, the attempts of Britain, France, and West Germany to impede progress in the negotiations had made the US government concerned that they would be successful in either preventing the negotiation of an agreement or pushing through a largely rhetorical treaty that would be meaningless in tackling the acid rain problem.

One of the reasons American diplomats cared about the dire prospects of acid rain diplomacy was undoubtedly because they believed the US would soon need to tackle the problem domestically as well as with Canada. Yet the primary motivation behind the sudden US intervention in the acid rain talks was to preserve its bargaining power in détente negotiations with the communist bloc. American diplomats feared that a poor outcome from the acid rain talks could reduce US leverage in discussions with the Soviets over the other Helsinki Accord issues of nuclear security and human rights.[118] In a State Department communication with government officials in London, Paris, and Bonn in February 1979, American diplomats told their allies in no uncertain terms that they needed to compromise with the Scandinavians or risk weakening the UN and allowing the Soviets to exploit a divided Western Caucus. As they explained in their letter:

> We are concerned, however, that if the public stalemate continues through the February 19 SAEP [UN Senior Advisers on Environmental Problems] meeting we will in April face the strong possibility of some members of the alliance vetoing a meeting. Another, and at present likely alternative would be western acceptance of a meeting with very little substantive content and not meeting our previously established criteria. This latter prospect could encourage the Soviets to politicize the UN ECE [UN Economic Commission for Europe] by using it in the future for essentially propaganda purposes and could set an unfortunate precedent.[119]

The memo strongly urged the governments of Britain, France, and West Germany to accept a compromise proposal drafted by the US State Department that contained binding commitments to continue negotiating on acid rain with specific deadlines for future amendments as well as strengthening of monitoring and research. The US acknowledged that it could not at present agree to the Nordic demands for reducing sulfur dioxide emissions because of the high costs, but stated that the Norwegian and Swedish governments had already conceded in private that they did not have the support to achieve their maximum position and would be willing to make concessions.[120]

Following the receipt of this memo, France suddenly appeared willing to consider signing a legally binding treaty and expressed its shift in position to Danish representatives and officials at the European Communities.[121] This change in the French stance gave new momentum to talks between the European Communities and Nordic countries at the end of February, and resulted in a revised agreement that nearly all countries supported.[122] The final version mirrored the contents of the proposal from the US State Department, consisting of broad pledges to periodically review pollution regulations and conduct further research on acid rain but with no firm commitments to reductions in fossil fuel emissions.[123] The final draft of the convention was approved at the UN in April 1979 and signed by all parties later that fall.[124]

Though this was perhaps a remarkable historical moment given that no legally binding international treaty had ever been negotiated on air pollution, Norwegian and Swedish delegates were left reeling from Britain's insistence that scientific uncertainties about acid rain stood in the way of government action to reduce fossil fuel pollutants. At the formal signing of the Convention in November 1979, the British Under Secretary of State for the Environment expressed his government's approval of the "balanced" approach taken in the Convention on matters for which "there are as yet no clear answers."[125] He asserted that it was still unknown whether air pollution was responsible for damaging ecosystems, and insisted that scientists must "first establish the facts" before discussions could continue on transboundary air pollution.[126] Only once governments were "armed with the facts," he concluded, could effective pollution policies be developed.[127] These sentiments were affirmed by the other major polluters, such as West Germany. While commending the Nordic countries for their moving testimonials of the environmental damages wrought by acid rain, the West German representative cautioned that there were still "considerable gaps in knowledge" that the scientific community needed to resolve.[128]

Undoubtedly anticipating such remarks, the Norwegian Minister of the Environment pleaded with those present to use scientific studies "for the purpose of making progress and not to postpone necessary decisions."[129] While the Norwegian government hoped that its willingness to compromise on emissions reductions in exchange for a legally binding

FIGURE 5.2 Delegates from the European Communities deposit ratified copies of the 1979 UN Convention on Long-range Transboundary Air Pollution in Geneva. Yutaka Nugata, "Common Market Countries Deposit Legal Instruments," United Nations, photo #263507, July 15, 1982.

convention would allow for more intricate, complicated discussions on lowering air pollutants to continue over the coming months and years, this did not come to pass. The major emitters, seemingly emboldened and united with the other European Communities member states, continued to insist that the science wasn't settled enough to schedule further negotiations on emissions reductions.[130] At first Norwegian representatives persisted in arguing that scientific research had proven acid rain's harmful effects "beyond all reasonable doubt," but by 1980, they began to frustratingly declare that no scientific proof could meet the standards of Britain and its allies.[131] While the politics of the Cold War may have given Scandinavians an opening to address acid rain across the continent and earned them American support, the doubt-mongering of coal interests had now swept Western European governments into an alliance against reducing fossil fuel emissions. Hope for further assistance from the US, however, was dashed as conservative governments rose to power in both America and Britain and quickly united against calls for cooperating on acid rain.

6 An Environmental Crisis Collides with a Conservative Revolution

Acid rain has the highest priority from the US perspective. Indeed it could be said that the single most important environmental policy objective of this Administration is to do nothing about acid rain![1]

Don Rolt, counselor for science, technology, energy and the environment at the British Embassy in Washington, DC, 1982

You may find it heartbreaking for Europe to invest an enormous amount in measures which turn out to be ineffective. Others will find it heartbreaking if nothing is done, and the consequences turn out to be as serious as feared not only for aquatic systems but also for vegetation and perhaps health.[2]

Hans Martin Seip, Norwegian Central Institute for Industrial Research, 1981

In a meeting held January 8, 1980, following the signing of the 1979 United Nations (UN) Convention on Long-range Transboundary Air Pollution, several scientists working for Britain's Central Electricity Generating Board (CEGB) briefed the organization's senior administrators about current air pollution complaints against the country's power plants and their ongoing research efforts. The topic of acid rain dominated much of the discussion because of the political momentum developing in Europe and North America on the issue.[3] But atmospheric scientist Ron Scriven, who had worked as a researcher for the CEGB for many years, reassured the attendees that there was increasing skepticism about the validity of acid rain studies among European scientists. Crediting the efforts of CEGB scientists for this shift, Scriven was emphatic that its laboratories would continue pursuing research "of a defensive nature" to repudiate potentially damaging evidence of Britain's role in causing acid rain.[4]

The CEGB's plan to continue combating international pressure to reduce fossil fuel emissions found a receptive ear in Britain's recently elected Prime Minister, Margaret Thatcher. Coming to power just after the 1979 UN Convention was negotiated, Thatcher's conservative, pro-business administration was committed to improving the economy and skeptical of environmental protections. On the other side of the Atlantic, a revolution was also happening in American politics with implications for acid rain and environmental policies more broadly. In the fall 1980 election, Ronald Reagan defeated then President Carter with a pitch to "make America great again" by cutting taxes and bolstering the economy. Environmental regulations were generally seen as anathema to the goals of Thatcher and Reagan, but acid rain particularly clashed with their views on how to catalyze economic growth. A series of energy crises during the late 1970s had contributed to recessions in Britain and the US, leading both Thatcher and Reagan to prioritize the revitalization of domestic fossil fuel industries. Reducing emissions from coal-fired power plants by installing expensive flue gas desulfurization technology was diametrically opposed to these aims. That acid rain seemed to affect a few fish populations in sensitive lakes only added to Thatcher and Reagan's commitment to do nothing about the problem.

But while the coal industries of both countries may have succeeded in characterizing acid rain as a minor nuisance, by the early 1980s ecological studies of its impacts were beginning to coalesce into an international consensus on the serious ecosystem damages caused by fossil fuels. Investigations into how increasing acidity could harm organisms had been ongoing since the early 1970s, most intensely in Norway and the US. Once the OECD project made it difficult for the British and American power industries to argue that their pollutants were not acidifying rainfall, these ecological studies came under attack by scientists working for the British CEGB and the Electric Power Research Institute (EPRI), the scientific arm of the American coal industry.

With conservative parties in power in both countries, the prospects for building off the 1979 UN Convention to implement pollution reductions seemed dire. Yet ecologists believed they could make a strong enough case to overcome the objections of coal industry scientists. In fact, they had designed their research studies in the late 1970s to do exactly that. Their work put these researchers and environmental officials in Europe and North America on a collision course with an ascendant politics of economic growth and energy independence. The Reagan and Thatcher administrations institutionalized the involvement of coal industry scientists in environmental politics in an unprecedented manner, raising questions about whether government environmental agencies were receiving unbiased reports on the state of knowledge about acid rain as well as complicating international negotiations on pollution reductions.

For a time, the deep influence of the coal industry in two of the world's largest polluters had a chilling effect on environmental research and forestalled progress on an international agreement to lower fossil fuel emissions. It corrupted scientific advising in both countries, forcing nations impacted by acid rain to reckon with whether to insert themselves into the science and politics of other states. With so much resistance to incurring any economic costs from pollution reductions, the bar for scientists studying acid rain's transport and ecological effects was also continually raised. No longer was industry concerned with whether studies had shown pollution could travel long distances in the atmosphere and be deposited in rainfall. Now researchers were told they needed to identify the exact power plants that damaged a certain region, a level of specificity nearly impossible to achieve with chaotic atmospheric processes. But the alignment of conservative politics with the coal industry eventually went even further, with government officials and industry scientists making dishonest representations of current studies and misleading assessments at home and abroad. These tactics were approved and sanctioned at the highest levels of the Reagan and Thatcher administrations. If a few ecosystems were harmed in the process, this was the price that needed to be paid to protect economic growth and the vitality of domestic coal production.

Ecology and the Question of Environmental Damage

When British coal industry scientists began criticizing the ecological research coming out of Norway in the mid-1970s, environmental scientists working on a Norwegian study of acid rain's effects took notice. Known as the "SNSF project" (Acid Rain's Effects on Forests and Fish, Sur nedbørs virkning på skog og fisk), it was the most expansive national science project in the country's history and the largest investigation of acid rain's environmental impact anywhere in the world. In 1976, the SNSF project's steering committee was already in the process of reevaluating research priorities and projects for the second and final phase of the program.[5] While the first phase of SNSF research from 1972 to 1976 had been carried out under the hypothesis that a decrease in pH in and of it-self damaged biological organisms, studies from the project's first four years had demonstrated that the mechanism was much more complex. For example, researchers found that the acidification of freshwater eco-systems involved the interaction of precipitation with chemicals in the soil leached into water runoff.

In the second phase of the project, the steering committee decided investigators should move beyond treating the ecological processes of acid rain as a "black box," correlating input of pH with environmen-tal damages. Instead, the second phase needed to more critically under-stand the causal relationship involving pH, soil components, and dam-age to freshwater life.[6] While research from the first phase of ecological studies demonstrated that decreasing pH in rivers and lakes reduced the survival of fish eggs and fry as well as killed adult fish, the exact physio-logical processes involved were still largely unknown. A similar revision in approach was also initiated regarding acid rain's impacts on forests. The first phase of the project had not produced clear evidence that acid rain damaged forest ecosystems, so subsequent projects were designed to look more closely at the interactions between vegetation and chemical substances in soils. Biologists also sought to document the broader eco-system effects from acid rain, such as loss of species diversity in fresh-water lakes and depletion of nutrients in soils and waterways.[7]

This shift in emphasis on precisely detailing the association between decreasing pH in rainfall, soil chemistry, and freshwater damages was motivated in large part by growing concerns among the SNSF project's leadership that supposed "alternative" causes of acid rain's damages, however ill-conceived, needed to be investigated in the next phase of research.[8] The goals for the second phase of the project explicitly in-cluded research evaluating the alternative mechanisms for acidification touted by the British and later American coal industries.[9]

The responsibility for examining possible alternative causes of acid rain's environmental damages primarily fell on the soil chemist Hans Martin Seip, who had served as a lecturer at the University of Oslo un-til 1976 when he joined Norway's Central Institute for Industrial Re-search as an environmental chemist.[10] Seip became acquainted with the SNSF program through his colleagues at the Institute who had been in-volved in the research, and in 1977 agreed to serve as the assistant direc-tor for research during the remainder of the project and oversee studies evaluating these claims.[11] A childhood friend of the new director of the SNSF program, Lars Walløe, Seip likely seemed an ideal candidate to mediate the dispute because of his university background, which gave him credibility with both the Ministry of Environment and academic scientists.[12]

Upon joining the SNSF project, Seip worked closely with two researchers from the Norwegian Institute for Water Research, Arne Henriksen and Richard "Dick" Wright, to investigate the importance of land use changes in acidification of the environment. Though his role was largely confined to analyses of ongoing fieldwork by SNSF researchers, Seip's ability to synthesize a diverse array of studies on rainfall, soil, and water interactions provided a powerful counterargument to contentions that something else was responsible for the observed damages. He developed what he termed "conceptual models" of possible causes of environmental acidification based on Norwegian ecological studies. These were (1) a direct relationship between acid rain and acidification of surface waters, (2) an indirect relationship, mediated by changes in the soil, or (3) a relationship between land use changes and acidification. While seemingly complicated, these "models" were simply an elaborate way to consider the weight of the evidence for each possible hypothesis, allowing Seip to use scientific terminology to point out the difference in support for the relationship between acid rain and damage to ecosystems compared with evidence suggesting land use changes were responsible. His reports concluded that acid rain was the major cause of environmental damage in Southern Norway while still acknowledging that land use changes could lead to environmental acidification in other areas.[13]

As a result of this work, by 1980 Seip had become one of the world's foremost ecological experts on acid rain pollution and took on the task of organizing rebuttals to the British coal industry about remaining scientific uncertainties.[14] The first such instance concerned a widely reported CEGB study on Norway's Tovdal River, discussed in chapter 3, that claimed to have found evidence for alternative causes of the observed damages despite SNSF studies showing otherwise. To evaluate why the results of the CEGB scientists' fieldwork differed so markedly from their own research, SNSF researchers conducted follow-up studies in both Southern Norway and Scotland during 1978 and 1979. Based on their measurements, Seip and Henriksen determined that acidity from land use changes did not meaningfully contribute to lowering the pH.[15] In an article they published in 1980, the two scientists systematically dismantled the CEGB research on the role of these "weak" acids in acidifying surface water ecosystems in Southern Norway and Southwestern Scotland, resulting in a notable disappearance of this claim in CEGB arguments on the scientific evidence for acid rain.[16]

With the completion of the SNSF project nearing, Norwegian ecologists decided to organize an international conference on acid precipitation's ecological effects. Held in Sandefjord, Norway, in the fall of 1980, it was overseen by Hans Martin Seip and Gene Likens, who had discovered acid rain in America, alongside Fred Last and Carl Olaf Tamm, top ecologists from Britain and Sweden, respectively. The conference included more than 300 participants, a third of whom hailed from the US. Although the majority of participants came from Scandinavia, the US, and Canada, researchers from east and west Europe attended as well.[17] Even scientists from both the British and American coal industry research institutes were in attendance, underscoring the importance they gave to the potential report from the proceedings.

In summarizing the current state of knowledge on acid rain, the conference report was careful to note that there was still only tenuous evidence of damages to vegetation and forests from fossil fuel pollution. But it emphasized that "it is evident that serious changes have been detected in aquatic ecosystems and there is no longer any doubt that most of these changes are related to incoming precipitation."[18] More than one hundred scientific papers submitted to the conference documented the impact of acid rain from the US to Norway to Sweden to the Netherlands. It was the most extensive compilation of environmental research on acid rain to date and highlighted the need to continue international negotiations on pollution reductions.[19]

At the same time, US and Canadian scientists were also amassing a plethora of environmental data documenting acid rain's harmful effects in North America and making progress in bilateral negotiations on the issue. Following the negotiation of the 1979 UN Convention, the US and Canadian governments committed to their own joint program of research and a formal diplomatic apparatus for continuing bilateral talks on reducing fossil fuel emissions that cause acid rain.[20] The next year, President Carter sought to implement a far-reaching government effort to study the transport of emissions across the US and examine their effects on freshwater and forest ecosystems. Known as the National Acid Precipitation Assessment Program (NAPAP), it was created through the Acid Precipitation Act of 1980 and expected to run for the next decade.[21] The Carter administration's new programs meant that the US government finally began spending more money on acid rain research than EPRI, the research arm of the coal industry, but the latter remained the second largest funder of acid rain studies in the country with $15 million spent per year.[22]

Though NAPAP promised to further document the existence and extent of acid rain in the US, there was already considerable evidence from university scientists and government researchers demonstrating the relationship between fossil fuel emissions and ecosystem harms. To synthesize the rapidly accumulating knowledge on acid rain, in 1979 the Environmental Protection Agency (EPA) asked the National Academy of Sciences to complete a report on the environmental consequences of fossil fuel combustion. Ellis Cowling, a biologist at North Carolina State University and chairman of the national acid deposition monitoring program for the US government, was the primary catalyst in requesting the academy report. The committee appointed to assess scientific studies in the area included top acid rain researchers from countries across North America and Europe, including Swedish scientist Svante Odén, who first discovered the problem, and Lars Overrein, one of Norway's top biologists in the SNSF project who investigated acid rain's effects on freshwater ecosystems.[23] While addressing fossil fuel pollution broadly, the report focused on the two main precursors to acid rain: sulfur dioxide and nitrogen oxides.

Published in 1981 on the heels of the international acid rain conference in Sandefjord, the National Academy of Sciences report underscored the growing consensus on the relationship between fossil fuel pollution and reduced populations of organisms in susceptible ecosystems. Noting that the vast majority of the world's freshwater lakes existed in areas sensitive to acid rainfall, the Committee stressed that additional controls on power plant emissions would be necessary to prevent the elimination of organisms "over substantial parts of their natural ranges." While terrestrial effects were still not conclusive, the Committee argued that chemical changes in soils from acid rain could prove harmful to forests, particularly in conjunction with other environmental stressors.[24] Overall, the scientists stated, the evidence connecting power plants with acid rain was sufficiently "overwhelming" to justify tightening emissions controls.

The report generated a media firestorm and drew the attention of government officials, particularly those in the newly inaugurated Reagan administration.[25] It responded by cutting off all funds to the National Academy of Sciences for work on acid rain the following year. The Reagan administration accused the National Academy of Sciences of being biased in its assessment, firing the first shot in what would become a highly politicized battle over acid rain science during the President's tenure.[26] The cessation of funding for the National Academy of Sciences was part of a larger effort to reduce research support for environmental pollution under the new administration and undermine the scientific case for regulatory controls. The EPA's budget was slated to shrink by 50 percent over the next several years, including massive cuts to grants for acid rain studies then underway among university scientists and government researchers.[27] The administration also decreased funding for the UN's environment program, a serious blow given America's prior leadership on international environmental issues throughout the previous decade.[28]

In Britain, government support for acid rain studies was looking equally dire. Prior to the late 1970s, little research had been done on acid rain that was not funded by the nationalized coal industry.[29] By 1981, just two organizations outside the CEGB had received government funding for studies on acid rain: the Institute for Terrestrial Ecology (ITE) and the Department of Agriculture and Fisheries in Scotland (DAFS). Founded in 1973, the ITE was an independent research group financed by Britain's Natural Environment Research Council to conduct basic biological and ecological research as well as carry out environmental studies under contract with the Department of the Environment, European Communities, and other governmental bodies as needed.[30] The ITE had begun researching the effects of sulfur dioxide on trees through a grant from the Department of the Environment in 1975 as part of the department's general research on air pollution.[31] The initial impetus for the work was to identify biological indicators of air pollution levels and to better discriminate between the effects of various pollutants.[32] The studies were overseen by Fred Last, who hired a young, recent PhD, David Fowler, to carry out the research. Fowler had completed his doctoral studies at Britain's Harwell laboratory as part of the OECD project on acid rain, and in 1976 he parlayed this work into investigations on whether rainfall in Scotland was also affected by acidic pollutants.[33] The DAFS studies were similar in scope, begun in the mid-1970s to provide an additional monitoring station in Scotland for the OECD project. When measurements at the Scottish station showed surprisingly high levels of acidity in precipitation, several members of the DAFS laboratory expanded their work to include studies of acid rain's impact on freshwater ecosystems compared with land use changes.[34]

By 1980, these two investigations were beginning to generate data showing that Scotland and Wales were also experiencing acid rain. The implications of the results were profound: acid rain was not just a problem Britain was exporting to other countries but a threat to its own environment.[35] Instead of leading to more support for further investigations, however, even these limited research endeavors were threatened by severe budget cuts to the Department of the Environment under the Thatcher administration.[36] Within a year, reduced funding for environmental research had resulted in a decision to eliminate much of the grants to government agencies for acid rain studies with the exception of the CEGB.[37]

The Reagan and Thatcher administrations' decision to undercut independent environmental assessments of acid rain set up a major conflict between American and British government officials and the majority of acid rain researchers. While ecologists in the field were achieving widespread consensus on acid rain's harmful effects, Reagan and Thatcher's departments of environment and energy were instead relying more and more on the outlier views of industry-backed scientists. As a result of CEGB and EPRI researchers gaining a strong foothold in these administrations, it became far more challenging for the consensus view to receive a fair appraisal in the British and American executive branches during the early 1980s.

Confronting Coal Industry Influence under Reagan and Thatcher

In the US, the arrival of the Reagan administration caused an uproar among environmentalists. Throughout the campaign, Ronald Reagan had made bizarre statements about environmental protection, perhaps most infamously claiming that trees cause more pollution than people. Many in the public held out hope that these were farcical remarks and need not be taken as genuine indications of Reagan's potential regulatory approach to pollution.[38] That changed quickly in the months following his inauguration after staffing decisions for the EPA and other government departments made it clear Reagan was quite serious about dismantling environmental protections.

President Reagan's most controversial appointments were Secretary of the Interior James Watt, a pro-business attorney who sought to open protected environmental areas to development, and EPA administrator Anne Gorsuch, another pro-business attorney and former state legislator with no background in environmental issues.[39] But while these two selections received the most public criticism, the hiring of industry-oriented staff went far beyond these two high-profile positions. The Reagan administration appointed industry-sympathetic officials as well as former industry scientists to key posts throughout the EPA. For example, university scientists on a special task force overseeing NAPAP, the main government research program for acid rain, were replaced by industry scientists. These included Ralph Perhac, then EPRI's director of the environmental assessment department, who had testified on behalf of the coal industry multiple times in previous years and managed their acid rain research.[40] Reagan also appointed James Mahoney, an atmospheric scientist and founder of a company called Environmental Research and Technology, who had previously received corporate contracts to assist with environmental work.[41] His firm soon became the largest employer of meteorologists outside the US government.[42]

Beyond the acid rain task force, several other presidential appointees at the EPA were staunch advocates for business interests and swiftly adopted the position of the coal industry on acid rain. John Hernandez, the deputy director of the EPA, oversaw extensive cooperation between the agency and EPRI on acid rain studies during the early 1980s, allowing them direct input on the government's research projects. EPRI soon began partnering with the EPA and Department of Energy on a series of studies designed to test the "linearity" of acid rain emissions, meaning whether a decrease in pollutants would result in a comparable decrease in acidic precipitation. These programs were first proposed by EPRI and were undertaken with the expectation that the Reagan administration would not act until their completion.[43] In the meantime, Hernandez reportedly banned use of the term "acid rain" in his presence and was highly critical of Canada's efforts to negotiate emissions reductions.[44] He eventually came under fire for forcing scientists at the agency to alter reports on the dangers of chemical pollutants at the request of the Dow Corporation as well as denying the involvement of lead smelters in causing increased lead pollution in Dallas, Texas.[45]

William Nierenberg, head of a controversial review of acid rain research for the Reagan administration, had also served on EPRI's advisory council during the 1970s and 1980s; even when not formally on the council, he participated in meetings at EPRI about environmental pollutants.[46] Nierenberg felt it benefited environmental scientists to develop relationships with the power industry, stating that those who served as EPRI advisors "have been sobered by their practical discussions with EPRI staff members."[47] He continued to receive advice from Chauncey Starr, the head of EPRI, while serving as chairman of the Reagan administration's Office on Science and Technology Policy Acid Rain Review Panel. The panel's eventual report on acid rain was highly criticized for minimizing the dangers of acid rain, and Nierenberg was later accused of watering down the scientific findings in its executive summary at the request of the White House.[48]

Other Reagan-appointed EPA officials with prior business ties took up the mantle of combating international pressure on acid rain from

Canada as well as Europe. Kathleen Bennett, an assistant EPA administrator, and Richard "Dick" Funkhouser, EPA's director of international
affairs, were especially influential. Prior to joining the EPA, Bennett had
worked as a government affairs coordinator for a paper manufactur-
ing company, while Funkhouser had worked for oil companies and in
the Foreign Service.[49] Together, they sought to undercut several international scientific reports on acid rain shortly after President Reagan took
office. As one example, in December 1981, Funkhouser instructed Bennett to ensure that scientific reports from the US-Canada working group
on acid rain included opening chapters and appendices that stressed
the "uncertainties, gaps, maxima-minima of work group data, in other
words, to place the studies into broadest perspective for policy makers."
Funkhouser feared that the scientific reports from the working group
would be used by Canada and the Scandinavian countries as well as
intergovernmental groups like the OECD and UN in order to pressure
the US government to make emissions reductions.[50] He also blocked the
release of a key OECD report on global environmental issues over its
discussion of acid rain, even though the report received approval from
other US agencies including the State Department.[51] Internal documents
show that Funkhouser along with the Department of Energy strongly
objected to any discussion of emissions controls in the OECD report
and wanted the document to call only for exchange of information and
further research on the problem.[52] Based on these actions, Scandinavian
environmental officials identified Funkhouser and Bennett as the two
key American obstacles to international action alongside Britain's Peter
Chester in the early 1980s.[53]

Career civil servants in the EPA were also put under pressure by their
counterparts in the Department of Energy to minimize research findings
that implicated fossil fuel emissions in the acid rain problem during the
early years of the administration. In conversations between the Department of Energy and EPA over issuing a 1981 government acid rain report, the Department of Energy made clear that it was "essential" for
EPA officials to amend certain language so that the regulatory impact
of the findings would be limited. For instance, they asked the EPA to
explicitly state that the report "should not be used as a basis for regulatory or legislative action."[54] They also wanted a reference inserted
about alleged uncertainties concerning whether reductions in sulfur dioxide would lead to comparable reductions in acid rain.[55] The controversy grew so heated that the EPA scientist on the government's acid
rain working group requested that a single Department of Energy scientist be appointed to discuss the findings, and expressed his frustration

that a number of the department's scientists would not sign-off on the aquatic effects section of the report despite his attempts to ensure it was "scientifically sound."[56] The exchange is emblematic of the growing desperation on the part of government scientists over the Reagan administration's suppression of acid rain research.

In Western Europe, the increasing boldness of the CEGB after Prime Minister Thatcher took office led to a confrontation between Norwegian ecologists and Britain's Department of the Environment that had lasting repercussions for the acid rain issue both domestically and abroad. The controversy started in the fall of 1981 when the director of the CEGB's research laboratories, Peter Chester, began lobbying Norwegian government officials on acid rain. He reached out to representatives he believed would be sympathetic to the coal industry position in order to discuss findings that he claimed challenged "the simple explanation that a reduction in SO_2 [sulfur dioxide] emissions in Western Europe would bring about a corresponding amelioration of fishery problems in Southern Scandinavia."[57] Chester wrote that gaps in knowledge could be cleared up by more studies over the coming years, and in the meantime "it would be heartbreaking for Europe to invest £ billions in measures which then turned out to be ineffective."[58] Unsure of what to do, the Norwegian officials conferred with the SNSF project steering committee and the Ministry of the Environment to discuss Chester's statements.[59]

After learning of the situation, Norwegian scientists decided to aggressively and directly respond. Hans Martin Seip was asked to draft an extensive rebuttal to each of Chester's claims, which was sent not only to Chester and the CEGB but Britain's Department of the Environment as well.[60] In his comments to Chester's assertions, Seip took great pains to identify portions of Chester's analysis with which he agreed, such as the difficulty in making quantitative predictions about the effect of reduced emissions. However, he argued that recent studies strongly supported a connection between sulfur dioxide emissions and acidity.[61] Seip reserved most of his criticisms for Chester's writings on the impact of acid rain on aquatic ecosystems, particularly Chester's claim that there was not a relationship between acid rain and acidity found in lakes. He pointed out that Chester's analysis had included lakes with markedly different geological features, and thus did not control for the ways in which various rock and soil types respond to acid rain. The importance of soil and rock types for the susceptibility of nearby freshwater ecosystems to acid rain had been known for many years. Indeed, it seems likely that Chester's decision to aggregate lakes of various geological types was a deliberate misrepresentation so that he could argue there was little correlation between acid rain and lake acidity without controlling for these confounding variables. Seip concluded his response by stressing that though many scientific details still needed to be worked out, if they all had to be cleared up before action was taken to reduce emissions, the environmental damage could be difficult to repair. "You may find it heartbreaking for Europe to invest an enormous amount in measures which turn out to be ineffective," Seip wrote, but "others will find it heartbreaking if nothing is done, and the consequences turn out to be as serious as feared."[62]

While it's unclear whether the British government ever responded directly to Seip or Norwegian government officials, his letter ignited wideranging discussions among British environmental officials, who began to more sharply question whether CEGB acid rain research was distorted to serve the political and economic interests of the coal industry. The Department of the Environment's chief scientist, Martin Holdgate, was particularly troubled by the exchange. After receiving a copy of Seip's reply he told his colleagues that he was concerned about the degree to which CEGB scientists had become the main source for government information on the acid rain problem. Though careful not to explicitly impugn their scientific integrity, Holdgate suggested that the coal industry's financial interest in avoiding emissions controls could be biasing their work.[63] In discussing a recent report published by Chester, Holdgate pointed out that the CEGB director had violated one of the most basic tenets of the scientific method in claiming there was insufficient evidence of acid rain's damaging effects. Chester had referenced only one source when there was an extensive body of literature on a particular topic. Yet although Holdgate and other officials in the Department of the Environment felt it would be wise to have an independent referee review the research, there were no government scientists outside the CEGB with extensive experience in acid rain studies.[64] The Thatcher administration's decision to defund acid rain research had left the department in an unfortunate bind: unable to rely on the CEGB and without other experts to turn to whom they felt they could trust.

It also left the British government woefully unprepared to deal with the explosive findings from the ITE and DAFS showing acid rain could also be affecting Northern Britain. Their studies were well known within the scientific community, having been presented at the SNSF's Sandefjord Conference back in 1980 as well as at several meetings on acid rain's environmental effects organized by the CEGB since then. However, officials in Britain's Department of the Environment were entirely unaware of this work. They only learned of the research as a result of a literature survey performed at the insistence of the Norwegian government, which refused to participate in a British proposed international study on acid rain's ecological effects unless Britain compiled an assessment of the studies done to date.[65]

Completed in the spring of 1982, the literature survey shocked members of Britain's Department of the Environment. Based largely on David Fowler's studies as well as those by the DAFS, the report showed that rainfall in areas of Scotland and Wales had experienced large declines in pH, which suggested that acid rain might be impacting Britain's environment. Perhaps most startlingly, the report stated that in some areas "the amount of H+ [a measure of acidity] deposited is of the same order as regions of Europe, Scandinavia and North America where inputs of acidity in rain are also relatively large."[66] Holdgate was stunned by the revelation that Britain could be experiencing a threat from acid rain similar to that faced by Norway, Sweden, Canada, and the eastern US, and told his colleagues that the Department of the Environment clearly needed to support further research in Scotland to determine whether there were any observable impacts on fish.[67]

Yet as weeks passed with little response from the Thatcher administration, someone—perhaps one of the authors of the report, one of its university-based scientific consultants, or even a member of the Department of the Environment—leaked a copy of the still unpublished draft to the *Observer*. It then published a harsh critique of Thatcher's cuts to environmental research in light of the recent findings concerning acid rain, sparking an enormous amount of attention from the British press, public, and Parliament.[68] The Department of the Environment was soon inundated with questions from legislators about the problem and what the government planned to do about it, first from members representing areas in Scotland but eventually including those from across Britain.[69]

With pressure mounting on the Thatcher administration to reinstate its funding for acid rain studies, Holdgate reached out to Fred Last at the ITE to discuss allocating resources for acid rain research to produce a scientific basis for policy decisions "independent of major vested interests."[70] His discussions with Last elicited an enthusiastic response about possible areas of collaboration, but also admonishments about the need for the Department of the Environment to acknowledge acid rain's effects on aquatic ecosystems and to keep an open mind about the potential impact on forest ecosystems.[71] Holdgate also received a strong rebuke from scientists at the Natural Environment Research Council, which had seen their programs on acid rain gutted in the previous years.[72] Upon learning of the sudden renewal of interest in financing such projects, they told Holdgate in no uncertain terms that the withdrawal of government funding for research outside the CEGB was having devastating effects both at home and abroad. The lack of support for their work had allowed CEGB scientists to assume an even greater international profile, they said, and the widespread mistrust of their neutrality in acid rain was damaging the British government's credibility with other nations.[73] Once these conversations were underway, someone leaked to the press that the Department of the Environment had agreed to approve further financing before a final budget had even been approved, apparently in the hopes of preventing government officials from reneging on their offer.[74]

Similar leaks of discord also started emanating from the US EPA a short time later. Civil servants who had worked for the EPA since before President Reagan took office gave a damning account of the corruption under the new administration to the *New York Times* in September 1982 following the British scandals over acid rain in Scotland and Wales. They described worsening relationships with Reagan officials over budget cuts, mismanagement, and their overall disdain for environmental science and regulation. But the acid rain issue seems to have especially vexed EPA workers, who were dismayed that the Reagan appointees continued to deny any link between fossil fuel emissions and acidic precipitation.[75] The question, however, was whether environmental scientists at the US EPA and Britain's Department of the Environment could overcome the growing influence of the coal industry over domestic legislation and international diplomacy.

International Pressure Meets Domestic Politics

One of the first opportunities to try to diminish the power of the coal industry in the Thatcher and Reagan administrations came in the summer of 1982 when Sweden hosted an international conference on acid rain in Stockholm. It invited all signatories to the 1979 UN Convention on Long-range Transboundary Air Pollution to attend with the aim of making progress on international negotiations for reductions in sulfur dioxide emissions. The conference had initially been proposed as a way to commemorate the ten-year anniversary of the first UN conference on environmental issues held in Stockholm in 1972, but soon assumed importance as an avenue to address acid rain since only four countries had ratified the 1979 UN Convention when planning began.[76] In cooperation with Norway and Canada, Swedish officials from the Ministry of Agriculture organized a broad campaign known as Miljö ' 82 (Environment ' 82) in order to "generate the interest of the media and environmental groups in the conference and exert pressure on major emitters."[77] They hoped that the meeting would maintain, and perhaps even further, the political momentum surrounding acid rain.[78]

The conference preparatory committee decided to first convene two parallel scientific expert meetings to discuss the ecological effects of acid rain and methods to control pollution emissions. These would be followed by a ministerial meeting of government representatives, with the scientific discussions intended to inform them about "the increased understanding of the mechanisms and effects of transboundary air-pollution-acidification and of possible options for control that have emerged since the signing of the Convention in 1979."[79] In addition to preparing their own background materials from Swedish scientists, the conference organizing committee asked each country's environmental officials to nominate its own scientists to attend the expert meetings.

As Britain's Department of the Environment prepared for the Stockholm Conference in the early months of 1982, its selection of representatives for the expert meetings proved to be a delicate issue because of the emerging concerns about the coal industry's influence in acid rain policy. After its chief scientific officer, Martin Holdgate, learned who would be attending from other countries, he cautioned the department that the scientific qualifications of these experts were beyond reproach. They included American ecologist Gene Likens, the German soil chemist Bernard Ulrich, and Swedish scientists such as Svante Odén, Henning Rodhe, and Goran Persson. These men could not be "dismissed as econuts" and would not take kindly to arguments about the lack of scientific evidence.[80] In Holdgate's view, it would therefore be crucial to ensure that all British scientific materials were "accurate and objective" and that the country's delegates held unimpeachable credentials.[81] However, revising the government's position on the issue would likely put them in a "head-on collision" with the CEGB.[82]

As a compromise solution, several environment officials suggested including nominees from universities or government research institutes, such as Fred Last and David Fowler, while still asking CEGB scientists like Ron Scriven or Gwyneth Howells to "ensure the UK expert delegation does not become CEGB dominated while recogising at the same time the important role CEGB scientists are playing in the acid rain issue."[83] On the matter of what these scientists should say at the conference about the evidence for acid rain's deleterious effects, Holdgate suggested the British delegates should openly accept that there had been an increase in acidity in Scandinavian lakes and rivers that was harming the environment. However, he felt they might nevertheless insist scientists needed to be able to "work backwards" from areas of damage to determine what degree of pollution control was necessary from individual power plants before reductions could be implemented. This was intended to provide a concession to CEGB scientists, since Department of the Environment officials feared they would balk at conceding that acid rain caused ecological damages if this implied "a readiness to contemplate emission reductions."[84] In general, Holdgate told his colleagues, it would be important to display a positive approach while still resisting pressure for "any gesture made for gestures sake, like some commitment to everybody to halve their sulphur dioxide emissions by the end of the century."[85] It was hoped that the Department of the Environment could convince the CEGB to agree on Holdgate's approach to the scientific discussions.

Unfortunately, as the British government entered its final preparations for the Stockholm gathering, it appeared that Thatcher's Cabinet planned to adopt a position on the scientific evidence in direct conflict with the Department of the Environment's recommended stance. Just a few weeks before the conference, Holdgate received minutes from a meeting of Cabinet officials with members of the Department of Energy and Department of the Interior. The notes indicated these officials were preparing to argue there was no scientifically proven link between fossil fuel emissions and acid rain.[86] Holdgate immediately rallied other environmental officials to tell Thatcher's Cabinet that no one in the scientific community except coal industry researchers disputed that these pollutants caused acid rain. He acknowledged there was still uncertainty over the "chemical details" but informed Thatcher's advisors that research indicated acid rain was the primary cause of observed damages in freshwater lakes and that some evidence was accumulating on acid rain's effects on forests. While the precise relationship between acid rain and an increase in acidity needed further quantification, denying a causal link altogether would put the British delegation "widely at variance with the general consensus of scientific opinion."[87] Given the disparity between the Thatcher administration's proposed line and the current scientific consensus, Holdgate then took the dramatic step of telling the Cabinet that as chief scientist of the Department of the Environment he would not support any such statements.[88]

Upon learning of Holdgate's objections, the Chief Scientific Officer of the Cabinet Office, Dr. Robert Nicholson, also contacted the Cabinet Office Deputy Secretary to state his agreement with Holdgate's position and analysis of the current literature on acid rain.[89] The Undersecretary for the Environment and upcoming British representative to the Stockholm Conference's ministerial meeting, Giles Shaw, lent his support to

Holdgate as well. Shaw informed the Secretary of State for the Environment that he would like to take a more positive approach to making emissions reductions at Stockholm rather than "referring the matter back to research."[90] Yet Holdgate's ultimatum received a tepid reply from officials at the Department of Energy, who insisted that scientific research was far from clear about the benefits of expensive reduction measures without addressing the more substantive points of his protests.[91] The Cabinet Office was equally dismissive and openly admitted that decisions about the British line at Stockholm had not been determined by the scientific evidence but rather by "tactical and political considerations."[92] They informed the Department of the Environment that all British representatives would be instructed to concede links between fossil fuel emissions and acid rain only as necessary. Though the Department of Energy insisted that these instructions would be limited to the ministerial meetings and not the scientific expert gathering, it seems clear that Thatcher's administration intended to adopt the CEGB tactics of using scientific uncertainty to undermine the Stockholm conference.[93]

The prospects for a successful meeting appeared equally hopeless to many American scientists and policymakers. Prior to the conference, EPA researchers contacted the British embassy in Washington, DC, to try to warn them about the Reagan administration's preparations for the meetings. One, James L. (Larry) Regens, confessed that he had been receiving "off the wall" questions from Dick Funkhouser on acid rain in advance of the Stockholm meeting and was concerned about the administration's approach to the expert meetings.[94] Regens' suspicions turned out to be well founded. Soon after informing the British embassy about the dire situation at the EPA, US diplomats told British environmental officials that the new US government tactic was to "exploit the uncertainties and apparent contradictions" to avoid pressures to reduce emissions.[95] The Chief Scientific Officer to President Reagan, Dr. George Keyworth, explicitly laid out their desired strategy to Martin Holdgate. He told the British scientist that he felt the Canadian "emotions outran science" and the administration would not agree to expensive action to reduce emissions with "no sound basis in science." Despite having received pushback from the Thatcher administration, Holdgate bluntly told Keyworth that the British Department of the Environment accepted that fossil fuel emissions caused acid rain and harmed freshwater ecosystems, and he strongly suggested that the Americans would be ill-served by a negative approach to the Stockholm conference.[96]

Despite efforts to improve their governments' positions, scientists at the EPA and Department of the Environment were unsuccessful in convincing either administration to alter its stance. Hard line experts touting the coal industry's perspective were sent to the scientific meetings from both Britain and the US, including Gwyneth Howells, Ron Scriven, and Kathleen Bennett. Bennett in particular received widespread criticism from the American scientific community for her statements in Stockholm. Acid rain researchers who attended the meeting as observers accused her of engaging in "specious and dangerous distortion" during negotiations and using "outdated research."[97] Notwithstanding the controversial positions taken by Bennett, Scriven, and Howells, the expert meetings achieved some notable points of agreement to transmit to the ministerial meeting the following week. At the ecological gathering, the scientific delegations reached a consensus that an overall reduction in emissions would lead to a similar decrease in acid rain, though certain areas might experience larger or smaller declines.[98] This was in stark contrast to the statements of CEGB and EPRI scientists, who claimed it was unclear whether lowering pollution levels would bring about any appreciable reduction in acid rain.

Progress was also made at the meeting on strategies for reducing emissions. British and American protests about the severe economic costs of a 50 percent reduction in power plant emissions were not supported by OECD analyses submitted to the conference. The organization's director of the environment, Jim McNeil, presented data suggesting that a 50 percent reduction could be achieved within a few years at a cost of just 2.5 million pounds. Britain's Leslie Reed, an official at the Central Unit on Environmental Pollution, had previously argued this sum would only achieve reductions of about 25 percent and raise British electricity bills by 15 percent.[99]

The scientific meetings at Stockholm thus plainly refuted the claims of coal industry scientists on both acid rain's environmental risks and the costs of controlling pollution emissions. Nevertheless, the British and American delegations returned from Stockholm largely pleased with the outcome of the conference.[100] Although the final declaration stated that future coal-fired power plants should have some form of pollution control, the delegations had achieved their aim of preventing any proclamations on the need to reduce emissions immediately.[101] At home, however, the fight over acid rain was only just beginning.

In Britain, the Department of the Environment continued its lobbying effort to alter government policy on acid rain in the wake of the conference. Officials decided to intensify their outreach to the Department of Energy, hoping to persuade the group that it would be wise for the administration to alter its stance. Writing to David Mellor, Thatcher's

Under-Secretary of State in the Department of Energy, they argued that the international pressure Britain received in Stockholm made it imperative for the country to reevaluate policies on acid rain.[102] Even the French, which had historically held similar views about the need for more scientific proof, had recently agreed to control emissions from any new power stations.[103] For these reasons, their letter stated, the British government should require the CEGB to review prospects for reducing emissions.[104]

The international backlash Britain received from its performance in Stockholm convinced the Department of Energy to finally join the Department of the Environment in putting the CEGB and National Coal Board "on public notice" that they may have to consider sulfur dioxide reductions in the future.[105] In January 1983, the Secretary of State of the Department of Energy, Nigel Lawson, wrote to the CEGB and National Coal Board to inform them that they needed to review their policies on sulfur dioxide emissions and acid rain in light of the 1982 Stockholm Conference and initiatives to control air pollution within the European Communities. However, the CEGB and its scientists were unmoved in their opposition. At meetings of the CEGB and National Coal Board following Lawson's warning, the head of CEGB research, Peter Chester, argued that the coal industry should maintain its stance that more research was needed. Perhaps, he noted, it would be wise to finance projects by outside scientists as a compromise with environmental officials.[106] However, any such project would still serve as a stalling tactic on acid rain.

In the US, the pressure from Stockholm was amplified the following year by a new acid rain report from the National Academy of Sciences. But environmental scientists had far less success capitalizing on it to earn the support of the American Department of Energy for acid rain reductions. Released in June 1983, the report specifically addressed the question of linearity, an issue EPRI had previously raised as a reason to delay reductions in fossil fuel emissions.[107] If there was not a "linear" relationship between emissions and acid rain, the argument went, then a 30 percent reduction in emissions might only reduce acid rain by 25 percent, 15 percent, or not at all. There was only one problem: by 1983, scientific studies had shown the relationship was, in fact, roughly linear. The National Academy of Sciences report explicitly stated that "there is no evidence of a strong nonlinearity in the relationships between long-term average emissions and deposition." While the scientists were careful to note that models were still unable to know which specific power plant was harming a certain area, they stated that an overall reduction in fossil fuel pollution would lead to a comparable reduction in acid rain.[108]

The report caused a stir among the public as well as Congress, which subsequently held a series of hearings in the fall of 1983 to discuss whether controls on coal-fired power plants were now warranted. Senator Max Baucus of Montana was instrumental in convening testimony on the report as part of the Senate Committee on the Environment and Public Works. He had expressed his support of acid rain legislation in the early years of the Reagan administration even though he represented a major coal-producing state. Baucus was frustrated that amendments to the Clean Air Act in 1977 required the installation of flue gas scrubbers in new power plants rather than allowing flexibility in meeting air quality standards. Montana was home to deposits of low-sulfur coal, which would likely have been in greater demand if regulations had offered other options in meeting the standards.[109] New acid rain legislation therefore had the potential to help producers of low-sulfur coal in his state. Baucus had also been an early critic of Kathleen Bennett's approach to acid rain work, believing that the EPA under her purview was acting in bad faith by calling for more research instead of trying to find policy solutions.[110]

Ahead of the Senate's hearings, Baucus reached out to Bernard Manowitz, chairman of applied research at Brookhaven National Laboratory, to ask the opinion of the government's national laboratories on the National Academy of Sciences report. Manowitz, who had overseen all Brookhaven's acid rain work since the mid-1970s, was an obvious point person to solicit comments on the report. The Department of Energy and its national laboratories had been conducting evaluations of linearity in acid rain emissions for several years, so they must have seemed ideal experts to assess the 1983 report. However, what was not disclosed to Congress or raised publicly at any time was that EPRI funded this work at the national laboratories, a relationship that seems to have impacted these scientists' assessment of the issue.[111]

Manowitz invited eight scientists to provide a critique of the report: Stephen E. Schwartz, Lawrence I. Kleinman, Paul Michael, and Fred Lipfert from Brookhaven laboratory; George N. Slinn, Terry Dana, and Richard C. Easter from Battelle Pacific Northwest laboratory; and Jack Shannon from Argonne laboratory. The funding relationships between these laboratories and EPRI stretched back years. For instance, in 1978, Brookhaven and Battelle laboratories began an EPRI funded project to study the chemical reactions involved in converting sulfur dioxide emissions to sulfate, the form of the pollutant in rainfall. The Department of Energy also collaborated with EPRI on the latter's acid rain monitoring network, known as the Sulfate Regional Experiment (SURE) program.[112]

In addition, Brookhaven and the US Department of Energy assisted EPRI with its ecological research on acid rain's impacts in the Adirondack mountains.[113] Scientists from the Battelle Pacific Northwest Laboratories who participated in the critique also had longstanding funding relationships with EPRI dating back to the mid-1970s, including on acid precipitation.[114]

EPRI funding was distributed to the national laboratories through grants, but the manager of the work was typically listed as an EPRI employee and the corresponding scientists at the institutions were not specified. It would therefore have been difficult at the time for an outsider to uncover their financial relationship. However, with hindsight it's evident that most if not all of these scientists were involved in the EPRI projects and staunch advocates of industry's position on acid rain. One of the Brookhaven scientists, Fred Lipfert, had recently made public statements that acid rain science was "scanty" and had a longstanding relationship with the power industry.[115] Before coming to Brookhaven, he had worked at the Long Island Lighting Company on pollution monitoring technology. In that role, he first began consulting for EPRI on acid rain in 1976 as they prepared to launch the SURE program.[116] Following his move to Brookhaven, he remained involved in EPRI research and soon became one of the leading proponents of the theory that fossil fuel emissions and acid rainfall had a nonlinear relationship.[117] After leaving Brookhaven to become an independent environmental consultant near the end of the decade, he worked on multiple contracts for EPRI on acid rain as well as other pollution problems; most of his publications supported the industry perspective on these issues, ranging from linearity in acid rain to the relationship between air pollution and public health.[118]

In addition, both Paul Michael and Lawrence Kleinman from Brookhaven worked with EPRI's SURE Program during the early 1980s, though Kleinman appears to have been much more involved in acid rain research.[119] Richard Easter, one of the three reviewers who prepared a joint response from Battelle laboratories, had worked for EPRI on the SURE program as well. Steven Schwartz, a scientist at Brookhaven, also collaborated on projects concerning the transformation of sulfur dioxide to sulfate in power plant plumes that were funded by EPRI.[120] Most of this work focused on atmospheric chemical reactions related to sulfate production, and none of these scientists had extensive experience modeling the long-range atmospheric transport of sulfur dioxide or other pollutants or was trained in atmospheric physics.

Their assessment of the National Academy of Sciences report was highly critical, so much so that congressional officials were startled by the harsh, combative tones in their reviews. The scientists unanimously argued that the report represented poorly conducted science and should not be taken seriously by policymakers. In his testimony to senators, Manowitz said that the report's results were inconclusive and inadequate to serve as anything more than a hypothesis for future research. As a result, he advised the committee that the US government should adopt the approach of the Nierenberg Committee, which acknowledged current "uncertainties" and was sensitive to possible costs of regulation.[121] His remarks were echoed by Jack Shannon from Argonne laboratory, who asserted that while the report's findings were reasonable they were not "conclusive."

Several senators, flummoxed by the testimony of Manowitz and Shannon, grilled the pair about their remarks and the severe critiques from the national laboratory scientists. Shannon received especially harsh treatment from the senators since he had actually been part of the National Academy of Sciences group and signed off on the report, leaving the committee members confused about his abrupt reversal in position. He had been personally quoted in the press just a year earlier saying that new computer models in fact showed "a 50 percent reduction in smokestack emissions would reduce the amount of downwind acid deposition by an almost equal amount."[122] Manowitz was eventually called out by a senator for perhaps letting self-interest guide his response to the report since the national laboratories were then heavily invested in studies of nonlinearity. Should the question be resolved, the senator noted, Brookhaven would perhaps lose a major funding source. This was as close as the Senate came to uncovering the financial connection between the national laboratories and the coal industry during the hearing.

Environmental scientists in attendance tried to remain positive about the intentions of the national laboratories scientists while still defending the report. Michael Oppenheimer, a leading atmospheric scientist who was one of the country's top researchers on linearity, testified that he was baffled by the critiques but believed the national laboratory scientists were honest, good people. He and Myron Urman, head of the National Research Council's Environmental Studies Board, attributed the disagreement to the national laboratories group looking at more microscopic issues in modeling transport and chemical reactions in the atmosphere rather than the larger question of whether a certain percentage reduction in emissions would result in a proportional reduction in acid rain. As they explained, when you look at small-scale atmospheric reactions there will always be some degree of "nonlinearity," but this did not necessarily apply to the overall impact of emissions reductions.

One startling aspect of the congressional debates over the National Academy of Sciences report is that models for predicting transport patterns had been under development in Europe for quite some time and had already demonstrated a linear relationship between emissions and acid rain.[123] Generating the necessary data to study this question was a major purpose of the OECD and UN acid rain research programs. Anton Eliassen, leading a group of Norwegian atmospheric scientists, produced the first models to predict dispersion patterns in the late 1970s, and they were continually refined as monitoring stations across the European continent generated new data.[124] Yet this work seems to have had little impact on the perspective of the national laboratory scientists toward the consensus reflected in the National Academy of Sciences report.

In fact, for the rest of the decade many of these scientists continued to argue against implementing pollution reductions because models could not perfectly predict which power plants were responsible for acidifying particular ecosystems. This was despite widespread acceptance of the linear relationship between emissions and acid rain within the scientific community. Steven Schwartz, for example, eventually conceded that regionally acid rain would be reduced proportionally with emissions reductions. He nevertheless maintained that because atmospheric physicists could not precisely model the dispersion of power plant pollution, regulators should be cautious in seeking control strategies.[125] His comments were criticized by two MIT atmospheric scientists in *Science* as giving "comfort to industrial and political leaders urging inaction on control of acid deposition precursors until a distant future when perfect scientific understanding might be reached," although Schwartz asserted this was not his intention.[126]

Were the national laboratory scientists who accepted EPRI grants for work on acid rain somehow biased by the funding? On one hand, this does not seem to have been the case for all scientists who received money from the organization. There were several university researchers who used grants from EPRI for acid rain work and yet were outspoken advocates for acid rain regulation. However, the relationship between the national laboratory scientists and the coal industry before and after the 1983 hearings certainly seems relevant in assessing their perspective on the issue. Most alarmingly, their ties were not disclosed to congressional officials or the public during their testimony. Would senators have evaluated their opinions differently had they known the source of money for their work on acid rain? It seems likely. And while it's impossible to know whether this would have changed the course of acid rain policy in the US in light of strenuous opposition to fossil fuel regulation among congressional representatives from mid-Western states, it absolutely had a negative impact on progress toward acid rain control in the coming years.[127]

By the mid-1980s, EPRI officials were pleased that "the complexities of the acid rain story were beginning to be appreciated in Congress," and felt they had largely succeeded in their campaign to prevent any requirements to reduce emissions.[128] The British CEGB was far less sanguine, since the Department of the Environment had convinced the Department of Energy of the need to reevaluate the country's acid rain policies. The contrast shows the significance of Norwegian environmental scientists' intervention with Britain's Department of the Environment to demonstrate that they were receiving biased information from the CEGB. Their efforts were aided by differences in the British civil service, which provides continuity in administrative positions regardless of who holds power in Parliament. That is why Martin Holdgate, the Department of the Environment's chief scientist, was able to retain his powerful position in the agency for nearly two decades. In the EPA, the voices of long-serving scientists were easily subsumed by the priorities of Reagan-appointed officials like Dick Funkhouser, Kathleen Bennett, and John Hernandez. The potential biases of the CEGB scientists were also plain for everyone to see as they were directly working for the British coal industry; those of the US national laboratory scientists and other EPRI contract recipients were not. A final crucial difference between the two countries would ultimately force the Thatcher administration to confront the dangers of fossil fuel pollution before its conservative ally. The country was party to a supranational union that, while weak in some respects, had enough authority to impose costs on Britain for breaking with the European Communities on acid rain.

7

Acid Rain and the Precautionary Principle

The most banal comment of all is that acid rain is "a £million problem with a £billion solution" . . . it plucks a "cost" figure out of the air.[1]

Anthony Tucker, science correspondent, *Guardian*, 1983

As regards the research policy question in my earlier note . . . I found the [acid rain] report of special interest. It essentially adopted a position that one should not entertain arguments of economic reasonableness because this could undermine the policy of "pure prevention." This, of course, is essentially the issue of policy choice by the Commission as a whole which I wish to clarify.[2]

Hywel Davies, Commission of the European Communities, 1983

In 1977 David Davies, a British geophysicist and the current editor-in-chief of *Nature*, publicly weighed in on the costs and benefits of reducing acid rain pollution. Responding to recently released figures on the amount of sulfur dioxide reaching Scandinavia from his home country, Davies came down squarely on the side of his countrymen who opposed lowering fossil fuel emissions. Calculating that the loss of fish in Southern Norway amounted to only one million dollars, Davies decried the "billion dollar" solution of reducing air pollution through costly control technologies for such paltry environmental benefits.[3] Erik Lykke, Norway's director general of the Ministry of the Environment, wrote a sharp rebuttal to Davies' editorial pointing out that the issue of acid rain was more than a question about the worth of salmon.

For one thing, although freshwater ecosystems may have been most vulnerable to acid rain, its impacts spread beyond lakes and rivers. Scientists suspected it was likely contributing to environmental damages in forests as well as corroding buildings and other infrastructure. But perhaps more troublingly, they believed that acid rain had the potential to react synergistically with other industrial chemicals in ways scientists were only beginning to understand. The long-term consequences of these emissions on the health of the environment therefore seemed impossible to quantify.[4]

Despite Lykke's counterarguments, Davies' characterization of acid rain as a million dollar problem with a billion dollar solution became a prominent trope during the 1980s in both Britain and the US.[5] Comparing the economic costs of controlling pollution against the economic benefits of reducing acid rain seemed to bolster the coal industry's argument that governments would be unwise to take action in the face of scientific uncertainty. Why require expensive flue gas desulfurization technology on power plants, when not only was the science of acid rain uncertain but any possible benefits minimal? While a seemingly straightforward argument on its surface, debates about the economics of acid rain were rooted in different ways of valuing the environment and calculating the economic cost of environmental damages. Was it only fish that needed to be included in a calculation of pollution damages from sulfur dioxide? Or should other types of organisms in aquatic ecosystems be incorporated into these assessments? Crucially, how should scientists and economists measure the possibility that emissions could cause future damages to the environment, and how could policymakers weigh the evidence for this risk against the more easily quantified economic costs of controlling pollution?

These questions came to dominate debates about acid rain in Europe during the mid-1980s as scientific research in West Germany found that entire forests were at risk of damage from acid rain, a troubling possibility that prior ecological studies had investigated without finding clear evidence one way or the other. However remote, the prospect that fossil fuel emissions could cause far-reaching environmental impacts across the continent provoked a massive public outcry and reshaped environmental diplomacy on acid rain. The British government would now have to fend off pressure not only from the Scandinavian countries but from West Germany as well, which was then a powerful force within the European Communities. Because of Britain's membership in the intergovernmental group, the Thatcher administration's policies on acid rain diverged sharply from those of President Reagan in this period—but not before the British coal industry made one final attempt to avoid emissions reductions by funding more scientific research.

Faced with the obstinacy of the coal industry about the state of scientific knowledge, European legislators were left with two options. They could continue debating whether sufficient proof had been amassed on acid rain's effects and whether reductions would improve the environment at a reasonable cost. Or, they could stop arguing about the science altogether and find a rationale for acting in the face of uncertainty given the risk of enormous environmental costs should the science turn out to be right. Environmental officials in West Germany and the European Communities ultimately chose the latter course. They turned the million dollar problem—billion dollar solution claim on its head and formally introduced the precautionary principle into an international environmental agreement for the first time.

Costs and Benefits of Precaution

The Norwegian spruce and pine forests of Scandinavia and Central Europe stretch for thousands of miles across fjords and mountain ranges. For centuries, the forests have been an enduring influence on German and Nordic cultures, and the possibility of their destruction from acid rain had long preoccupied Scandinavian ecologists.[6] Yet when Norway's massive national study on how acid rain affected ecosystems concluded in 1980, no definitive findings had been produced about the impact of acid rain on forest ecosystems even though the evidence for damages on freshwater ecosystems was extremely robust. Within a year, however, the potential for acid rain to damage Europe's forests saw renewed interest from scientists, government officials, and the public thanks to a series of articles published in *Der Spiegel*, which were based on the work of the scientist Bernard Ulrich and several of his colleagues at the University of Göttingen in West Germany.

Ulrich, originally trained as a soil chemist, began studies of forest ecosystems in Solling, West Germany, through the International Biological Program (IBP), which coordinated large-scale biological research projects across Europe and North America from 1964 to 1974.[7] The Solling area, a mountainous region far from industries in West Germany, was chosen as the IBP's site for research on a forest biome.[8] After several years of measurements, Ulrich and his colleagues discovered that samples of rainwater and soil runoff showed increases in acidity despite the location's distance from polluting sources.[9] At the same time, trees in the West German Black Forest began showing signs of serious damage. Conifers stood like barren skeletons without needles on their branches; beech trees turned yellow and lost their leaves.[10] Ulrich and several of his

FIGURE 7.1 Forest damages in West Germany suspected to be caused by acid rain. Swedish and Norwegian NGO Secretariats on Acid Rain, "Acid News," January 1984, UK Parliamentary Archives, HC/CP/9445, p. 1.

colleagues believed that the observed changes in the forests were likely connected to precipitation acidity and hypothesized that acid rain was releasing toxic metals from the soil and damaging tree roots. Over time, they argued, it had the potential to cause the "collapse of the entire forest ecosystem," though earlier research in Scandinavia had not found reason to suspect this was an imminent danger.[11] To address the problem, Ulrich called for reducing sulfur dioxide emissions to protect areas in East and West Germany, Poland, and Czechoslovakia as well as adding lime to affected soils as a stopgap measure in the meantime.[12]

Sensitive to the policy implications of this research, Ulrich sent his work to newspaper reporters immediately after publishing the findings in a scientific journal.[13] *Der Spiegel*, a leading German press, soon began a series of articles on the possible loss of the country's forests from acid rain that attracted international attention. Though his efforts to agitate the public earned him a controversial reputation even among like-minded researchers in Scandinavia, Ulrich said he felt a moral obligation to speak directly to the German public about the potential threat of acid rain even though he believed there was only an 80 to 85 percent chance his theory was correct.[14] While Ulrich's dire pronouncements also contradicted the prevailing scientific consensus on the ability of acid rain to harm forest ecosystems, the massive public uproar over his findings spurred the West German government to quickly formulate domestic policies to reduce the country's sulfur dioxide pollution.[15] If acid rain was truly to blame for the observed damages, however, it would not be enough for the country to act alone. Given the dispersion patterns of pollution over Europe, West Germany would need to convince its neighbors to reduce their fossil fuel emissions as well. Yet Britain, the largest polluter in Western Europe, was unlikely to be moved to action by pleas about the fate of Germany's forests or by Ulrich's uncorroborated research, especially considering the insistence of Britain's coal industry that acid rain science was too uncertain.

Fortunately for West Germany, a newly emerging economic rationale for reducing the pollutants that caused acid rain coalesced around the same time as Ulrich's dire predictions were put forward. Economists and policymakers at the Organisation for Economic Cooperation and Development (OECD) had begun researching the economics of environmental pollution as far back as the early 1970s. OECD officials sought to pioneer ways of accounting for "externalities" like pollution, particularly out of concern that different environmental policies in various countries could lead to trade imbalances as well as the export of polluting industries to places with lax regulations.[16] As it became clear during the 1970s that an expected rise in energy demand would increase reliance on coal and oil by 1980, the OECD's economic research on the costs and

FIGURE 7.2 Photograph of Bernhard Ulrich, taken by Fred Pearce and later featured in "The Strange Death of Europe's Trees," *New Scientist*, December 4, 1986, p. 41.

benefits of pollution control assumed even larger importance within the organization.[17]

It also attracted the interest of several renowned European economists, including the Nobel prize winner Wassily Leontief, who assisted the OECD Secretariat with evaluating the economic costs and benefits of reducing fossil fuel pollution. In 1980, their analysis found that the cost of abatement in member states currently stood at about 1–2 percent of each country's gross domestic product (GDP), while the damages from pollution were 3–5 percent of GDP.[18] The OECD subsequently released several reports arguing that the economic costs of lowering fossil fuel pollutants were equal to, if not less than, the economic value of better environmental and human health as well as preservation of property values.[19] The OECD's calculations were widely reported in the media throughout the early 1980s, particularly in the West German press regarding the implications for the possible costs and benefits of reducing acid rain.[20] West German delegates to the European Communities subsequently advocated for using these cost-benefit analyses as a basis for further international negotiations on acid rain, and pledged that the European Communities would strengthen its role in acid rain diplomacy under its coming presidency of the group.[21]

Although the OECD's economic work was a significant step forward, it did not resolve the ongoing conflict about whether the science of acid rain's harms was robust enough to take action. Previously, the OECD had tried to examine possible ways industrialized nations could tackle problems of scientific uncertainty and environmental externalities. Together with its work on the economic costs of pollution, in the 1970s the OECD had investigated what it called "anticipatory environment policies" to address the issue of external harms not reflected in the current energy markets.[22] The idea of acting in advance to prevent environmental damage was certainly not a new one. Political arguments about the need for careful planning and risk management in environmental protection have a long history, including the earliest efforts to cope with radioactive contamination and DDT.[23] But using the possibility of severe costs from future damages as a justification for limiting fossil fuel pollution was a more recent phenomenon.

Fortuitously, the earliest attempts to do so had already begun in West Germany shortly before acid rain became a major scientific and political issue. In the 1970s, West German officials introduced the term "vorsorge" into environmental legislation for the first time, which means "precaution" in English. It is the earliest known use of the now famous word in environmental regulation anywhere in the industrialized world.

When employed by West German officials in these years, it often referred solely to the need for long-term planning and foresight regarding the environment.[24] The West German idea of a "precaution policy" for air pollution, for instance, was intended to maintain the current air quality in areas with little pollution as well as reduce total pollution from more heavily affected parts of the country.[25]

As acid rain became a prominent issue in West Germany, the idea of precaution took on a new meaning. Rather than focusing solely on managing the environment in a way that would protect it for generations to come, precautionary policies came to mean acting to protect the environment when faced with imprecise or uncertain scientific evidence.[26] This novel framing of precautionary measures eventually became the modern precautionary principle still widely used in environmental policymaking today. It places the "burden of proof" on industries to show that their products are safe instead of requiring environmental scientists to demonstrate that they are harmful before limiting or removing substances from use.

West Germany, in consultation with the European Communities, redefined precaution in this way to try to overcome Britain's opposition to acid rain regulations following Ulrich's research. In 1982, West Germany began pressuring the European Communities to discuss implementing regulations to reduce sulfur dioxide emissions, including inviting the president of the European Communities, Gaston Thorn, to examine the damage to West Germany's Black Forest allegedly caused by acid rain. The West German government's concerted efforts to bring attention to acid rain led many European Communities' officials to consider whether the organization should reevaluate its approach to environmental legislation.[27] There was significant concern among officials at the European Communities that fossil fuel regulations introduced by one member state could undermine the group's economic integration, particularly if major polluters like Britain chose not to impose similar limitations on power plant emissions.[28]

Though not specifically using the word "precautionary," the Commission responded to West German fears about acid rain by drafting a new environmental program that explicitly endorsed what officials dubbed a "pure preventative" approach to pollution.[29] Over the next several years, the European Communities put forward a series of policy proposals under this rubric that made it clear the organization would no longer wait for supposed scientific certainty before acting to address acid rain. For example, in June 1983 the European Communities adopted a resolution calling for urgent steps to reduce acid rain before irreversible destruction of forests occurred.[30] It simultaneously introduced a directive requiring authorization for the building of new power plants likely to cause air pollution and started drafting legislation specifically to limit the emissions of power plants responsible for acid rain.[31]

In a subsequent 1984 resolution passed by the European Parliament on acid rain, representatives explicitly acknowledged that there were still scientific uncertainties pertaining to the exact causal mechanisms of acid rain's environmental damages. However, their submission requested that the European Council nevertheless draw up measures to halve the levels of air pollutants responsible for acid rain and other environmental impacts within the next five years.[32] It contended that the ecological effects were dire enough that steps needed to be taken before the environment was subjected to an unacceptable amount of degradation.[33] Together, the measures sent a clear message that trying to block progress on acid rain by claiming the science was still uncertain would be a problematic strategy going forward.

The irony, of course, is that scientific consensus on the dangers of acid rain had been building for more than a decade. Aside from the coal industry, the vast majority of researchers in Europe and North America agreed that acid rain was a serious threat to freshwater ecosystems and that reducing fossil fuel emissions would help protect these areas. Environmental officials in Scandinavia, Canada, and the US had argued for years that enough knowledge had been gained to act. The precautionary approach to acid rain was not formulated because current knowledge was suddenly called into question. It was introduced because West Germany and officials at the European Communities needed a way to move beyond the coal industry's demands for more and more research. The precautionary principle, as they redefined it, gave them a way to act with "uncertain" knowledge and undercut the ability of industry to further delay the imposition of regulations.

As the European Communities developed its precautionary approach to acid rain, British energy and environmental policymakers began to discuss the implications of supranational regulations for the country's power plants. William Waldegrave, the Minister of the Environment under Margaret Thatcher, made it clear to other British department heads that the country would not be able to stonewall the European Communities' proposals altogether. Nor, he believed, would it be acceptable to "argue that legislation must wait until the final scientific proof of causation," in light of the European Communities' new precautionary principles.[34] Despite this warning, members of Britain's Central Electricity Generating Board (CEGB) and National Coal Board were already in the process of putting together a strategy to do just that under the guise of obtaining true scientific objectivity on acid rain research.

A Scientific "Bribe"

As the European Communities' regulations were being drafted throughout 1982 and 1983, the leaders of Britain's National Coal Board and CEGB feared that the country could soon find itself forced to regulate emissions as a member of the intergovernmental body.[35] Officials at the CEGB believed that they had at best five years before they would be forced to introduce controls at power plants to reduce sulfur dioxide.[36] The National Coal Board in particular feared that these directives, combined with the CEGB's mandate to begin promoting nuclear power stations, could result in a sharp reversal in current government policy surrounding pollution emissions.[37]

Britain's coal industry leaders were highly critical of the new emphasis on cost-benefit analyses incorporated into the precautionary legislation, arguing that the directives did not adequately address the specifics of getting "value for money" spent on reductions.[38] Yet the National Coal Board and the CEGB agreed that the political winds of Europe had dramatically shifted with these new measures in the European Communities, making it essential to prepare for the possibility that some type of reductions would need to be made in the near future. In the meantime, members of both groups felt the best course of action was to try to buy another five years before having to limit emissions. Their hope was that this delay would allow the British power industry to find emission controls less expensive than flue gas desulfurization, which the country would now be forced to import at great cost because of its decision to stop investing in the technology in the early 1970s under the assumption that they would prove victorious in the scientific debates on acid rain.[39]

The question then was how to engineer such a delay in introducing pollution control technologies while not appearing outwardly obstructionist. In February 1983, Peter Chester, director of the CEGB's research laboratories, pitched a possible strategy to the leadership of the CEGB and National Coal Board. Chester proposed that a five-year research project, jointly financed by the two organizations but overseen by an independent scientific agency, could give Britain sufficient time to devise less expensive means of reducing pollution and "persuade the right people at the right time" that other approaches were preferable to flue gas desulfurization.[40] Within a few weeks, the chairman of the CEGB, Sir Walter Marshall, had contacted the president of the British Royal

Society to float the idea of having the Society oversee and manage a potential five-year research project on acid rain in Scandinavia.[41] Marshall told him that ideally the Royal Society could reach out to the Norwegian Academy of Science and Letters as well as the Royal Swedish Academy of Science to participate in the study. Indeed, their inclusion seems to have been suggested at the last minute as a way of avoiding accusations that Britain was isolated in its lack of acceptance of the scientific evidence for acid rain's environmental damages.[42] Although members of Britain's National Coal Board noted in meetings with the CEGB that Scandinavian scientists had grown impatient with arguments that more research was needed and didn't seem likely to join Britain's Royal Society in the work, they nevertheless agreed to provide financial support for the potential project as a way of further stalling the introduction of control technology.[43] In the meantime, the CEGB and National Coal Board would launch an intensive, undisclosed effort to study American flue gas desulfurization technology and other possible technical means of lowering sulfur dioxide concentrations.[44]

The supposed independence of the British Royal Society in such a scientific project, however, is murky. The CEGB was highly involved in creating and overseeing the research project from the outset of its proposal to the group in June 1983. For example, the CEGB wanted the Royal Society to come to a joint decision with the Board about the objectives of the research if the project were to move forward. And although the CEGB told the Royal Society that a management team of independent scientists would ultimately have authority over the direction and content of the studies once they began, the CEGB stipulated that members of its own research laboratories would be represented on the management team overseeing the work in a "non-voting" capacity. Marshall claimed that this arrangement was intended to demonstrate the "independence" of the work, and yet the difference between "voting" on research matters versus having their opinions heard was never precisely defined.[45]

The CEGB's proposal to the Royal Society also included an appendix on possible appropriate questions for the project that made it apparent that the organization had certain unstated expectations for the research. It claimed ecological studies carried out over the previous decade had confirmed that the original scientific hypotheses about acid rain's damaging environmental effects were "overly pessimistic" and that the "costs to the European energy consumer" of regulating fossil fuel emissions demanded more rigorous review given these changing perceptions.[46] The document then laid out the key questions it hoped the study would answer, which were seemingly designed to exonerate the coal industry. The

CEGB asked the Royal Society to determine what other factors could be responsible for the loss of biological life, what made certain lakes more vulnerable to damage than others, to what extent these susceptible lakes are adversely affected "by acid deposition itself," and most importantly, what amount of environmental improvement could realistically be expected by a given amount of pollution reduction in Britain.[47]

Despite the blurry lines surrounding the CEGB's degree of control over the research project as well as the British Royal Society's lack of prior experience in environmental research, the President ultimately agreed to the CEGB's offer with no suggested changes or additions. Why the Royal Society's leadership decided to take on the project is not entirely clear, but they were likely influenced by the close ties of the CEGB to the society. Marshall himself had been a fellow of the Royal Society since 1971, and the CEGB's former research director, John S. Forrest, had been elected a fellow of the Royal Society in 1966 and served as its vice president after his retirement from the CEGB from 1972 through 1975.[48] A few weeks after Marshall formally approached them, the Royal Society reached out to its Norwegian and Swedish counterparts to request their involvement with the research project, officially titled the Surface Water Acidification Program (SWAP).[49]

Immediately after receiving the British Royal Society's invitation, the leadership of the Norwegian Academy of Science and Letters contacted Hans Martin Seip and Lars Walløe, both professors at the University of Oslo, as well as Gunnar Abrahamsen from Norway's Institute for Forest Research to discuss the possibility of Norway joining the study. The three scientists had been intimately involved in Norway's SNSF project (Acid Rain's Effects on Forests and Fish, Sur nedbørs virkning på skog og fisk), the largest ecological study of acid rain in Scandinavia.[50] Seip, Walløe, and Abrahamsen worked closely with the Academy in nominating the Norwegian scientific representatives to SWAP, which was originally intended to have four delegates from Britain, two from Norway, and two from Sweden as well as a British director of the project. Crucially, the Norwegian Academy advocated for the appointment of Seip as a Scandinavian "consultant" to the program's recently chosen British director, John Mason, who had just retired as director of the British Meteorological Office and had no background in acid rain studies.

At the same time, the Norwegian Academy of Science and Letters as well as the Royal Swedish Academy of Sciences established a working relationship with Norway's Ministry of the Environment to discuss the program and share their concerns about the research being usurped by the British power industry.[51] While the independent scientific academies appear to have been enthusiastic about the project at the outset, many Norwegian and Swedish scientists with longstanding experience in acid rain research reacted skeptically to the CEGB's proposal.[52] They believed that the CEGB and National Coal Board were likely using the research as a way to avoid making emissions reductions, which internal documents show was indeed the purpose of the study.

Their suspicions were shared by many environmental officials in the Norwegian and Swedish governments, notably Erik Lykke, Norway's director general at the Ministry of Environment. Writing to his colleagues in August 1983, Lykke lamented that while a large majority of British university scientists had accepted the considerable evidence linking acid rain to environmental damages, the CEGB and its scientists were continuing their "spirited campaign based on the uncertainty that will always be associated with complicated research conditions. The clear objective has been to postpone as long as possible all decisions on reducing emissions."[53] Although acknowledging that the five-year research program could deepen understanding of the relationship between air pollution and aquatic effects, Lykke pointed out that nearly all Norwegian ecologists with expertise in acid rain research did not expect further studies to produce dramatically different results.[54] Nonetheless, the Ministry of the Environment concluded that it would not be appropriate to interfere with the decisions of the Norwegian Academy of Science and Letters nor the Royal Swedish Academy of Sciences regarding their participation in the study, though Lykke felt it would be worthwhile to at least express the opinion to these organizations that the Ministry felt the study was "politically motivated."[55]

Despite these criticisms of SWAP, the Norwegian and Swedish academies agreed to sign on to the project. This decision reveals a certain naiveté about the British intentions, as representatives from the societies privately insisted to government officials that the British Royal Society had made it clear their concerns about the CEGB's degree of involvement were unfounded.[56] Yet it also appears that the leadership of the societies may have been tempted by the money itself, noting in their meetings with members of the Norwegian government that studying sulfur dioxide emissions need not be limited simply to "acid rain" but could potentially include a wide range of other pollution mechanisms.[57] Perhaps they envisioned the project taking shape around more basic research interests; it certainly appears that the British Royal Society gave them reason to believe they would have some input in designing the research program over the next year. In conversations with the two academies, the British Royal Society and Peter Chester explicitly denied that the research was intended to stave off emissions reductions, but rather was a necessary follow-up to the 1979 UN Convention's requirement to promote research on acid rain.[58] Following these consultations, the three societies scheduled meetings that fall to begin discussions of the work program and publicly announced the start of SWAP in September 1983.

Environmentalists responded to the announcement with protests outside the British Royal Society headquarters in London, depositing a coffin of dead fish at its doorstep the week the five-year study was launched.[59] In Norway, the media reported on the concerns of the Norwegian Ministry of the Environment that the program was in danger of being biased or controlled by the CEGB, as well as used to postpone emissions reductions. In an interview with *Aftenposten*, Erik Lykke was critical of the fact that the funds for the study came directly from industry as opposed to general government funds, which had been provided for acid rain research in most other countries; he nevertheless expressed the hope that the British Royal Society would ensure that the CEGB did not distort the research program.[60] Several Norwegian government scientists spoke out publicly against the study as well, including Lars Overrein, the director of Norway's Institute for Water Research. Overrein had overseen a great deal of Norway's acid rain research and told British reporters that he feared the Royal Society intended to "rediscover the wheel" given the extensive studies that had already been conducted on the problem.[61]

FIGURE 7.3 Environmental groups protest outside CEGB headquarters on December 15, 1983, shortly after SWAP was launched. The crowd included six Santas (front and center) and forty dead trees, including a Christmas tree in a coffin. Swedish and Norwegian NGO Secretariats on Acid Rain, "Acid News," January 1984, UK Parliamentary Archives, HC/CP/9445, p. 4.

A few weeks later, the first meeting of the three Royal Societies about the research program did nothing to ease the doubts of Scandinavian scientists and environmental officials regarding the objectivity of the studies. Although there were many British ecologists who had worked on the acid rain problem for years, only one of the five British researchers appointed to manage the project had any prior experience with acid rain studies.[62] Lars Walløe, the former director of Norway's SNSF research program, attended the gathering as one of the Norwegian scientific delegates and reported on the discussions to the Ministry of the Environment. He told the Ministry that he believed the leadership of the Norwegian and Swedish academies were naïve about the British scientific academy's ability to be unbiased in the research. After Walløe conferred with Professor Carl Olaf Tamm, a Swedish researcher who had been asked to join the management team, the two agreed that they would stay on and try to steer the research program in the most objective direction possible, while committing to resign if they should fail to do so.[63]

As the scientists attempted to come to an agreement about the research direction of SWAP, it became evident that representatives from the three countries had vastly different ideas about how to proceed. One of the early points of disagreement was the negligible amount of money and resources set aside for studying acid rain in Britain. At its outset, nearly all of the study's funds was allocated for research in Scandinavia despite the recent findings from ecologists in Scotland and Wales about ongoing acidification in these regions.[64] Scandinavian researchers believed that British damages should also get attention; focusing on only their countries suggested the program was politically motivated. In addition, the Norwegian and Swedish scientists felt that the British scientific representatives had adopted the now familiar line promoted by Chester and the CEGB's scientists during discussions of the current state of the field.[65] As one of the two Norwegian attendees reported to Norway's Ministry of the Environment:

> There are indications of water acidification in Scotland, English Lake district and Wales. But it is quite clear that the British were not particularly keen to learn more about the situation in the UK. Royal Society money would be used in Scandinavia, and the means to do acid rain research in the UK was very limited. The discussions followed a familiar pattern otherwise. The British constantly pointed out uncertainties in the relationship between deposition and acidification of freshwaters in line with P.F. Chester's arguments.[66]

Other developments in the program during the fall of 1983 served to further validate the Norwegians' fears. In total, eight British delegates, including selected scientific delegates as well as members of the Royal Society, CEGB, and National Coal Board, were initially represented on the management team with only two scientists from Norway and two scientists from Sweden. The distribution of power led Hans Christensen, Norway's Secretary of the Royal Norwegian Council for Scientific and Industrial Research, to warn that the research program smacked of British "environmental colonialism."[67] More ominously, the CEGB representative was Chester himself.[68]

After the SWAP management team held its first meetings in the fall of 1983 and spring of 1984 to review grant proposals and discuss what projects to support, Seip was finally brought onboard to assist Mason and perhaps avoid the impression that the British members were completely dominating SWAP decisions. In an attempt to reduce tension among the participants, Seip suggested that he and Mason work on publishing a "consensus" paper outlining the current knowledge of acid rain as agreed to by the three academies.[69] However, shortly after Seip became involved, one of the two Swedish scientific delegates, Carl Tamm, wrote a letter to the Royal Swedish Academy of Sciences and Norwegian Academy of Science and Letters threatening to pull out of the project. He told the academies that British scientists on the management team had given "short and shallow" consideration to the project's implementation and was highly critical of John Mason's paltry background in acid rain research.[70] A few days later, Tamm turned down a request to appear on behalf of the project at a US symposium on acid rain. He explained to the organizers that they would be better off talking to one of the British representatives from the program since he was no longer sure what his involvement would be, and the British would be better positioned to explain "what the Royal Society expects to get out of the cooperative project."[71]

In a last-ditch effort to salvage his role in the research, Tamm directly contacted the chairman of the project, Richard Southwood, an Oxford University biologist who also led Britain's Royal Commission on Environmental Pollution at the time. Encouraged by Southwood to express his frank views on the disagreements hampering SWAP, Tamm wrote a scathing critique of Mason's management of the research, statements Mason had made about the unreliability of research data on acid rain, and especially the problem of achieving consensus on how to proceed with the research.[72] As Tamm explained to Southwood:

> Fully objective science is a goal we must strive for, but probably
> never achieve. Even in so called "hard" science (as distinguished

from e.g. social science or humanities) the choice of problems and of the methods to solve them includes subjective moments. Consensus can thus not be a prerequisite for cooperation, instead the existence of different opinions within and between groups of scientists is a factor stimulating scientific progress. However, if the representatives of the various schools do not feel respect for the opinions and the methods used by their dissident colleagues, it would be better to work separately and let the scientific world give the final judgement [sic], than to force scientists into an unwilling cooperation.[73]

The inability to agree on the project's direction or achieve consensus on the current state of the field severely hindered the execution of the project. Just three grants were awarded in the year after the project launched, and only two percent of the 5 million pounds budgeted for the studies was spent.[74] Seip and Mason's "consensus" paper, published in *Ambio* in early 1985, offered little reason to believe that the project might serve as an avenue in building upon previous and existing research efforts in Scandinavia. Much of the document reiterated objections previously addressed by Seip and others in Norway about the need to investigate whether land use changes could be contributing to acidity and whether enough long-term, standardized studies existed to draw conclusions about the role of acid rain in causing loss of life in freshwater ecosystems.[75]

A year later, a sharp disagreement over releasing SWAP's data publicly dealt a final blow to any hope that the project would result in productive collaboration between the three countries. At a meeting of the management team to review the current status of the research in 1986, the five British scientists on the board used their majority status to overrule a request by the Scandinavian scientific representatives to release annual reports of the study's results. The action was dubbed a "gag on research" in the media and led Lars Walløe to publicly protest that the British decision made it impossible for the results to receive appropriate attention from governments and the public.[76]

After the incident, the group elected to meet only once per year until the conclusion of the study in 1990. Richard Southwood and John Mason were given broad authority to oversee the program in the interim after consulting with "one or two" other members, both of whom could presumably be British scientists. It is difficult not to interpret this as freezing out the Scandinavians from further serious involvement in the oversight of the program. There is little record of any substantive discussions of ongoing research except for the preparation of a midterm and final report on the project in 1987 and 1990.[77]

Despite the obvious reservations held by many of the Scandinavian scientists participating in the project, notably Tamm and Walløe, none of the Norwegian or Swedish members of the management team opted to withdraw from SWAP in protest. Their intentions may have been good; SWAP did provide generous resources for further work on the problem, and it's clear that some of the British representatives, such as Southwood, were at least attempting to work with the Scandinavians in good faith.

And yet, their decision to remain in the project provided continuous fodder for members of Britain's CEGB and National Coal Board to claim that they were not obstructing progress but simply waiting for the final, definitive results from the program. Both at home and abroad, SWAP became a frequently touted initiative of Prime Minister Margaret Thatcher's administration, supposedly a testament to the country's commitment to addressing acid rain.[78] A moment of honest, public accounting seems to have occurred only once, when the CEGB and National Coal Board were questioned in a Parliamentary hearing on acid rain the year after SWAP began. At the hearing, members of the House of Commons openly accused the groups of using the program as a scientific "bribe" to the Scandinavians so they could avoid addressing the problem politically. British parliamentary officials who visited Norway and Sweden in the early 1980s to educate themselves about acid rain had been told by Scandinavian government officials that the SWAP study was a ruse intended to get the country out of making pollution reductions. When confronted with these charges, CEGB and National Coal Board officials at first denied them. But when questioned further about the true objectives of the research, they retorted that it served the two countries right if their scientists could be bribed.[79] Yet while the CEGB may have been optimistic about the efficacy of SWAP in staving off pollution regulations, instead the project soon became a means for the Thatcher administration to save face as it reversed its acid rain policies under pressure from the European Communities.

Britain Joins the Acid Rain Club

In December 1983, members of Britain's House of Commons Environment Committee voted to begin an inquiry into acid rain, a momentous break with the Thatcher administration.[80] The investigation was led by Sir Hugh Rossi, a member of Thatcher's own Conservative Party, who had joined the Environment Committee just a few months before. It was motivated partly by reports of acid rain occurring in Scotland and Wales, which had galvanized members of Parliament over the previous year, but also by the European Communities' efforts to regulate power plant emissions. With growing mistrust in Prime Minister Thatcher's treatment of the problem, the inquiry was designed to settle the question of what was known scientifically about acid rain and whether Britain should join its neighbors in implementing pollution regulations to address it.[81] Rossi and the other representatives gathered background information on acid rain studies over several months and scheduled in-person visits to Norway, Sweden, and West Germany to study the damage caused by acid rain as well as meet with environmental scientists to hear their expert opinions.[82]

As Britain's House of Commons was ratcheting up its attention to acid rain, Britain and the US were faced with renewed lobbying to reduce air pollution from an international meeting of environmental ministers held in Ottawa, Canada, in March 1984. Delegations from ten countries met to commit their nations to reducing sulfur dioxide emissions by 30 percent from 1980 levels before 1993.[83] Since the majority of these countries had already implemented domestic reductions equivalent to those agreed upon in Ottawa, the meeting was held less to advance the policies of those in attendance and more to persuade Britain and the US to consent to similarly lowering their air pollution levels.[84] Dubbed the "Thirty Percent Club," the group included the Scandinavian countries, Canada, West Germany, the Netherlands, Switzerland, Austria, and France.[85] The conference was deliberately scheduled to take place just prior to a joint submission by Canada and Norway to the United Nations (UN) of a binding annex to the 1979 UN Convention on long-range transboundary air pollution, which would require countries to reduce sulfur dioxide emissions by 30 percent from 1980 levels before 1993.[86] These developments placed Britain in an even more isolated position among Western European nations and threatened to further strain relations between the country and fellow members of the European Communities.

On the heels of the UN submission, Britain's House of Commons began its acid rain hearings in May 1984 with pointed questions about whether Britain should join the Thirty Percent Club and agree to the proposed European Communities directive on power plant emissions.[87] Witnesses from the CEGB, National Coal Board, Department of the Environment, environmental groups, and government laboratories were all called to give their assessment. Central to the hearings was the question of whether Britain should adopt a precautionary approach similar to the

European Communities or if the country should instead wait for SWAP to provide "definitive" evidence on the matter.

Parliamentary officials stressed this issue in their questioning of the CEGB in particular, which took place at the outset of the hearings. Peter Chester, director of CEGB research, and Walter Marshall, chairman of the organization, appeared before the Environment Committee along-side several other scientists and top officials from the organization. The Environment Committee representatives opened the dialogue by presenting some of the most recent scientific findings concerning the impact of acid rain on the environment and Britain's culpability in the problem. Yet when asked about evidence showing the role of CEGB emissions in causing acid rain, Chester and Marshall reiterated their frequent arguments that there was no scientific consensus on whether there was "genuine environmental damage."[88] In addition, they sidestepped the committee's questions about how much longer scientific studies would have to continue before enough information was available to make policy decisions. As Chester put it:

> The scientific process takes the time that it takes. If you want to use the answers you go with that. If you do not have the time for whatever reason you make a decision that is not based on science.[89]

Marshall added that he hoped the SWAP studies would have the answer by the end of its program of research in 1988 but could not guarantee this timeline. He reiterated that the involvement of the Royal Societies should produce "a consensus scientific opinion" and firmly denounced charges that the study was being used to procrastinate on the issue.

The members of the House of Commons Environment Committee paused their questioning here, however, to openly discuss the difficult decision they faced based on the CEGB's testimony. The chairman, Sir Hugh Rossi, explained that what most concerned the committee was that the environmental devastation from acid rain could be severe by the time the necessary studies had been carried out to the satisfaction of the CEGB. Parliament did not want the responsibility, he said, of further delaying emission reductions in case the existing research on environmental damages turned out to have substance. Another member of the Environment Committee chimed in to say that the West German legislature had obviously concluded that the probability of severe environmental harm from acid rain was high enough to justify immediate action. [90] Marshall countered that he felt politicians were not the best judge of such matters and that the West German officials were not as logical as the representatives were making them out to be.[91]

The Thatcher administration's Secretary of State for the Environment, William Waldegrave, affirmed the CEGB's remarks in his department's formal submission to the hearings. Waldegrave wrote that the Prime Minister did not believe the expense of installing flue gas desulfurization in existing power stations could be justified "while scientific knowledge is developing and the environmental benefit remains uncertain."[92] His response emphasized the complexity of the scientific problems, the unique geography and soil types that made some areas in Scandinavia more susceptible to acid rain, and the potential for other factors, whether different pollutants like ozone or land use changes, to contribute to acidification of ecosystems.[93]

Despite the claims of the CEGB and several top Thatcher administration officials, the Committee came to the unanimous conclusion that Britain needed to take immediate steps to combat the effects of acid rain after receiving further testimony from other government departments and visiting affected areas in West Germany, Norway, and Sweden.[94] In a final report published shortly after the conclusion of the hearings, they argued that Britain should join the Thirty Percent Club and begin the process of reducing acidic pollutions from CEGB power plants through the installation of point source control technologies. Most shockingly, the Environment Committee asserted that the CEGB had deliberately misled Parliament and the country by insisting that there was no scientific consensus on the issue.[95] These conclusions were largely influenced by the evidence submitted from scientists at the Natural Environment Research Council, which the Committee noted had no ties to industry interests. While the House of Commons' report was careful to acknowledge that the causal mechanisms linking pollution emissions to specific environmental damages were still imprecise, it argued that the risks of inaction were too great not to reduce these pollutants.[96] The House of Commons recommendations were followed shortly thereafter by similar proposals from the House of Lords European Communities Committee, which suggested that the country cut emissions by thirty percent given the acceptance of this target by a growing number of European nations.[97]

The CEGB response to Parliament's recommendations was swift. It issued a press release claiming that the House of Commons report on acid rain distorted its testimony and came to mistaken conclusions on the scientific evidence. Rossi was so angry over the public lashing that he wrote Marshall a sharp rebuke explaining that all the CEGB's evidence had been vetted through several independent scientific experts and "found wanting."[98] Rossi also accused the CEGB of misinforming the public about these issues by quoting higher figures on the costs of reducing sulfur dioxide and nitrogen oxides publicly than they had given in their written and oral testimony.[99] He then published his letter so that the CEGB's charges could be clearly and unequivocally debunked.[100]

Members of the European Communities as well as representatives from governments throughout Western Europe hoped that the sparring between Parliament and the CEGB might herald a shift in Britain's position on acid rain.[101] But at least publicly, Thatcher's government continued to refuse to join the Thirty Percent Club or to discuss flue gas desulfurization installation in power plants.[102] British delegates subsequently declined to sign the 1985 UN Helsinki Protocol on acid rain, the first diplomatic accord requiring specific reductions in emissions of an air pollutant.[103] All signatories agreed to lower their sulfur dioxide emissions by 30 percent from 1980 levels. Out of the original parties to the 1979 UN convention on transboundary air pollution, only Britain, the US, and Poland abstained from signing the new protocol.[104]

Privately, however, the leadership of the CEGB and National Coal Board suspected reductions in the country's air pollution were imminent given the confluence of domestic pressures and the European Communities measures. At the time the House of Commons report was released, other Communities' member states were calling for even speedier reductions in emissions alongside five proposed directives on combating air pollution and acid rain.[105] Though outwardly still opposing any such protocols, by the spring of 1985, the CEGB and National Coal Board began working behind the scenes to influence possible European Communities regulations. Officials from the two organizations told British delegates to the European Communities that political pressures from other member governments could result in an agreement anytime in the near future, and expressed their hope that the potential reductions could be implemented through overall emission reductions rather than individual plant controls.[106] The rapidly moving consensus within the European Communities on regulating emissions from large power plants also led members of Britain's Department of the Environment and Department of Energy to begin discussing the need for Britain to establish some form of domestic policy before the European Communities moved to implement its air pollution directives.[107]

These covert attempts at preparing a domestic response became even more hurried in the spring of 1986 when the European Council passed a declaration stating that the Commission would create a regulatory framework for lowering fossil fuel emissions. Member states needed to agree on detailed requirements no later than December 31st of that year.[108] Dutch delegates put forward an initial proposal for reductions to serve as a starting point for negotiations that was more aggressive than the proposal made by the Thirty Percent Club just two years prior. The Dutch plan stipulated that reductions were to be made by 1995 based on 1980 emissions levels; member states would need to lower sulfur dioxide levels by 60 percent before 1995 as well as lower nitrogen oxide emissions and dust emissions by 40 percent that same year.[109] West Germany, Denmark, and France all lent their support to the proposal. While Italy, Ireland, and Greece originally sided with Britain in opposition to the proposed cuts, members of Britain's Department of the Environment feared that the Commission would appease the latter two through special exemptions and that Italy was open to amending its position.[110] These predictions were borne out during further Council deliberations on a compromise agreement several weeks later, which resulted in a commitment by all European Communities' member states except Britain to reduce their fossil fuel emissions.[111]

Internal discussions among British Cabinet officials reveal that the Prime Minister's top advisers saw the European Council decision as having dire consequences for Britain's broader foreign policy objectives and integration into the European Communities. A month after the organization drafted its directive on power plant emissions, Thatcher convened a meeting of Cabinet officials responsible for European policy to review the impending regulations. At the gathering, members of the Department of the Environment made it clear that the legislation was "not going away" and argued that although Britain could continue to oppose all attempts to reduce emissions it would be extremely difficult to do so in the face of unanimous support for the plan from the rest of the member states. More than that, however, Britain's strident opposition could inhibit its ability to work with other member states on a variety of European-wide policies. Officials in the Department of the Environment noted that refusing to negotiate on acid rain would prove particularly problematic once Britain took over the European Council Presidency in the second half of 1986. Since it was unlikely that the group would finalize legislation before then, Britain would be in the awkward position of trying to lead the Communities while in the midst of a split with all other member states on a major environmental and energy initiative.[112] Martin Holdgate, the Department of the Environment's chief scientist, thus strongly recommended that Britain agree to negotiate.[113]

Thatcher's Cabinet officials, while anxious to discuss the matter with the CEGB, gave Holdgate permission to initiate conversations with other department heads on how the country might be able to agree to some kind of control measures for the first time.[114] Their continued sensitivity to the CEGB's input initially frustrated Holdgate and other environmental officials, who feared that Britain's "environmental control policy [was] being completely dominated by the views of CEGB, Energy and Treasury."[115] Nonetheless, they forged ahead with drafting possible domestic legislation on air pollution control from power plants in May 1986 to provide a negotiating position for talks with the rest of the European Communities, though they were not optimistic about the likelihood of the administration following through on their plans.

In fact, the CEGB appears to have harbored a much different view of the administration's intentions after speaking with Cabinet officials. In June 1986, members of the CEGB and National Coal Board agreed that the Thatcher administration seemed more and more inclined to come to an agreement with the rest of the European Communities member states because of the "political need to do something substantive."[116] The objective for the coal industry, then, would be to ensure they could minimize the amount of pollution Britain would need to reduce as well as the speed at which it was required to do so.

They would also need to amend their position on the science of acid rain to reflect the government's decision to move forward with reductions. Writing in July 1986 during these internal policy debates, Peter Chester composed a treatise on the "Evolution of Understanding" regarding acid rain that was a blatant attempt to revise the CEGB's stance on acid rain science in light of the impending regulations. While recycling some of his now familiar criticisms of acid rain research, for the first time he gave a subtle acknowledgment to the need for future regulation of British power plants. Chester nevertheless argued that since it would likely take a long time for the environment to recover, countries should not commit themselves to "any arbitrary target reduction or deadline."[117] It was in this vein that he justified the continued importance of the SWAP research, stating that its purpose—"To determine what changes would be brought about in water chemistry and fishery status in Norway and Sweden by given levels of reduction of man-made sulphur deposition"—remained as necessary as ever.[118]

SWAP, however, was about to become a convenient front for the Thatcher administration to claim that the "science" had finally proven convincing rather than appearing to capitulate to pressure from the European Communities. In July 1986, Prime Minister Thatcher suddenly told members of her cabinet that "new evidence" on acid rain needed to be urgently examined by the chief scientific advisers of the Department of the Environment and the Department of Energy as well as by John Mason in his role as head of SWAP.[119] What exactly constituted this new evidence was never stated. In fact, the SWAP scientists were not even initially involved in drafting the report on the supposed new evidence. Instead, the chief scientific advisers of the Cabinet Office, Department of Energy, and Department of the Environment alongside the director of technology, research and planning at the CEGB composed an internal report on the state of acid rain science. The report's findings were only then reviewed by John Mason and Richard Southwood, who gave their consent to its contents.[120]

In many places, the report still emphasized uncertainties in scientific research on acid rain in ways that mirrored the CEGB's arguments to Parliament. It stressed that the interaction between acid rain and ecological harms was "complex" and lamented the lack of a quantitative relationship between sources and receptors. However, the report acknowledged that acid rain had a "major" effect on freshwater ecosystems, a remarkable admission for the British government.[121] Crucially, the report concluded by stating that emissions reductions would be necessary to ensure the recovery of affected areas.[122]

After internally reviewing the consensus report from the three departments, in September 1986 Prime Minister Thatcher announced that in light of new research findings from SWAP her government would begin the process of installing flue gas desulfurization on power plants.[123] Precisely timed to coincide with a visit of Thatcher to Norway, the statement marked Britain's first commitment to lowering power plant emissions to combat acid rain.[124] Though vague on specific deadlines, the government pledged that the CEGB would "play a part" in solving the problem by lowering its emissions over the next several decades.[125] When questioned on the timing of her decision, the Prime Minister vehemently denied that it was "political pressure" that caused a reversal in policy and attributed her decision solely to new information gleaned from the SWAP project.[126]

Of course, there were no such new findings from SWAP itself, which had been largely dead in the water because of tensions between British and Scandinavian scientists. The "new evidence" was constructed from members of the Department of the Environment, Department of Energy, and CEGB per the orders of the Prime Minister. But by using the program as a pretense for the country's capitulation to demands from the European Communities, Thatcher's government was able to maintain its image as a paragon of scientific objectivity and economic prudence with the media and the public. In the end, SWAP had been a master political manipulation of the scientific process, used first to delay making emissions reductions under the guise of needing more data and eventually deployed as cover for the administration's decision to bow to international pressure.

Britain's acceptance of emissions controls at the behest of the European Communities was monumental, paving the way for real progress on acid rain after years of stalled diplomacy. Following Margaret Thatcher's announcement that Britain would install flue gas desulfurization technology in several of its power plants in 1986, the European Communities successfully negotiated the 1988 Large Combustion Plant Directive. The legislation required all member states to lower emissions of sulfur dioxides beyond the requirements of the 1985 UN Helsinki Protocol, which had stipulated signatories would reduce these pollutants by 30 percent from 1980 levels before the year 1993. While Britain was able to bargain for a slightly less onerous schedule of reductions compared to the other member states, it nonetheless committed to reducing emissions by sixty percent from 1980 levels before 2003.[127]

Despite these achievements, the Thatcher administration's elevation of coal industry science to the highest levels of government advising had serious consequences for environmental diplomacy. At the end of the decade, European environmental officials were not inclined to rely on the scientific community for specific guidance in formulating pollution policies. Whether in the European Communities or UN, government representatives instead utilized precautionary approaches in order to divorce political decision-making from debates about the state of scientific proof between industry scientists and environmental researchers.[128]

The precautionary principle would eventually transform the way government officials in Europe and around the world understood their responsibilities to act when faced with regulating pollution. In 1992, the precautionary principle was finally written into a global environmental accord known as Agenda 21 during the Rio Earth Summit. The declaration from the conference states that the precautionary approach should be widely applied to protect the environment and that "lack of full scientific certainty shall not be used as a reason for postponing cost-effective measures to prevent environmental degradation."[129] In many ways, Agenda 21 and other international environmental agreements that followed were codifying the lessons from acid rain and shifting the burden of proof to industry on whether their products were environmentally damaging. Yet by pitting industry scientists against the rest of the scientific community, the precautionary approach still left unanswered how governments could best use expertise in developing future regulations.

worst, however. At the urging of the coal industry, President Reagan announced soon after that if reelected for a second term he would merely boost research funding for scientific studies of the problem and the development of new technologies to reduce emissions.[3]

Following Britain's capitulation to the European Communities in 1986, the US became the only country still refusing to agree to acid rain reductions in the industrialized world. The coal industry was emboldened by Reagan's staunch support of their position and continued their attacks on acid rain science at the start of his second term. Some of the most blatant were directed at Gene Likens, the ecologist who had originally discovered acid rain in the eastern US with his colleague F. Herbert Bormann. In 1984, the president of the Ecological Society of America asked Likens to write a piece on acid rain for the *Wall Street Journal*. The issue had received misleading coverage in the paper and Likens hoped to have an opportunity to share the current scientific consensus with its readership. What he didn't know was that the *Wall Street Journal* planned to run his article alongside commentary from Alan Katzenstein, a consultant for the Edison Electric Institute, the major lobbying arm of the coal industry. According to Likens, the publication edited his piece without allowing him final review and weakened parts of the original. A colleague sent the articles to Likens along with two aspirins and a note saying he might need to take the pills before reading them.[4]

The conservative publication's side-by-side, equal treatment of a coal industry consultant and one of the leading researchers on acid rain is emblematic of the hardening stalemate over the issue between right-leaning groups and environmental advocates during Reagan's second term. As far back as 1980, the *Wall Street Journal* would pair comments from environmental scientists or members of the EPA concerning acid rain with those of the Electric Power Research Institute (EPRI), the coal industry's scientific arm.[5] Yet their lopsided treatment at that time could be excused by the fact that EPRI was then the largest funder of acid rain studies. By 1984, despite widespread consensus among environmental scientists in Europe and North America that acid rain was causing ecological damages, conservative leaning publications like the *Wall Street Journal* and the *Economist* were continuing to toe the industry line on the state of knowledge about the problem.[6] To make matters worse for environmental scientists, President Reagan had seemingly suffered little over his position, winning a resounding reelection that year. With no supranational authority that could exert pressure on the US like the European Communities did with Britain, the prospects for a change in US policy on acid rain appeared slim.

FIGURE 8.1 Gene Likens briefing President Ronald Reagan, Cabinet officials, and staff on acid rain at the White House in 1983. Photograph from Gene Likens.

The reelection of President Reagan also did not bode well for the government's major research program on acid rain, the National Acid Precipitation Assessment Program (NAPAP). Its ability to produce unbiased, objective findings about acid rain would be severely tested in his second term. Scientists like Gene Likens fought strenuously in the scientific community and in the public arena against attempts to discredit research on acid rain's harms, but they were often stymied by an administration that continued to view denying science as a convenient way to avoid regulating pollution. It would take a transformation in the American public's concern about the Reagan administration's environmental policies on acid rain to push the US government toward concrete steps to lower fossil fuel emissions. The backlash was partly a response to local environmental problems that had suffered from mismanagement and neglect thanks to Reagan's philosophy of deregulation. But it was also heavily motivated by the specter of even more severe planetary threats from fossil fuels, which made acid rain appear to be a warning signal of direr environmental consequences on the horizon.

The Last Holdout

While the British power industry responded to fears of regulation by trying to bribe Scandinavian scientists with research funding, EPRI took the opposite approach. The scientific arm of the coal industry amplified its attacks on the leading scientists studying acid rain in the middle of the decade, and not just in the popular press. EPRI-funded researchers also eviscerated ecological studies in scientific journals, accusing longstanding leaders in the field like Gene Likens, Thomas Butler, and Charles Cowgill of mistakes and bias in tying acid rain to fossil fuel emissions.[7] Though these scholarly takedowns did not meet some of the typical standards of original research and had little impact on the rest of the field, they gave power industry scientists a veneer of respectability that allowed them entry into conferences and, crucially, the halls of Congress.[8] During 1984 hearings on acid rain, representatives from the coal industry argued EPRI employed "some of the most capable and credible scientists and engineers in the nation" before going on to question the connection between coal emissions and acid rain. Their proof of EPRI's credibility was the simple fact that its scientists' publications had been cited by others in the field.[9] Despite their questionable objectivity, EPRI continued to be a major sponsor of government-run studies on acid rain, notably research from the national laboratories into the effects of acid rain on forests, and it maintained a presidentially appointed seat on the government's acid rain task force.[10]

The Reagan administration's manipulation of government-sponsored acid rain research also continued throughout the second term of his administration. The most glaring incidents occurred with NAPAP, which was the largest acid rain study then underway and expected to conclude in 1990. NAPAP came under fire repeatedly for misrepresenting research results and minimizing the severity of the problem, and there were reports that scientists working for the program feared if they spoke out against its findings they would lose federal grant money or their jobs.[11] After several scientists who participated in the project claimed a 1987 interim report distorted their research results to support the position of the Reagan administration, NAPAP's director, Lawrence Kulp, resigned mere days after its publication.[12] Under pressure from Congress, the new director of NAPAP, James Mahoney, was forced to recall the interim assessment and pledged to employ more outside reviewers for the final report.[13] Kulp nevertheless continued to stand by the report's conclusions that acid rain did not cause serious environmental damage and that reductions in sulfur dioxide were not worth the cost.[14] He went on to become an environmental consultant specializing in energy issues for EPRI and the coal industry.[15]

If scientific and political debates on acid rain were still mired in the familiar impasse between the coal industry and environmental scientists, the American public was starting to question whether environmental deregulation was a wise course for the country. Opinion polls showed a marked rise in concern about environmental pollution over the course of the Reagan administration, particularly regarding issues such as toxic waste and water contamination.[16] The 1986 midterm elections drove home the political consequences for the Republican party, as Democrats took control of both houses of Congress and won a number of close races by placing environmental issues at the center of their campaigns.[17] In their opening salvo against President Reagan, the Senate and House of Representatives passed new legislation on water pollution and were successful in overriding Reagan's veto of the bill.[18]

The following year, fears of public animosity because of the administration's environmental policies led President Reagan to embrace a United Nations (UN) agreement curbing nitrogen oxide emissions, one of the main contributors to acid rain and largely produced by automobiles. Known as the Sophia Protocol, the agreement included both Britain and the US among its signatories and required all parties to freeze emissions of nitrogen oxides at 1987 levels by the year 1994.[19] Environmental activists and observers from the scientific community were quick to point out that election politics were the likely cause of President Reagan's abrupt decision to participate in the international accord, and hoped it might signal a new approach to acid rain within the US government.[20] Republican senators had urged Vice President George H. W. Bush to convince Reagan to sign the accord after the brutal midterm elections, fearing further public ire.[21] That year, President Reagan also instructed the EPA to begin a series of joint projects with the Soviet Union on global environmental pollution, ranging from acid rain to ozone depletion.[22] But despite these tentative hints at an American shift on acid rain, President Reagan still refused to implement cuts to sulfur dioxide, the main acid rain pollutant and a byproduct of coal-fired power plants. Negotiations with Canada remained stalled over US intransigence, while members of Congress continually failed to garner enough support to pass legislation in the face of industry lobbying.[23]

For environmental issues to move to the forefront of American consciousness, it would take an even more potent threat than acid rain: global warming. Scientific interest in the issue stretches back more than a century, but public attention started to coalesce in the late 1980s. In 1988, the country experienced the hottest summer on record, and concern about environmental issues skyrocketed. James Hansen, a noted climatologist, gave powerful testimony on the dangers of climate change to

Congress in June, captivating lawmakers and building political momentum for environmental regulation. Scientists, environmental activists, and the media began connecting acid rain to the climate change problem more and more in public discourse about the country's dependence on fossil fuels.[24] Environmental groups reported two to three times more calls and letters from the public as well as increasing donations. Polling by the *New York Times* and CBS News found nearly two-thirds of Americans agreed with the statement that "protecting the environment is so important that requirements and standards cannot be too high," with only 22 percent disagreeing. This was a remarkable difference from the outset of the decade, when polling showed an even split with 45 percent of Americans in favor of environmental protection and 42 percent disagreeing that it was important.[25]

The resulting backlash against the Reagan administration's environmental policies became a major election issue during the summer and fall of the 1988 Presidential campaign. Then–Vice President Bush was lambasted for Reagan's record on acid rain and an overall weakening of the EPA.[26] Even before securing the Democratic Party's nomination, Michael Dukakis criticized President Reagan for doing nothing about the problem.[27] Faced with falling poll numbers against Dukakis in the summer before the November election, Bush made a decisive break with President Reagan on environmental issues and acid rain in particular. Calling himself an "environmentalist," he promised to act on acid rain and endorsed a number of other clean air and water initiatives.[28]

After Bush's successful election to the presidency, he was then faced with the challenge of whether and how to control acid rain emissions. Though it had been a major campaign promise, many officials in the Bush White House argued vociferously that the president should abandon the issue because it would anger business interests. Unsurprisingly, President Bush faced enormous pressure from the coal industry not to move forward with emissions reductions. At a minimum, they argued, he should wait for the results of NAPAP before enacting new regulations. For President Bush, however, keeping his campaign promise to address acid rain appears to have fortified his commitment to formulate a proposal despite opposition from fossil fuel interests as well as former Reagan officials.[29]

With the looming issue of global warming hanging over these deliberations, President Bush and EPA officials partnered with the Environmental Defense Fund to come up with a novel way of regulating fossil fuel emissions that could serve as a possible test run for controlling greenhouse gases. The heart of the plan was a "cap-and-trade" mechanism that set a limit on total emissions for the country and then issued industry permits to pollute up to a certain amount. Power plants could choose to reduce their own emissions and sell any unused permits to others, or purchase permits from plants that were lowering pollution to reduce their obligation to cut emissions. The basic idea of cap-and-trade had been around for decades, but this was the first time it was used in environmental regulation anywhere in the world.[30] Prior environmental legislation usually specified either uniform limits for emitters or the use of a particular kind of control technology, called the "command and control" approach.[31]

A revolution in clean coal technology also made the prospect of reducing acid rain emissions more appealing to the Bush administration. While flue gas desulfurization was still expensive to install in existing coal power plants, new innovations in rebuilding these facilities with cleaner burning technologies promised greater reductions for reduced cost. They were developed out of a 1984 program with industry called the Clean Coal Demonstration Program, and had the potential to not only cut sulfur dioxides but also reduce nitrogen oxides as well.[32] In the proposed acid rain regulations, power plants that opted to use this technology were even granted an extended deadline for making the cuts.[33] Entities like EPRI, while not supportive of the administration's push to enact regulations on acid rain, appreciated that they would have options for how to achieve the necessary reductions.[34]

In contrast to industry, some legislators and environmentalists felt the Bush administration proposal did not go far enough in controlling emissions. Congressmen such as Representative Edward Markey from Massachusetts argued that the Bush administration's proposal passed up a crucial opportunity to address the country's larger dependence on fossil fuels and the problem of global warming by focusing only on the sulfur dioxide pollution responsible for acid rain.[35] Environmental groups, even including the Environmental Defense Fund that helped develop the cap-and-trade system, were also critical of the final proposed program and feared the regulations would not go far enough to curb the country's fossil fuel pollution.[36] Nevertheless, the legislation passed with overwhelming majorities in both houses of Congress and was signed into law by President Bush in 1990. A long-awaited agreement with Canada on acid rain followed shortly thereafter. Since the Canadian government had already commenced the process of halving the country's sulfur dioxide emissions in 1984, the treaty largely codified existing domestic legislation in both countries with a few exceptions.[37] It did, however, broaden the two country's commitments to further cooperation on

FIGURE 8.2 President George H. W. Bush signs proposed legislation to amend the Clean Air Act with Secretary James Watkins and William Reilly before it's sent to Congress. George Bush Presidential Library and Museum, July 21, 1989, photo P05226-19A.

transnational pollution beyond acid rain, including mandatory consultations and scientific cooperation on monitoring pollution.

Although US legislation intended to mimic the freedom and simplic-
ity of the market, in practice the 1990 Clean Air Act amendments on
acid rain were incredibly complex. While allowing greater choice for
emitters in comparison to previous regulations, it nevertheless involved
considerable government control over implementation. Forty different
formulas were developed to determine how many pollution allowances
would be allotted to power plants, which attempted to take into account
everything from prior reductions to especially high costs for certain facilities.[38] Utilities, unsure what price their allowances would fetch on the
market, were hesitant to participate. It took the government-run Ten-
nessee Valley Authority to make the first trade with Wisconsin Power
in 1992 in order to launch trading of the permits, and the EPA did not
release the final rules for allowances, monitoring, and penalties until the
following year.[39] While having a bit of a rocky start, the acid rain legislation saw some success in just the first two years of its implementation.
Emissions from coal-fired power plants fell by nearly half, well below
the government's cap.[40]

The US approach to acid rain reductions through cap-and-trade was
hailed as a promising model for Europeans, and perhaps even for tackling future fossil fuel problems like climate change. Yet as a model of
using science in policymaking, the American program mirrored the European and especially British experiences with acid rain. President Bush
may not have used justifications of precaution in the same explicit manner as the European Communities, but his administration moved ahead
on regulation before the government's own enormous research project,
NAPAP, was completed. There is little in the historical record to suggest that there was a definitive scientific breakthrough that helped push
through acid rain legislation. Instead, it was pressure from the American people, frightened over the prospect of climate change and angered
by the Reagan administration's malfeasance in managing environmental
problems, that brought about a change in policy. At the close of the decade, then, the relationship between environmental science and politics
on both sides of the Atlantic was in disarray, with scientific experts playing a minimal role in setting emissions caps, the timetable of reductions,
or which plants should make cuts.[41] Research was viewed as important
in identifying threats but unable to give governments guidance in weighing future environmental risks against the continual demands of industry
for more certainty.

A Pyrrhic Victory for Scientific Expertise

By the mid-1990s, scientific, governmental, and public attention to acid rain plummeted from levels seen in the late 1970s and 1980s. With governments beginning to make reductions, research funding for further studies from both government and private industry lessened in Europe and North America. Rising concerns over global threats to the environment from depletion of the ozone layer and climatic change also reduced popular environmental agitation over acid rain. Compared to these environmental issues, acid rain seemed a minor problem that was already in the process of being resolved.

But while acid rain may have started to fade from public view, environmental officials continued to raise questions about whether government approaches were optimal from an economic and scientific perspective. Members of the British government were especially dubious about the wisdom of using across the board emissions reductions and insisted that there were more economical and environmentally beneficial means of addressing the problem. Many Scandinavian scientists were wary of the reduction approaches for a different reason. The UN and European Communities' agreements simply required cuts of a set percentage from emissions levels at a selected baseline year, which did not account for the latest scientific research on ecosystem vulnerability. One method for ascertaining acid sensitivity that began receiving attention in the early years of the decade was known as the "critical loads" concept. Critical loads are quantitative thresholds for exposure to pollutants; if pollution exceeds these limits, harmful ecosystem effects will occur. Defined as the "tolerable level of acid deposition," they consider the sensitivity of different environments to acid rain.[42] The idea of creating such a mechanism to identify the most susceptible environmental areas was not new.[43] Norwegian scientists, such as Brynjulf Ottar, had envisioned doing precisely this through combining ecological and atmospheric research as far back as the mid-1970s.

Norwegian representatives first raised the possibility of using the critical loads concept in future international agreements on acid rain during negotiations for the 1985 UN Helsinki Protocol on sulfur dioxide.[44] Scandinavian ecologists proposed using critical loads as an alternative to percentage emissions reductions out of concern that the current reduction scheme would not prevent long-term environmental damage to the most sensitive ecosystems impacted by acid rain.[45] Data from the UN's European monitoring network, which had finally begun amassing measurements from countries in both Western and Eastern Europe after 1980, were crucial to the ability to utilize the method. While the early data from the network had been prone to errors, eventually atmospheric physicists at the Norwegian Meteorological Institute became significantly better at locating the offending power plants responsible for specific pollution deposition through acid rain. This evolving data set, combined with ecological studies in Scandinavia, led the signatories of the 1988 UN Sophia Protocol on nitrogen oxides to agree to develop critical loads as the basis for future diplomatic negotiations on air pollution in the 1990s.

Central to this recommitment to incorporating scientific expertise directly into the policymaking process was the formulation of a mathematical model based on a newly emerging discipline of global environmental systems analysis, which combined social science methods with those of the natural sciences. Known as the Regional Air Pollution Information and Simulation (RAINS) model, it was created by scientists at the International Institute for Applied Systems Analysis (IIASA) in Austria. Founded in 1972 after six years of negotiations between the US and the Soviet Union, the IIASA was a nongovernmental research institution created for the express purpose of stimulating scientific cooperation across the iron curtain.[46] Its work was focused on applying the tools of system analysis developed through the RAND Corporation after the Second World War to global issues such as energy, population, and environmental problems. Following the successful establishment of the institute in the early 1970s, its director Roger Levien created programs of research on energy, food, and agriculture that brought together scholars from both the natural and social sciences.[47] The IIASA's work on environmental issues began with analyses of marine natural resources leading up to the UN Convention on the Law of the Sea in 1982. Led by the esteemed scientist and public statesman Harvey Brooks, a former Harvard professor and member of the US Presidential Scientific Advisory Committee during the Eisenhower, Kennedy, and Johnson administrations, the IIASA's environmental researchers were committed to using the tools of systems analysis to provide policy-relevant scientific advising to government officials.

Beginning in the early 1980s, Brooks and other IIASA scientists were invited to attend the first meetings of the UN Executive Committee tasked with implementing the 1979 UN Convention on transboundary air pollution, particularly coordination of the European monitoring network.[48] Through these meetings, scientific representatives of the IIASA from the Netherlands and Poland came up with a model that tried to anticipate the costs and benefits of possible emission reduction scenarios.[49] The RAINS model worked by combining information on fuel usage and anticipated emission levels with data on the atmospheric transport and environmental effects of acidic pollutants.[50] It was comprised of six "sub-models" for sulfur dioxide as well as nitrogen oxides, and attempted to project the environmental impact of various changes in energy policies on soil acidification, lake acidification, groundwater acidification, and forest health.[51] IIASA scientists first presented the model at a conference of acid rain researchers from Western Europe, the US, and Canada in November 1986, many of whom were hopeful it could eventually be used in policymaking circles to perform cost-benefit analyses of different reduction options.[52]

Although this work received encouragement and support from scientists with longstanding involvement in acid rain studies, it was ultimately spearheaded by a group of researchers very different than those historically involved in studying the problem. For instance, although he continued to support acid rain research through the Norwegian Institute for Air Research, Brynjulf Ottar largely turned his attention away from acid rain to studies of arctic pollution in the mid-1980s, seeing this as the new frontier for assessing global environmental dangers.[53] Anton Eliassen remained involved in the UN's European monitoring network through the 1990s, but most of his British counterparts withdrew from the international acid rain debates. After the privatization of Britain's power industry in the late 1980s and early 1990s, its researchers left for academic appointments or new, quite different positions in the energy companies that emerged out of the dissolved Central Electricity Generating Board (CEGB). None of the private companies would mount a research apparatus similar in scope to the CEGB laboratories. While many ecologists in Scandinavia who had pioneered studies on the environmental effects of acid rain continued with their work and assisted with the RAINS model, they were not at the forefront of its eventual use in the negotiation of further emissions reductions.

Instead, scientists from countries that had largely been outside the fray of the intense political debates surrounding acid rain took up the challenge of bringing policy-relevant science and systems analysis into discussions at the UN.[54] The lesson some seemed to learn from acid rain's political and scientific history was that policymakers would be better served by the creation of independent groups of scientists that could work toward consensus reports. This impulse was part of the rationale for creating bodies like the Intergovernmental Panel on Climate Change (IPCC), whose express purpose was to produce consensus among the scientific community and give policy advice on environmental pollution.[55]

After several additional years of refining their model, in 1993 IIASA scientists gathered in Oslo, Norway, with environmental delegates from the US, Canada, and European nations to discuss the possibility of using their scientific model to optimize further emissions reductions. In its final iteration, the RAINS model relied on the idea that each ecosystem had a "critical load" of pollution tolerance that would lead to serious damage once reached. In combination with data on atmospheric transport and energy usage, the model sought to identify the power plants most responsible for causing damage from acid rain and apply cuts directly to these stations.[56] Some of the government officials were confident that once such locations were identified, participating countries could be persuaded to pool their resources to finance reductions at sites outside their national borders at cheaper costs rather than making equal reductions among all countries.

And yet, one of the most striking and provocative findings of the RAINS model was that for a considerable number of locations, no amount of reductions would be enough to prevent harmful effects on the environment. Some ecological systems in Scandinavia were so vulnerable to acid rain that only completely shutting down the offending power plants would protect them. In fact, when IIASA scientists ran the models to try to optimize environmental benefits based on different reduction strategies, they found that avoiding critical loads for all affected areas would cost over five times as much as the current reduction plans.

Because of these and other problems, eventually the meeting's participants agreed to use an interim "target" load. Instead of trying to prevent severe ecological damage outright, target loads instead dealt with the question of what amount of protection could be realistically achieved given the various economic, social, and political issues facing each of the negotiating countries.[57] The resulting UN agreement from the meeting, the Oslo Protocol of 1994, consequently adopted target loads as the appropriate benchmark and agreed to a diverse range of percentage cuts designed to distribute the reductions in the fairest, most environmentally and economically reasonable way. However, European countries were forced to accept that ten percent of affected ecosystems would continue to be damaged by acid rain despite signatories lowering emissions by an average of nearly sixty percent.[58]

In some respects, the use of the RAINS model represented a long-awaited triumph for environmental scientists and officials who had worked on the acid rain problem for decades. Scientific experts had finally earned a seat at the negotiating table and were able to help craft the mechanisms of a reduction scheme to optimize environmental benefits and economic costs. There would no longer be a need to justify taking steps out of a sense of precaution; researchers could quantify the exact damage levels expected and how to address them. On the other hand, the IIASA's modeling showed more starkly than ever the environmental tradeoffs inherent in fossil fuel dependence. There was no solution that would allow continued use of these energy sources without serious environmental damage in some areas. In time, continued studies of acid rain's effects would call into question whether critical loads underestimated how susceptible ecosystems were to acid rain, while a deepening impasse over global warming challenged the wisdom of relying on consensus reports to move policy. Like much of the acid rain story, the ascension of IIASA scientists to a seat at the negotiating table was not an ultimate solution to the problem, but another example of how environmental damage from fossil fuels defied straightforward fixes from science and technology.

The Environmental Legacy of Acid Rain

As the first international fossil fuel pollution problem to be "solved," acid rain has been hailed as a major success story in environmental science and diplomacy.[59] When it comes to achieving the goals set out by policymakers, there are good reasons for such a rosy view. Following the negotiation of international treaties to reduce sulfur dioxide and nitrogen oxides, emissions levels fell sharply across Europe and North America. Within Europe as a whole, levels of sulfur dioxide pollution fell by 67 percent between 1980 and 2000, from about 55 million tons in 1980 to less than 20 million tons in 2000. The continent's worst polluter, Britain, reduced its pollution by 90 percent over that period, as did a reunified Germany. Yet the reductions were not entirely the result of lowering pollution emissions from coal-fired power plants. A large proportion occurred because of switching to less polluting fuels, whether oil, natural gas, nuclear, or renewable sources. Improved energy efficiency also contributed to declines, especially after the oil shocks of the late 1970s. In addition, part of the reductions came about because of economic stagnation, particularly in Eastern Europe after the fall of communism in 1990.[60] For European environmental officials, the intervening years have demonstrated that government regulation of a polluting industry can go only so far in making progress toward a cleaner, more sustainable society.

In the US, the 1990 Clean Air Act regulations were comparably successful in terms of reductions achieved and even more so in terms of economic costs. Sulfur dioxide emissions from power plants were halved by the year 2000 from 1980 levels of 17.5 million tons, and at far less expense than had been anticipated.[61] The EPA predicted the emissions cuts would total about $10 billion a year, with estimates far higher from industry groups like EPRI. Instead, the cap-and-trade program cost a mere $1 billion a year in the 1990s, a paltry sum compared to the health and environmental benefits calculated to be over $10 billion a year. The innovative use of a trading system, which allowed for cuts to be made in plants where it was cheaper to do so, deserves credit for much of the savings. But the reduced economic burden was also the result of lower prices for installing flue gas desulfurization units than industry had expected.[62] Once the full cap took effect in 2000, the cost of the plan eventually went up slightly to $3 billion a year but remained much lower than the $25 billion estimated by the coal industry and lower still than over $100 billion in total environmental and health benefits accrued from the cuts.[63]

Environmental research on acid rain has continued since controls went into effect, although funding and support has declined considerably from earlier periods when the issue had greater political salience. Although the EPA continues to monitor acid rain pollution, NAPAP's conclusion in 1990 left many ecologists struggling for adequate resources to continue research on acid rain's effects. European ecological studies have received support through intergovernmental programs at the UN and European Communities but are also well below funding levels in the 1970s and 1980s.[64] Nevertheless, the legacy of the environmental research programs for acid rain can be felt in both Europe and North America. The European monitoring program originally created to study acid rain pollution across the iron curtain remains in operation today and is the largest source of data on a range of air pollutants affecting populations across the continent. Many of the scientists involved in its development, such as Norwegian atmospheric physicist Anton Eliassen, have maintained their involvement in the network as it has sought to provide guidance to policymakers on reducing pollutants such as ozone and particulate matter. It has also served as a model for the creation of environmental monitoring networks around the world, particularly in Asia as countries like China have struggled to control their fossil fuel emissions.[65]

Among scientists with ties to the coal industry, many parlayed their experience with acid rain into climate change debates. There was a sharp difference, however, that demarcated their involvement in the two issues.

In the case of acid rain, organizations like EPRI and the CEGB sought to maintain at least the veneer of scientific credibility. The American and British industries funded actual research projects and amassed an army of experts who could plausibly claim expertise in the field. While the representations of findings from these programs were problematic, there was a sense that industry needed to participate in scientific studies to influence political debates on the issue. Not so with climate change, which has seen industry groups deploy researchers with no involvement in environmental science to represent their views. Organizations like the George C. Marshall Institute, whose founders included scientists heavily involved with EPRI and other fossil fuel companies, have published papers critical of the science of climate change despite doing scarcely any research of their own on the issue.[66] Rather than engaging directly in scientific debates through conferences, publications, and their own work, these industry-funded experts have become little more than shills for companies willing to pay them as long as they tout the fossil fuel industry line. Oil companies such as Exxon Mobil have received especially sharp criticism in recent years for misleading investors about the dangers of climate change and mounting a multimillion dollar campaign to misinform the public.[67] If anything, fossil fuel companies have been emboldened by the coal industry's experience with using denials of science to forestall progress on acid rain.

The legacy of acid rain for the environment itself has been no less challenging. With attention turning to other pressing pollution problems like climate change, countries suffering from acid rain damages began the slow process of trying to assist ecosystems in their recovery. In areas throughout the US, Canada, Scandinavia, and Germany, environmental agencies tried to engineer stopgap measures to reduce the impact of acid rain by adding alkaline solutions to soils, rivers, and waterways. Known as "kalking" or "liming," several experiments during the 1970s found it could reduce acidity in affected lakes and soils to a certain degree, though it was unclear whether adding these chemicals could bring back freshwater species to lifeless lakes. In the 1980s, Norway and Sweden began using helicopters to disperse these substances over regions of their countries most at risk from acidic pollutants in the hope that these measures would buffer their ecosystems while waiting for the diplomatic accords to go into effect. These two countries undertook the most extensive liming efforts, covering vast areas compared to the limited use along the eastern US.[68]

However, despite these interim measures and the rapidly decreasing emissions from power plants, ecological recovery from acid rain emissions has been far from straightforward. The use of critical loads as a metric to protect ecosystems from damage has become controversial in Europe as data have shown several problems with the method since the 1990s.[69] After the negotiation of the Oslo Protocol using the RAINS model, many ecosystems failed to achieve the target loads set by the international agreement despite governments making the necessary emission reductions. More troublingly for many environmental scientists, even those areas that did meet the target loads set in the Protocol often had signs of damage from acid rain, raising questions about whether there was some problem with the models themselves or the data used to generate them.[70]

To this day, lakes and rivers in North America and Europe continue to show the deleterious effects of acid rain, and thousands of them will no longer support aquatic organisms.[71] Some areas heavily affected by acid rain may take many decades to fully recover.[72] In the US, recent reports from the EPA and Government Accountability Office underscore that acid rain's ecological impacts are still being felt. Nearly 90 percent of the Great Lakes region, for example, remains in danger from acid emissions.[73] Many areas sensitive to acid deposition may even suffer further deterioration without more cuts.[74] Given the extensive reductions countries have already made in fossil fuel emissions, the most promising avenue for further environmental protection is likely to come from a wholesale revolution in the energy industry rather than the installation of further pollution controls.

The issue of forest damage has remained a challenging area of study, but ongoing research efforts confirmed acid rain can act as a harmful stressor and contribute to the decline of tree species in both North America and Europe. We now know that by depleting nutrients from the soil, acid rain can weaken trees and make them vulnerable to drought, disease, or other climatic stressors. Trees exposed to a combination of acid rain, ozone, and local pollution sources appear to be particularly at risk, such as the region known as the "black triangle" within Germany, Poland, and the Czech Republic.[75] Long-term studies on both continents have demonstrated overall declines in productivity for many forest stands, though massive die-offs were thankfully avoided.[76] The absence of worst case scenario decimation has led some to call the "forest death" fears an example of environmental overreach, dramatizing dangers unnecessarily. There is an argument to be made that the media did sensationalize some of the research results, but it's misleading to suggest that there was no evidence to support concerns or that they have been debunked by additional research.[77] In fact, some studies have underscored the extent to which acid rain can threaten forest productivity even faster than expected.[78]

The looming issue of climate change threatens to upend the progress made on combating acid rain's environmental effects. Warming temperatures may create dislocations of tree species, while disruptions to the water cycle could exacerbate effects in forest ecosystems. Some research has suggested that a warmer climate might actually be protective against damages by increasing weathering rates of sulfur dioxide and nitrogen oxides from soils. At the moment, there is simply not enough scientific research to conclude one way or the other how climatic changes will interact with acid rain, whether for good or ill.[79]

However, it is clear that the need to address fossil fuel pollution globally will only grow in the coming years. Since the early 1990s, there has been a deepening realization that many fossil fuel pollutants do not simply travel through the atmosphere on a regional scale, as was discovered with acid rain, but may have a global circulation pattern. These include sulfur dioxide as well as pollutants with more direct impact on human health, such as particulate matter and persistent organic pollutants.[80] In combination with concerns over climate change, scientists and policymakers are now faced with thinking about the multiplying effects of many different pollutants over increasing temporal and geographical scales. Regardless of whether climate change has a beneficial, harmful, or neutral impact on acid rain's ecological effects, it will be difficult for affected areas to completely recover if humanity continues to rely on coal, oil, and natural gas for energy, not to mention the future harms developing nations could suffer as they industrialize. The environmental legacy of acid rain is therefore not yet able to be confined to a history book. It remains in the soil, the air, and the water, an invisible reminder of humanity's power to harm the natural world and the price we pay for our fossil fuel dependence.

Epilogue: The Climate Change Reckoning

In 1979, at a US conference on chemicals in the environment held in Rochester, New York, Norwegian delegate Erik Lykke warned those in attendance that it would be imperative to mount a broad international response to the dangers of fossil fuel pollution. Acid rain, then the primary subject of attention for industrialized countries, was only the first of what were likely to be many environmental dangers caused by burning coal, oil, and natural gas. The potential long-term effects of increasing levels of carbon dioxide, he said, would pose an especially grave threat through changes to the climate. Emission controls would be essential for humanity to avoid further harms on a regional or global scale, and acid rain regulations would provide a first step in attempting to wean ourselves off fossil fuels.[1]

Lykke was neither the first nor the last environmental official to argue that acid rain held important lessons for addressing climate change. Many environmental scientists involved in acid rain also spoke out about how we could learn from the problem to better address greenhouse gas pollution. As far back as 1970, Bert Bolin, one of the Swedish scientists who helped prepare a special report on acid rain for the United Nations' first conference on international environmental issues, made the case for why acid rain should serve as a warning and possible guide for countries concerned about climate change.[2] Bolin would go on to become the driving force behind the Intergovernmental Panel on Climate Change (IPCC), the leading scientific body tasked with assessing current evidence for global warming and its impacts. The IPCC's formation reflected a desire for a more robust and coordinated effort to achieve broad scientific consensus that could be funneled directly into the policymaking process and avert the impasses that occurred with acid rain.[3] In its mission to coordinate thousands of scientists to produce comprehensive, clear assessments of the dangers from global warming, the IPCC was largely successful. After its founding in 1988, the group published numerous reports synthesizing the latest climate science and its implications for the planet.[4] In 2007, it was awarded a Nobel Peace Prize for these efforts.[5]

Despite the incredible achievements of the IPCC, the translation of its scientific achievements into concrete policy or diplomatic accords has been thwarted time and time again. Regardless of the merits of the decision, the US contributed to a serious global impasse over its refusal to join the 1997 Kyoto Protocol on climate change. Even European countries who sought to meet the targets of the accord have been criticized for not doing enough. Portions of the reductions they achieved did not result from direct emissions cuts, but rather from lowered energy usage following the 2008 financial crisis as well as the movement of polluting industries to China and other developing countries.[6] While recent international efforts seemed to finally bear fruit with the 2015 Paris agreement, the current Trump administration's withdrawal from the treaty has left nations around the world wondering how to move forward without one of the biggest polluters onboard.[7] Compounding the situation, fossil fuel corporations and their paid experts have continued to try to discredit the scientific research behind climate change.[8]

Historians aren't necessarily the most equipped to provide advice on how to move forward. Our eyes are trained toward the past, and an enormous amount has changed in just a short span of time on the issue of climate change. Had I finished this book a few years ago, I would have felt compelled to write a very different epilogue than the one here. It seemed that the world was moving in a positive direction on climate change issues; however flawed, the 2015 Paris agreement represented an important step in addressing the problem. But perhaps because of the longer perspective history provides, it can offer us a port in the storm. Even when the path forward looks bleak, we have come through similarly dire straits before. It's no guarantee we can do it again, but it is reason to have hope.

EPILOGUE   194

So what are the lessons the history of acid rain might offer us, as we try to tackle the next big environmental problem of our time? The first factor to examine is the role of science, and particularly "scientific consensus," in paving the way for international agreements and domestic regulations. Unfortunately, many scientists and policymakers who have worked on climate change as well as ozone depletion failed to fully comprehend the role of scientific expertise in creating political momentum for reducing pollution. For acid rain, there is little evidence to suggest that scientific studies convinced polluters like the US, Britain, France, and West Germany to lower their emissions; the same is true of ozone depletion.[9] Many environmental scientists who worked on acid rain during the 1970s and 1980s were not shy in talking to government officials and the media about the problem. They made every effort to disseminate their work and were outspoken about the need for governments to reduce fossil fuel pollutants or risk ecological damages. But although their actions helped to exert pressure on industrial emitters, their arguments did not convince governments in major polluting nations to change their policies. Privately, many officials in these nations acknowledged that these scientists were right. They simply didn't believe it was in their national interest to do anything about a problem that would have a far bigger impact on other countries and cost their industries a great deal of money to solve.

We are faced with a similar situation on climate change. The countries that will suffer the most are developing nations, especially areas of Africa, Asia, and Latin America with large populations in severe poverty. Meanwhile, the countries that are the worst polluters have the greatest capacity to withstand adverse environmental conditions that might result from climate change. Further emphasizing the state of scientific knowledge is unlikely to persuade the political establishments or the public to take action. Nor does it seem fair to expect scientists to become more vocal about the dangers in the hopes of generating political will.[10] There is little evidence from acid rain to suggest this would be effective.

Instead, we should look to three key moments in the history of acid rain where deadlock was broken internationally and domestically for future guidance on climate change, none of which involved persuading key actors of the scientific merits for implementing fossil fuel reductions. The first was the successful negotiation of the 1979 UN Convention on transboundary air pollution, which involved all European nations as well as the US and Canada. During the initial discussions between the communist and capitalist blocs, the US State Department exerted pressure on Britain, France, and West Germany to agree to the treaty. American diplomats believed that failure to negotiate an environmental accord, which the Soviet Union supported, would seriously cripple the country's efforts to seek agreements with the Soviet Union on nuclear armaments and human rights. At that time, negotiations to reduce environmental pollution were seen as far less contentious than discussions over the other two issues, and Washington hoped a treaty on acid rain could build momentum to achieve the country's other diplomatic objectives with the Soviet Union. The politics of détente consequently helped motivate reluctant polluters to cooperate on the environment to ensure their other foreign policy goals would be met.

Climate change could similarly benefit from the international community linking negotiations on reducing greenhouse gases with other diplomatic efforts, particularly matters of national security and trade. At a minimum, it makes practical sense given the far-reaching implications of a warming world. Specialists in economic development have long noted that efforts to reduce poverty will be much more difficult as global warming intensifies. Military analysts have also argued that climate change could have serious security implications by further destabilizing volatile regions in the Middle East and Africa. But unlike the UN acid rain negotiations, which were part of larger talks about the Cold War, climate change diplomacy has been separated from other foreign policy objectives. And so far, European countries have proven unwilling to exert serious pressure on a reluctant US by leveraging their cooperation on other foreign policy matters.[11] There are some indications that German Chancellor Angela Merkel may be taking steps along these lines with the cooperation of countries in the European Union as well as Russia and Saudi Arabia, but much stronger actions will be necessary for them to succeed in convincing the now isolated US to change its position.[12]

The second and third events that brought about important changes in addressing acid rain were the decisions by Britain and the US to lower their emissions. For Britain, their reversal in position came about only because of pressure from the European Communities, which presented a unified front on reducing fossil fuel emissions after West Germany realized its own environment might be threatened by acid rain. Without harmonization of regulations within the Communities, many economic experts feared that industries in countries with lower environmental standards would have an unfair advantage, possibly creating trade distortions among member states. The British government, which had previously refused to reduce the country's pollution to combat acid rain, eventually agreed to West German demands because of its desire to remain in the European Communities and strengthen the political and economic unity within Western Europe. The drive for European integration and a willingness to concede authority to a supranational group were thus instrumental in combating acid rain.

The world is facing a much different situation with climate change, as Britain's recent decision to leave the European Union has demonstrated. Nationalist impulses in countries like Britain and the US will make it far more difficult to utilize a similar strategy in the years ahead.[13] However, Britain's position with the European Communities on acid rain nevertheless underscores how important it will be for other countries to band together and exert economic pressure on countries that might not otherwise lower their fossil fuel emissions. The British government's desire to maintain the economic advantages it received through its membership convinced it to agree to the European Communities' pollution reduction scheme. Proposals to force outliers on climate change to pay a tax on exports to other countries would be one way to achieve similar pressure without requiring a supranational authority seeking to motivate compliance with recalcitrant member states.[14] While this path could unleash a trade war and should be used only once all other avenues are exhausted, the long-term risks to the environment of rogue states that refuse to reduce emissions should keep this strategy on the table.[15]

For the US, a shift in the government's approach to acid rain came about after an intense public outcry, which itself was a product of heightened media attention to environmental issues during the summer heat wave of 1988. The massive surge in support for acid rain legislation was especially influential in that year's presidential race and helped convince President George H. W. Bush to make the problem a centerpiece of his early days in office. After the election of President Donald Trump and little mention of climate change during the 2016 campaign, it may seem like sparking a similar movement today would be near impossible.[16]

Yet it was partly the mismanagement of national and local environmental issues under the Reagan administration that helped swing public opinion toward support for pollution controls. With the Trump administration also dismantling a host of environmental regulations unrelated to climate change, it may be possible for environmental groups to capitalize on more personal, local issues to make pollution problems a top concern once again.[17] If the acid rain case is any indication, public attention to climate change is likely to rise because of the broad assault on environmental regulation under the current President.

Resistance to moving away from fossil fuels will remain so long as those who stand to benefit from their use have special influence with elected officials. We will all need to work together—domestically and internationally—to ensure the support of unbiased environmental research and make our voices heard in the halls of government. If not, many years from now, historians will be writing a retrospective about climate change far different than the story of acid rain.

Acknowledgments

The process of writing this book began nearly a decade ago, and I've benefited enormously from many people and organizations that supported me along the way. My earliest interest in understanding the history of environmental science and pollution problems began at Princeton University, and I am grateful to have worked with several professors there who nurtured both my mind and spirit while on campus. Outside of the history department, Anne Matthews, Joyce Carol Oates, and Susan Choi were instrumental in honing my ability to craft a compelling narrative and have significantly influenced the way I think about writing history. Gregory van der Vink introduced me to the use of scientific and technical data in policymaking, and my ability to understand the complex geophysics involved in environmental problems is thanks to his legendary abilities as a teacher.

Among the history faculty, Anthony Grafton, D. Graham Burnett, and Helen Tilley were incredible sources of support and counsel during and after my time on campus. I am most indebted to Angela Creager, who served as my advisor and has remained a beloved mentor in the years since. It was through working one on one with Angela that I learned how to conduct original historical research, and her example as a scholar is one I continually strove to emulate.

The bulk of my work on this project took place at Yale University, where I was fortunate to join an exceptional community of scholars in the history department and the program in the history of science and medicine. I want to express my sincere thanks to Paola Bertucci, Bettyann Kevles, Naomi Rogers, Paul Sabin, Jenifer van Vleck, Jay Winter, and John Warner. Patrick Cohrs, William Rankin, and Bruno Strasser deserve special mention for their tireless guidance and feedback. I'm also grateful for the camaraderie of my fellow students, including Debbie Doroshow, Courtney Thompson, Jenna Healey, Kate Irving, Kelly O'Donnell, Helen Curry, Heidi Knoblauch, Robin Scheffler, Mary Brazelton, Joy Rankin, Tom Reznick, Jed Gross, Todd Olszewski, Ying Jia Tan, Gerardo Con Diaz, Justin Barr, and Brendan Matz.

Many of my peers from other institutions gave thoughtful comments on my work. I am particularly thankful for encouragement I received from Leah Aronowsky, James Bergman, Connemara Duran, Cara Kiernan Fallon, Evan Hepler-Smith, Whitney Laemmli, Mary Mitchell, Timothy Nunan, Alistair Sponsel, and Roger Turner, who have formed part of my wider community of young scholars. There are also faculty members who have provided feedback from locations far and wide. James Fleming gave me advice even before I embarked on graduate study in the field and has had an enormous impact on my understanding of the earth sciences. At various stages of this project, Peder Anker, Carin Berkowitz, Janet Browne, Deborah Coen, Gene Cittadino, Michael Egan, Jacob Darwin Hamblin, Toshihiro Higuchi, Sheila Jasanoff, Arne Kaijser, John Krige, John McNeill, Barbara Naddeo, Naomi Oreskes, Richard Revesz, Jody Roberts, Nils Roll-Hansen, Pamela Smith, Matt Stanley, Spencer Weart, and two anonymous reviewers have each offered helpful suggestions on my research, and I thank them for their genuine interest in my work. My editor, Karen Merikangas Darling, deserves special mention for her guidance and encouragement through the publication process, as does Mary Corrado for her detailed attention to the manuscript.

The most auspicious event in the development of this book was my decision to work with Daniel Kevles, who has not only been a supremely dedicated adviser but a true mentor to me in every sense of the word. He challenged me to push myself farther than I ever thought possible, and I owe him the deepest of intellectual debts. I will be forever thankful for finding a kindred intellectual guide to see me through my work. I am grateful he never balked at the ambition of this project, as it was his quiet confidence that propelled me to take the many leaps involved in pulling it together. His suggestions and comments have helped me immensely, and I cannot thank him enough for all the time and energy he invested in my intellectual growth.

I have been extremely fortunate to receive funding from several institutions to pursue research for this book. A National Science Foundation Graduate Research Fellowship supported my doctoral studies and allowed me an extra year of writing and research time, which I greatly appreciated. Grants from the American Meteorological Society, the American-Scandinavian Foundation, the Yale MacMillan Center for International and Area Studies, the Chemical Heritage Foundation, and the Society for the History of Alchemy and Chemistry are the reason I was able to spend considerable time in archives abroad, and this project would not have been possible without their generosity.

There were countless archivists and librarians who helped me access materials for this work, and I would like to express my gratitude to those at the British Parliamentary Archives, Bundesarchiv Berlin, Bundesarchiv Koblenz, European Commission Archives, Nasjonalbiblioteket, the British National Archives, the US National Archives and Records Administration II, Norsk institutt for luftforskning, Organisation for Economic Co-operation and Development, Norsk Riksarkivet, Svensk Riksarkivet Arninge, Svensk Riksarkivet Marieberg, Royal Society Archives London, Surrey History Centre Archives, United Nations Economic Commission for Europe Archives, and World Health Organization Archives. Several people deserve special mention for going above and beyond the call of duty to help me locate historical documents. My sincere thanks to Maria del Mar Sanchez at the United Nations; Jan-Anno Schuur, Patricia Fiore, and Francis Pellerin at the OECD; Kari Kvamsdal at the Norwegian Institute for Air Research, Magnus Sollid at the Norsk Riksarkiv, as well as Prime Minister Kåre Willoch for special permission to see closed records of the Ministry of Environment and Maren Esmark to see closed records of Naturvernforbundet; Lievan Baert at the European Commission archives; and Øystein Hov, Ronald Barnes, Bernard Fisher, Tony Kallend, Alan Webb, Richard Skeffington, and Peter Collins for allowing me to either keep or copy their personal papers and materials concerning acid rain.

Finally, I want to thank my family for their continued faith in me throughout the long process of finishing this book. My parents, Tony and Judy, and sister Amanda, have always believed in the project, and I count myself unbelievably fortunate to have them as a sounding board for my ideas. While completing this work, I also joined a new family who lent their support to these efforts. I sincerely appreciate the ways in which Peter, Peggy, Scott, and Cameron Bielski have welcomed me into their lives and shown interest in my research.

My greatest debt is to my husband, Craig, who has been my closest confidant for nearly half our lives. It is impossible to put into words how our relationship has shaped the person I am today, but I have looked forward to the opportunity to try. Craig has unwaveringly encouraged me every step along the way, even during the times that meant having to part for extended archival travels. My work is immeasurably better because of our discussions, whether over dinner or over the ether from two different continents. His wry sense of humor and quick wit has made even the most challenging days in the making of this project happy ones. I dedicate the book to him and to our son, Theo—to infinity, plus one.

Notes

INTRODUCTION

1. Henrik Ibsen, *Brand: A Dramatic Poem in Five Acts*, translated by Charles Harold Herford (London: W. Heinemann, 1898).

2. Interestingly, as journalist Fred Pearce has noted, many of the English translations of *Brand* omitted this passage. See Fred Pearce, *Acid Rain* (New York: Penguin, 1987), p. 15; James Lovelock, *The Revenge of Gaia: Earth's Climate in Crisis and the Fate of Humanity* (New York: Basic Books, 2006), p. 116.

3. J. R. McNeill, *Something New under the Sun: An Environmental History of the Twentieth-Century World*, reprint edition (New York: W. W. Norton, 2001), pp. 31, 64.

4. Some scientists did suggest that on certain occasions soot deposits from Britain could have reached portions of Norway in the nineteenth century, though these were not extensively studied. See Peter Brimblecombe, Trevor Davies, and Martyn Tranter, "Nineteenth Century Black Scottish Showers," *Atmospheric Environment* 20, no. 5 (1986): 1053–1057, p. 1057.

5. Compare, for instance, the term as it was originally coined by Robert Angus Smith in 1872. See Robert Angus Smith, *Air and Rain: The Beginnings of Chemical Climatology* (London: Longmans, Green, 1872).

6. Scientists were involved in government advising as early as the Progressive Era in the US, but their role became much greater in both Europe and the US following the world wars. See A. Hunter Dupree, *Science in the Federal Government: A History of Policies and Activities*, revised (Johns Hopkins University Press, 1986); John Krige, *American Hegemony and the Postwar Reconstruction of Science in Europe* (Cambridge, MA:

MIT Press, 2008); Stuart W. Leslie, *The Cold War and American Science: The Military-Industrial-Academic Complex at MIT and Stanford* (New York: Columbia University Press, 1994); Daniel J. Kevles, *The Physicists: The History of a Scientific Community in Modern America* (Cambridge, MA: Harvard University Press, 1995); Gregg Herken, *Cardinal Choices: Presidential Science Advising from the Atomic Bomb to SDI*, revised and expanded edition, Stanford Nuclear Age Series (Stanford: Stanford University Press, 2000); Benjamin P. Greene, *Eisenhower, Science Advice, and the Nuclear Test-Ban Debate, 1945–1963*, Stanford Nuclear Age Series (Stanford: Stanford University Press, 2007); Zuoyue Wang, *In Sputnik's Shadow: The President's Science Advisory Committee and Cold War America* (New Brunswick, NJ: Rutgers University Press, 2008).

CHAPTER ONE

1. John Roberts, Minister of the Environment, Canada, "Notes for an Address by the Honourable John Roberts, Minister of the Environment, to Fund-Raising Dinner" (Canadian Coalition on Acid Rain at the Albany Club, Toronto, Ontario, October 19, 1981), RA-PA-0641-E-Eg-Loo13, Norwegian National Archives, pp. 1–2.

2. Gro Harlem Brundtland, "Concluding Statement by Mrs. Harlem-Brundtland," November 1974, RA-S-2532–2-Dca-Lo409, Norwegian National Archives, pp. 2–3. The dispersion of mercury and PCBs on a regional and global scale also became matters of international concern at the time acid rain was identified. See Erik Lykke. "Statement by Erik Lykke, Head of Norwegian Delegation. Meeting of the Environment Committee at Ministerial Level, 13th November, 1974," November 13, 1974, RA-S-2532-2-Dca-Lo409, Norwegian National Archives, p. 1.

3. On the history of energy use and pollution since the Industrial Revolution, see Alfred W. Crosby, *Children of the Sun: A History of Humanity's Unappeasable Appetite For Energy* (New York: W. W. Norton, 2007) and J. R. McNeill, *Something New under the Sun: An Environmental History of the Twentieth-Century World*, reprint edition (W. W. Norton, 2001). The three famous instances of documented deaths from smog pollution were in 1930 in Liège, Belgium, in 1948 in Donora, Pennsylvania, and in 1952 in London, England. See Organisation for Economic Co-operation and Development, *Methods of Measuring Air Pollution: Report of the Working Party on Methods of Measuring Air Pollution and Survey Techniques* (Paris: Organisation for Economic Co-operation and Development, 1964), p. 7.

4. For a historical overview of air pollution in the Ruhr valley from coal, see Franz-Josef Brüggemeier, "A Nature Fit for Industry: The Environmental History of the Ruhr Basin, 1840–1990," *Environmental History Review* 18, no. 1 (April 1, 1994): 35–54, pp. 40–44. For a historical overview of air pollution surrounding London, particularly regarding smoke, see Peter Brimblecombe, *The Big Smoke: A History of Air Pollution in London since Medieval Times* (Methuen, 1987).

5. C. Donald Ahrens, *Meteorology Today: An Introduction to Weather, Climate, and the Environment* (Boston: Cengage Learning, 2007), p. 490.

6. Indur M. Goklany, *Clearing the Air: The Real Story of the War on Air Pollution* (Washington, DC: Cato Institute, 1999), pp. 14–21; David Stradling, *Smokestacks and Progressives: Environmentalists, Engineers, and Air Quality in America, 1881–1951* (Baltimore: Johns Hopkins University Press, 2002), pp. 99–104; Joel A. Tarr and Bill C. Lamperes, "Changing Fuel Use Behavior and Energy Transitions: The Pittsburgh Smoke Control Movement, 1940–1950: A Case Study in Historical Analogy," *Journal of Social History* 14, no. 4 (July 1, 1981): 561–588.

7. Though deposits of sulfur have been used by many ancient civilizations, sulfur was first identified as a chemical element by Antoine Lavoisier in 1777. The gaseous form of "sulfur dioxide" was used in a number of technological applications in the nineteenth century for public health, including as a disinfectant. See Cyrus Edson, "Disinfection of Dwellings by Means of Sulphur Dioxide," *Public Health Papers and Reports* 15 (1889): 65–68. "Sulfurous" smelling smoke had been attributed to the fogs around London during the nineteenth century, earning it the nickname "pea-soup." However, scientific research into the dangers of sulfur dioxide as a pollutant, as opposed to smoke and smog more generally, began only after the first documented pollution "disaster" in the Meuse Valley of Belgium in 1930. It increased progressively after the Second World War as a result of continuing investigations into pollution disasters. See D. B. Keyes, "Pure Air for Our Cities," *Scientific Monthly* 35, no. 5 (November 1, 1932): 427–430; E. F. A., "Atmospheric Pollution," *Journal of the Royal Society of Arts* 85, no. 4430 (October 15, 1937): 1041–1043. On the impact of the 1948 Donora incident and concerns about Los Angeles pollution on sulfur dioxide research, see Morris Katz, "Symposium on Analytical Methods and Instrumentation in Air Pollution Introduction," *Analytical Chemistry* 27, no. 5 (May 1, 1955): 692–694; "London's New Fog Suffused by Gases: Medical Journal Confirms Warning of Lethal Content from Smokeless Fuels," *New York Times*, November 21, 1953, sec. Business Financial, p. 30; Carroll E. Williams, "Air Pollution Is Discussed: Communities Face New Problems, Control Group Warned," *Sun*, May 28, 1953, p. 13; "Victory by Science Over Smog Seen by Hitchcock: Air Pollution Can Be Banished, but It Will Take Time and Money, Research Chief Says," *Los Angeles Times*, December 31, 1954, p. 4.

8. Jerome Alexander, "The Fatal Belgian Fog," *Science* 73, no. 1882 (January 23, 1931): 96–97.

9. Ann House, "Brussels Is Terrorized," *Chicago Daily Tribune*, December 6, 1930, p. 1; Edmond Taylor, "68 Die in Europe's 'Gas Fog': 'Death Cloud' Strikes at Belgium, France; Strange Malady Hits Many; Hospitals in Paris Fill Up. Bulletin," *Chicago Daily Tribune*, December 6, 1930, p. 1; "Belgium's Poison Fog Cases Likened to the 'Black Death,'" *New York Times*, December 6, 1930, p. 1; "Scores Die, 300 Stricken by Poison Fog in Belgium," *New York Times*, December 6, 1930, p. 1; "64 Belgians Killed by Gas in Heavy Fog: Victims Die in Same Manner as in War—Various Theories Held as to Cause," *Hartford Courant*, December 6, 1930, p. 1; "64 in Belgium Killed by Gas Borne on Fog: Mysterious Poison Mist Rolls along Meuse, Reaping Toll; German War Stores at Liege Recalled; Panic-Stricken Villagers Believe Them Source of Fatalities; Others Hold Fumes of Factories Cause; London Professor Advances Theory 'Black Death' Has Returned,"

Washington Post, December 6, 1930, p. 1; "Black Death Theory Held: 64 Die in Belgium in Mysterious Fog," *Washington Post*, December 6, 1930, p. 1; "Black, Death Is Feared as Mysterious Malady Slays Belgian Peasants: Fifty-eight Die Suddenly While Groping Through Heavy Fog Which Envelops Country for Three Days—Animals Also Victims. Poison Gas Also Is Suggestion. British Professor Says Conditions Resemble Those Attending Mediaeval Plague Which Carried Off One-third of Western Europe's Population," *Globe*, December 6, 1930, p. 1; "Deadly Fog in Belgium: Fifty People Die in 24 Hours in Meuse Valley Poison Gas or Factory Fumes?," *Manchester Guardian*, December 6, 1930, p. 11; "Mysterious Vapour: Panic-Stricken Population's Ordeal," *Scotsman*, December 6, 1930, p. 11; "Mystery Fog Brings Death to 64 Belgians: Villagers Are Asphyxiated as Poisonous Fumes Invade Meuse Valley Animals Perish; Region in Panic; Abandoned German War Gases Believed Possible Source," *Sun*, December 6, 1930, p. 2; "Poison Fog Kills 64 in Belgium: Villagers of Meuse Valley in Panic as Mysterious Mist Brings Death as Did Lethal Gas in War; Liege Vicinity Is Affected; 'Black Death' Theory Held, Disputed; Health Minister Starts Probe; 14 Dead and Hundreds Ill in Town of Engis," *Daily Boston Globe*, December 6, 1930, p. 1.

10. Somer H. Ann, "Belgian Queen Visits Victims in 'Fog Death' Belt," *Chicago Daily Tribune*, December 8, 1930, p. 4.

11. "Poison Fog Kills Scores in Terrorized Belgium: Deadly Black Mist Creeps on as Men and Animals Die Horribly; Noxious Vapors Invade Paris," *Los Angeles Times*, December 6, 1930, p. 2; "Prof J. B. S. Haldane Leans to 'Black Death' Theory," *Daily Boston Globe*, December 6, 1930, p. 2; "Similar to Black Death, Thinks British Scientist," *Sun*, December 6, 1930, p. 2; "Expert Discusses Probable Causes of Fog Deaths," *Science News-Letter* 18, no. 505 (December 13, 1930): 383; Alexander, "The Fatal Belgian Fog."

12. Keyes, "Pure Air for our Cities"; C. F. Talman, "Death-Dealing Fogs A Scientific Puzzle; Two Aspects of the Strange Effects of the Ice Fog," *New York Times*, December 14, 1930; J. Firket, "Sur les causes des accidents survenus dans la vallée de la Meuse, lors des brouillards de décembre 1930," *Acad. roy. Méd. Belg.*, 11, no. 683–741 (1931); André Allix, "A propos des brouillards lyonnais. 4. Le brouillard mortel de Liège et les risques pour Lyon," *Les Études Rhodaniennes* 8, no. 3 (1932): 133–144; "Air Pollution by Smoke," *Journal of the American Medical Association* 97, no. 18 (October 31, 1931): 1312–1313; "Belgium," *British Medical Journal* 2, no. 3756 (December 31, 1932): 1207–1208; "Atmospheric Pollution and Pulmonary Diseases," *British Medical Journal* 1, no. 3658 (February 14, 1931): 277; J. S. Haldane, "Atmospheric Pollution and Fogs." *British Medical Journal* 1, no. 3660 (February 28, 1931): 366–367; " 'Poison Fog' Report Submitted in Belgium: Board Says Abnormal Weather Added to Factories' Pollution of Air Caused Deaths," *New York Times*, November 13, 1931, p. 10; " 'Death-Fog' Mystery Is Explained: Belgian Report Reveals Fatal Pall Was Caused by Cold, Damp and Sulphur Gases," *Sun*, November 29, 1931, p. 56; T. K. C., "More About the Meuse Valley Catastrophe," *Journal of the Franklin Institute* 214, no. 2 (August 1932): 214.

13. The work of Alice Hamilton on industrial toxic gases is a notable exception for including sulfur dioxide. See Alice Hamilton, *Industrial Poisons in the United States* (New York: Macmillan, 1929). For an in-depth history of the

smoke abatement movement in America from the nineteenth century through the Second World War, see Stradling, *Smokestacks and Progressives*. For a discussion of exposure to hazardous chemicals in the workplace, see Christopher C. Sellers, *Hazards of the Job: From Industrial Disease to Environmental Health Science* (Chapel Hill: UNC Press Books, 1999).

14. William West, "On the Proportion of Sulphur in Coal," *Proceedings of the Geological and Polytechnic Society of the West Riding of Yorkshire* 1 (January 1, 1839): 48–52. On the harmful effects of sulfur pollution on buildings and property, see for example "Committee for the Investigation of Atmospheric Pollution: First Report (April 1914 to March 1915)," *Lancet* 187, no. 4826 (February 26, 1916): i–iii, p. iii; W. F. M. Goss, "Smoke as a Source of Atmospheric Pollution," *Journal of the Franklin Institute* 181, no. 3 (March 1916): 305–338, pp. 310, 333–334.

15. H. B. Meller, "Clean Air, an Achievable Asset," *Journal of the Franklin Institute* 217, no. 6 (June 1934): 709–727; Alfred Romanoff, "Sulfur Dioxide Poisoning as a Cause of Asthma," *Journal of Allergy* 10, no. 2 (January 1939): 166–169; "Our Industrial Air," *Lancet* 246, no. 6383 (December 29, 1945): 857. It should be noted that many works still cited building damages as the major concerning effect of sulfur pollution. See for example "The Measurement of Atmospheric Pollution," *Public Health* 47 (October 1933): 215–216; T. K. C., "A Quick Test for Air Pollution," *Journal of the Franklin Institute* 213, no. 3 (March 1932): 349–350.

16. T. K. C., "Toxic Fogs," *Journal of the Franklin Institute* 213, no. 3 (March 1932): 344–345. R. H. O., "City Air Searched for Sulphur Fumes," *Journal of the Franklin Institute* 226, no. 6 (December 1938): 830.

17. For these earlier positive views on smoke, see Stradling, *Smokestacks and Progressives*, and Peter Thorsheim, *Inventing Pollution: Coal, Smoke, and Culture in Britain since 1800* (Columbus: Ohio University Press, 2006). This is not to say that such views were held by all citizens of industrialized countries, as smoke control advocacy did become a potent political force in many urban areas during the late nineteenth and early twentieth centuries.

18. For an account of the Donora smog episode, see Lynne Page Snyder, " 'The Death-Dealing Smog over Donora, Pennsylvania': Industrial Air Pollution, Public Health Policy, and the Politics of Expertise, 1948–1949," *Environmental History Review* 18, no. 1 (April 1, 1994): 117–139.

19. While initially the fatalities were estimated to be 4,000 in the Great London Smog, new analyses have concluded that the death toll was likely as high as 12,000. Michelle L. Bell, Devra L. Davis, and Tony Fletcher, "A Retrospective Assessment of Mortality from the London Smog Episode of 1952: The Role of Influenza and Pollution," *Environmental Health Perspectives* 112, no. 1 (January 2004): 6–8; "200 Døde Etter Skodda i London," *Firda Folkeblad*, December 11, 1952, National Library of Norway, Oslo, Norway, p. 2.

20. Brimblecombe, *The Big Smoke*, pp. 166-169.

21. Author interview with Ronald Barnes, retired, former Principal Environmental Adviser, Esso, March 4, 2013, Wantage, England.

22. E. C. Halliday, "A Historical Review of Atmospheric Pollution," in *Air Pollution* (Geneva, Switzerland: World Health Organization, 1961), p. 11. The

word "smog" was first used by British scientist Henry Des Voeux in 1904 but was soon adopted in the US to describe severe pollution episodes in Los Angeles. See Christine L. Corton, *London Fog: The Biography* (Cambridge, MA: Harvard University Press, 2015), p. 27.

23. OECD, *Air Pollution in the Iron and Steel Industry* (Paris: Organisation for Economic Cooperation and Development, 1963), p. 86; M. W. Holdgate, *A Perspective of Environmental Pollution* (Cambridge: Cambridge University Press, 1980), pp. 183–184.

24. In addition to studying the meteorological aspects of air pollution and smog, researchers also looked at the effects of air pollution on the general health of urban populations. "Scientists Declare War on Air Pollution," *Science News-Letter* 64, no. 6 (August 8, 1953): 85; Holdgate, *A Perspective of Environmental Pollution*, pp. 183–184; James P. Dixon, Norton Nelson, John R. Goldsmith, et al., "National Conference on Air Pollution: Conference Report," *Public Health Reports* 74, no. 5 (May 1, 1959): 409–427; "Death Trap from Smog," *Science News-Letter* 66, no. 18 (October 30, 1954): 279; John Pemberton and C. Goldberg, "Air Pollution and Bronchitis," *British Medical Journal* 2, no. 4887 (September 4, 1954): 567–570; M. Neiburger, "Weather Modification and Smog," *Science* 126, no. 3275 (October 4, 1957): 637–645.

25. "Smog Study Set in West Germany: Government Investigates Health Danger after Public Protests; Ruhr Hardest Hit," *Los Angeles Times*, March 31, 1957, p. A12.

26. The scientists were drawn from the Verein Deutscher Ingenieure (German Engineers Association), which had been studying dust problems for thirty years. Special Committee for Iron and Steel of the OECD, *Air Pollution in the Iron and Steel Industry* (Paris: Organisation for Economic Cooperation and Development, 1963), pp. 77–78.

27. "Grand Jurors Quiz Mayor in Smog Inquiry," *Los Angeles Times*, October 21, 1954, p. 1; "Diners Wear Masks in Protest on Smog," *Los Angeles Times*, October 15, 1954, p. 1.

28. "President Signs $15,000,000 Smog Study Bill," *Los Angeles Times*, July 15, 1955, sec. Part I, p. 8; Robert T. Hartmann, "House OKs $25,000,000 Bill for Smog Research: Senate Expected to Drop Its Own Version and Back 5-Year Plan Passed by Representatives," *Los Angeles Times*, July 6, 1955, sec. Part I, p. 8; Bess Furman, "$25,000,000 Study of Smog Proposed: Five-Year U.S. Survey Asked— Federal Aid to Localities Envisaged in Program," *New York Times*, June 12, 1955, p. 68; "Senate Passes $15,000,000 Bill for Smog Study," *Los Angeles Times*, June 1, 1955, sec. Part I, p. 2.

29. D. A. Layne, "A Review on Smog," *Journal of the Royal Society for the Promotion of Health* 75, no. 2 (February 1, 1955): 171–192, pp. 172–174; S. H. Richards, "The Measurement and Survey of Atmospheric Pollution," *Journal of the Royal Society for the Promotion of Health* 71, no. 3 (May 1, 1951): 216–226, pp. 216–217; "London's New Fog Suffused by Gases: Medical Journal Confirms Warning of Lethal Content from Smokeless Fuels," *New York Times*, November 21, 1953, sec. Business Financial, p. 30; "Smog Masks," *British Medical Journal* 2, no. 4846 (November 21, 1953): 1145–1146; "Victory by Science over Smog Seen

by Hitchcock," p. 4; Williams, "Air Pollution Is Discussed," p. 13; P. W. West and G. C. Gaeke, "Fixation of Sulfur Dioxide as Disulfitomercurate (II) and Subsequent Colorimetric Estimation," *Analytical Chemistry* 28, no. 12 (December 1, 1956): 1816–1819.

30. For a discussion of the role of sulfur dioxide in the smog, see for example Layne, "A Review on Smog," p. 171, and Richards, "The Measurement and Survey of Atmospheric Pollution," p. 217. For the quotation, see "Smoke and Sulphur," *Lancet* 265, no. 6857 (January 29, 1955): 242–243.

31. Quotation from Lauren B. Hitchcock, president of the Southern California Air Pollution Foundation, Los Angeles, from an address to the first symposium on air pollution held by the American Association for the Advancement of Science. "Victory by Science over Smog Seen by Hitchcock."

32. Clarence A. Mills, "The Donora Episode," *Science* 111, no. 2873 (January 20, 1950): 67–68; West and Gaeke, "Fixation of Sulfur Dioxide as Disulfitomercurate (II) and Subsequent Colorimetric Estimation."

33. H. H. Schrenk et al., "Air Pollution in Donora, Pa.: Epidemiology of the Unusual Smog Episode of October 1948; Preliminary Report," no. 306 (1949); Lester Breslow, "The Epidemiologist Looks at Smog," *Public Health Reports* 70, no. 11 (November 1, 1955): 1140–1143; Central Unit on Environmental Pollution, "Emissions of Sulfur Dioxide in the UK. Doc B," August 1978, RA-S-2532-2-Dca-L0294, Norwegian National Archives, Oslo, Norway.

34. Regarding the controversy over sulfur dioxide's role in the Donora and London smog deaths, see Norton Nelson, Hugh E. C. Beaver, Arthur C. Stern, et al., "Air Pollution Control: A New Frontier," *Public Health Reports* 70, no. 7 (July 1, 1955): 633–661, pp. 634–635; Leslie Silverman and Philip Drinker, "The Donora Episode—A Reply to Clarence A. Mills," *Science* 112, no. 2899 (July 21, 1950): 92–93; Clarence A. Mills, "Dr. Mills' Rejoinder," *Science* 112, no. 2899 (July 21, 1950): 93–94; A. E. Martin, "Mortality and Morbidity Statistics and Air Pollution," *Proceedings of the Royal Society of Medicine* 57, no. 10 Pt 2 (October 1964): 969–975; World Health Organization, "WHO Environmental Health Criteria for Oxides of Sulfur and Particulates," 1974, BAC 2627/97 No. 457, European Commission Archives, pp. 1–2.

35. Central Unit on Environmental Pollution, "Emissions of Sulfur Dioxide in the UK. Doc B," p. 5; S. R. Craxford, "Draft. Clean Air in Britain: The Goal in Sight" (Warren Spring Laboratory, November 1976), AT 34-90, UK National Archives, Kew, UK, pp. 2–3; Hessell Tiltman, "Lessons of Bikini Explosion: Serious Effects of Seaborne Radio-Activity Japan's Growing Anxiety," *Manchester Guardian*, March 24, 1954, p. 1.

36. Craxford, "Draft. Clean Air in Britain," p. 2.

37. Few such solutions were immediately available, with perhaps the exception of coal washing. Craxford, "Draft. Clean Air in Britain," p. 2.

38. Lauren B. Hitchcock and Helen G. Marcus, "Some Scientific Aspects of the Urban Air Pollution Problem," *Scientific Monthly* 81, no. 1 (July 1, 1955): 10–21.

39. Tall stacks were used, at least as an interim measure, by the vast majority of these countries. For discussions of their initial adoption in the US, see Joel A. Tarr, *The Search for the Ultimate Sink: Urban Pollution in Historical Perspective*

(Akron: University of Akron Press, 1996), pp. 18–20; "Air Pollution Fight Spurs Trend to Taller Chimneys," *Sun*, July 21, 1968. On their extensive adoption, see Air Management Sector Group, "Summary of Reports by Delegates on Major Events of Relevance in Countries since the 10th Meeting of the Air Management Sector Group. Annex III: Proposed Examination of the Economic and Administrative Options for Control of Sulphur Oxides and Particulate Matter," April 25, 1973, NR/ENV/73.17, OECD Archives, Paris, France, p. 41; Environment Committee, "Recommended Guidelines for Action: Proposals by the Air Management Sector Group Arising from the Ad Hoc Study on Air Pollution from Fuel Combustion in Stationary Sources and from Other Considerations," May 15, 1973, ENV (73) 27, OECD Archives, Paris, France, p. 7. On the widespread use of high stacks by Britain and France, see OECD Environment Directorate, "Report of Joint Ad Hoc Group on Air Pollution from Fuel Combustion in Stationary Sources," May 19, 1972, PAC/70.7, OECD Archives, Paris, France, pp. xi–xii, 30; OECD, "Conclusions of the Ad Hoc Study on Air Pollution from Fuel Combustion in Stationary Sources," April 18, 1972, PAC/70.8, OECD Archives, Paris, France, p. 12; Erik Lykke, "Tiltak for å begrense transport av luftforurensninger og sur nedbør," March 3, 1976, RA-S-2532-2-Dca-L0004, Norwegian National Archives, Oslo, Norway; Charles Schaeffer, "Soviet Air Is Cleaner: Russian Experts Unimpressed by U.S. Anti-pollution Laws," *Jerusalem Post*, July 30, 1964, p. 5.

40. Daniel J. Kevles, *The Physicists: The History of a Scientific Community in Modern America* (Cambridge, MA: Harvard University Press, 1995), p. 382; "Radioactive Fallout in the Marshall Islands," *Science* 122, no. 3181 (1955): 1178–1179.

41. "The Ashes of Death," *Time* 63, no. 13 (March 29, 1954): 19; John Davy, "Experiments with the H-Bomb," *Observer* (London), March 28, 1954; Tiltman, "Lessons of Bikini Explosion."

42. Helen M. Davis, "Hazards of Smog," *Science News-Letter* 67, no. 19 (May 7, 1955): 298–300, p. 299; Walter Sullivan, "Plan to Avoid Radioactive Smog Prepared for Nuclear Reactor," *New York Times*, January 27, 1959, p. 7.

43. "Doctor Links Tax to End of Smog," *Sun*, February 4, 1957, p. 4; "Doctor Withholds Tax, Demands Curb on Smog," *Hartford Courant*, February 4, 1957, p. 7.

44. Allan M. Winkler, *Life under a Cloud* (Chicago: University of Illinois Press, 1999), p. 102.

45. Ibid., p. 132.

46. Rachel Carson, *Silent Spring* (Boston: Houghton Mifflin Harcourt, 2002).

47. Ibid. Carson compared the hazards of radiation to those of manmade chemicals numerous times in her work. See especially pp. 6, 8, 37, 39, 61, 208, and 211. DDT was first synthesized in 1874 in Germany, but its effectiveness in killing insects was not recognized until 1939, when a Swiss scientist applied the compound to flies in his laboratory. DDT's superiority over other insecticides was soon recognized, and within a year from the Swiss discovery, DDT was distributed in large quantities to the US armed services to control body lice, which transmit typhus, and to eradicate mosquitoes, which transmit malaria. For a history of DDT's health and environmental impacts, see David Kinkela, *DDT and the American*

Century: Global Health, Environmental Politics, and the Pesticide That Changed the World (Chapel Hill: University of North Carolina Press, 2013).

48. Fears about the impact of human activities on the environment can be traced back to the late nineteenth century, with the emergence of conservation groups in both North America and Western Europe. These organizations were largely concerned with the encroachment of railroads into previously undeveloped areas and urbanization. The wave of environmentalism that emerged in the 1960s was different in important respects, such as the makeup of its participants, and it responded to problems previously unknown to industrialized societies: nuclear energy, depletion of natural resources, toxic wastes, and, importantly for this study, acid rain. As one example of the divergence in these groups, the modern environmental movement had important links to student activists of this period. For an examination of the emergence of environmental movements in Western Europe, see Russell J. Dalton's *The Green Rainbow: Environmental Groups in Western Europe* (New Haven, CT: Yale University Press, 1994). Dalton examined hundreds of environmental advocacy groups in Western Europe and argues that there were important distinctions between the conservation and the modern environmental movements. He locates the impetus for the latter in Rachel Carson's *Silent Spring*, which had a significant impact on Western Europe in addition to the United States. Her work spawned a host of similar books by European authors, including *Avant que nature meure* (Before Nature Dies), *Spillran av ett moln* (On the Shred of a Cloud), *Der Tanz mit dem Teufel* (Dance with the Devil), *Im Wurgegriff des Fortschritts* (The Stranglehold of Progress); see Dalton, *The Green Rainbow*, pp. 33–36. On the modern American environmental movement, see Scott Hamilton Dewey, *Don't Breathe the Air: Air Pollution and U.S. Environmental Politics, 1945–1970* (College Station: Texas A&M University Press, 2000); Samuel P. Hays, *Beauty, Health, and Permanence: Environmental Politics in the United States, 1955–1985* (Cambridge: Cambridge University Press, 1989).

49. Bill L. Long, *International Environmental Issues and the OECD, 1950–2000: An Historical Perspective* (Paris: OECD Publishing, 2000), p. 13; and US Congress, *1972 Survey of Environmental Activities of International Organizations* (Washington, DC: US G.P.O., 1972), p. 104.

50. Organisation for Economic Co-operation and Development, *Methods of Measuring Air Pollution*, p. 7.

51. For other examples of this phenomenon, see Amy L. S. Staples, *The Birth of Development: How the World Bank, Food and Agriculture Organization, and World Health Organization Have Changed the World 1945–1965*, annotated edition (Kent, OH: Kent State University Press, 2006).

52. For example, the International Atomic Energy Agency assisted in coordinating research on atmospheric pollution from radioactive fallout, while several intergovernmental organizations including the OECD and Council of Europe worked to facilitate scientific cooperation on pesticides in the environment. On the need for international coordination of scientific policies, see Elinor Langer, "Science and Government: OECD Ministers for Science Compare Experiences on National Policies," *Science* 142, no. 3590 (October 18, 1963): 372–373.

53. Voluntary international cooperation among scientists outside state apparatuses has a much longer history. Scientists had exchanged publications and

held meetings with one another across borders for centuries, cooperating on research projects concerned with geological and cartographic surveying. In the nineteenth century the organization of collaborative projects became much more common. Beginning in 1847, a number of international scientific congresses were convened for the first time, including the International Health Congress in 1851 and the International Congress of Geodesy in 1862. Joint experiments and observations became more and more widespread in the late nineteenth century, including arrangements for an "international year" of research around a common scientific problem, such as the International Polar Year during 1882 and 1883 during which scientists from thirty-five countries carried out simultaneous geophysical studies. The first nongovernmental science organizations with permanent standing committees or bureaus arose out of these meetings after the Second World War. These groups rarely had their own laboratories or research centers and were more preoccupied with generating contacts among their members for joint projects to be done at other institutions. Exceptions were the Naples Zoological Station and the Jungfraujoch Scientific Station. See Organisation for Economic Co-operation and Development, *International Scientific Organisations: Some Aspects of International Scientific Co-operation* (Paris: Organisation for Economic Co-operation and Development, 1965), pp. 11–17.

54. The major intergovernmental organizations working on scientific issues became the UN Educational, Scientific and Cultural Organization (UNESCO), OECD, and NATO. Other specialized scientific associations included the International Atomic Energy Agency, the European Space Agency, the World Health Organization, and the World Meteorological Organization. The numbers given in this chapter include organizations in the fields of health, hygiene, and agriculture that were engaged in scientific studies as well. Clearly defining "international scientific organisations" is difficult for this period, as there was a huge variety of aims, activities, and methods of cooperating between groups. Ibid., pp. 20, 271–272.

55. Ibid., p. 20.

56. Kevles, *The Physicists*; Robert Buderi, *The Invention That Changed the World: How a Small Group of Radar Pioneers Won the Second World War and Launched a Tech Revolution* (New York: Touchstone, 1998); Robert Bud, *Penicillin: Triumph and Tragedy* (Oxford: Oxford University Press, 2008).

57. On scientific cooperation among European nations in the period after the Second World War and its roots in greater national investment in research and the growing costs of technology and equipment, see Alexander King, "Science and Technology in the New Europe," *Daedalus* 93, no. 1 (January 1, 1964): 434–458.

58. Christopher Freeman, *Science, Economic Growth, and Government Policy* (Paris: Organization for Economic Cooperation and Development, 1963), p. 36.

59. Ibid., p. 7.

60. John Krige, *American Hegemony and the Postwar Reconstruction of Science in Europe* (Cambridge, MA: MIT Press, 2006), p. 18. The Marshall Plan was proposed after Britain was forced to withdraw troops and assistance from Turkey and Greece because of economic hardship after the war. The US government felt that if it did not intervene, the Greek Communists would capitalize on

the unrest, seize power, and align the country with the Soviet Union, with potentially catastrophic results for other weak governments in Europe.

61. For more detail on the incorporation of science into American foreign policy in the years after the Second World War, see ibid., pp. 15–39.

62. Because of security concerns initially the US steered Europeans toward nonmilitary projects, such as work in high energy physics at CERN. Ibid, pp. 61–72. On the importance of scientific cooperation for US foreign policy toward Europe after the Second World War, also see Clark A. Miller, " 'An Effective Instrument of Peace': Scientific Cooperation as an Instrument of U.S. Foreign Policy, 1938–1950," *Osiris* 21 (January 1, 2006): 133–160, p. 156.

63. Charles S. Maier, *The Cold War in Europe: Era of a Divided Continent* (Princeton: Markus Wiener Publishers, 1996), p. 205; Organisation for European Economic Co-operation, *At Work for Europe: An Account of the Activities of the Organization for European Cooperation*, 3rd ed. (Paris: OECD, 1956), p. 11.

64. The Soviet Union's launch of Sputnik in 1957 caused US officials to evaluate whether the number and expertise of Europe's scientists were adequate in comparison to Soviet manpower. See *Background Documents Relating to the Organization for Economic Cooperation and Development* (Washington, DC: US G.P.O., 1961), p. 6. These worries were shared by NATO, which played an active role on issues of scientific manpower. See Krige, *American Hegemony and the Postwar Reconstruction of Science in Europe*, pp. 198–208.

65. Organisation for European Economic Co-operation, *At Work for Europe*, pp. 85–96.

66. United States Congress, *Organization for Economic Cooperation and Development Hearings Before the United States Senate Committee on Foreign Relations, Eighty-Seventh Congress, First Session, on Feb. 14, 15, Mar. 1, 6, 1961* (Washington, DC: US G.P.O., 1961), pp. 5–6.

67. Ibid., p. 7.

68. Ibid., p. 11.

69. *Background Documents Relating to the Organization for Economic Cooperation and Development*, p. 6. The OEEC/OECD was one of many intergovernmental and nongovernmental organizations formed during this time period. See Staples, *The Birth of Development*, p. 4.

70. Ibid., p. 307.

71. Christopher Freeman, *Science, Economic Growth, and Government Policy*, (Paris: Organization for Economic Cooperation and Development, 1963), p. 7.

72. On the growing role of the OECD in science policy formation, see Langer, "Science and Government."

73. Organisation for Economic Co-operation and Development, *International Scientific Cooperation* (Paris: Organisation for Economic Co-operation and Development, 1965), p. 62. While the London smog disaster was the most severe occurrence, pollution episodes in major cities continued to threaten public health and cause mortality spikes over the course of the 1950s and 1960s in both Europe and the US. See Frederick W. Lipfert, *Air Pollution and Community Health: A Critical Review and Data Sourcebook* (New York: Van Nostrand Reinhold, 1994), pp. 120–124.

74. Organisation for Economic Co-operation and Development, *Methods of Measuring Air Pollution*, p. 9.

75. Ibid., pp. 7–10.

76. US Congress, *1972 Survey of Environmental Activities of International Organizations*, pp. 75–99.

77. Ibid. In 1969, on the recommendation of President Nixon, NATO had also gotten involved in investigating water pollution and the possibility of creating clean automobile engines but did not look at air pollution specifically.

78. International Atomic Energy Agency, "The Activities of the International Atomic Energy in Air Pollution Problems," 1966, GX 33/1/1. Air Pollution—General—Air Pollution arising from Various Domestic Commercial and Industrial Sources, United Nations Archive, Economic Commission of Europe.

79. "Enclosure. Letter from Dr. P. Dorelle, Deputy Director-General, to Mr. V. Velebit, Executive Secretary, Economic Commission for Europe," January 23, 1967, GX 33/1/1. Air Pollution—General—Air Pollution arising from Various Domestic Commercial and Industrial Sources, United Nations Archive, Economic Commission of Europe. Also see US Congress, *1972 Survey of Environmental Activities of International Organizations*, pp. 75–99.

80. The networks included observatories intended to reveal long-term changes in atmospheric chemistry. However, these stations were built in places remote from pollution sources, which made them impractical for use on regional studies of air pollution. See US Congress, *1972 Survey of Environmental Activities of International Organizations*, pp. 75–99.

81. H. Harcourt, Représentant du Conseil de l'Europe, "Intervention," in J. G. ten Houten, ed., *Air Pollution: Proceedings of the First European Congress on the Influence of Air Pollution on Plants and Animals, Wageningen, April 22 to 27, 1968* (Wageningen: Centre for Agricultural Publishing and Documentation, 1969), p. 9.

82. Council of Europe, "Committee of Experts for the Conservation of Nature and Landscape: Fourth Session, Strasbourg, 2nd–6th November 1965," December 20, 1965, GX 33/1/1. Air Pollution—General—Air Pollution arising from Various Domestic Commercial and Industrial Sources, United Nations Archive, Economic Commission of Europe, p. 19.

83. G. Adinolfi, "First Steps toward European Cooperation in Reducing Air Pollution. Activities of the Council of Europe," *Law and Contemporary Problems* 33, no. 2 (April 1, 1968): 421–426, p. 422.

84. "Enclosure: Air Pollution. Letter from Michael Harris, Deputy Secretary General, OECD, to Vladimir Velebit, Executive Secretary, UN ECE," January 11, 1967, GX 33/1/1. Air Pollution—General—Air Pollution arising from Various Domestic Commercial and Industrial Sources, United Nations Archive, Economic Commission of Europe.

85. US Congress, *1972 Survey of Environmental Activities of International Organizations*, pp. 75–99.

86. "Enclosure: Air Pollution. Letter from Michael Harris, Deputy Secretary General, OECD, to Vladimir Velebit, Executive Secretary, UN ECE.".

87. Ibid., p. 5.

88. Brynjulf Ottar, "An Assessment of the OECD Study on Long Range Transport of Air Pollutants (LRTAP)," *Atmospheric Environment* 12, no. 1 (1978): 445–454, p. 447.

89. The possibility of convening a seminar to study background concentrations of air pollutants in areas remote from emissions sources was first proposed in the OECD during 1968. See Organisation for Economic Co-operation and Development, Directorate for Scientific Affairs, "Air Pollution: Background and Pilot Stations Joint OECD/Swedish Seminar Stockholm, 7th and 8th December 1967," December 21, 1967, BAC 70/1984 No. 370 Année(s): 1963–1967, European Commission Archive.

90. "Pauling Asks End of Nuclear Tests: Winner of Nobel Prize Cites Effects of Fallout," *Sun*, March 18, 1955, p. 1; G. V. Ferguson, "Fallout Is Worst in North Canada," *The Washington Post, Times Herald*, September 24, 1961, p. E2; "Russian Blast Fallout Found by U. S. in Food," *Chicago Daily Tribune*, October 13, 1961, p. 3; "Radioactive Cloud Misses Ireland," *Irish Times*, November 8, 1961, p. 11; "Fallout from Russian Blast May Appear In U.S. by Mid-Month; Milk, Air Sampled," *Wall Street Journal*, August 6, 1962, p. 3; Committee on Meteorological Aspects of the Effects of Atomic Radiation, "Meteorological Aspects of Atomic Radiation," *Science* 124, no. 3212 (July 20, 1956): 105–112; E. A. Martell and P. J. Drevinsky, "Atmospheric Transport of Artificial Radioactivity," *Science* 132, no. 3439 (November 25, 1960): 1523–1531.

91. John Davy, "Insecticide Traces in Rain Water," *Observer* (London), August 1, 1965, p. 3; J. L. George and D. E. H. Frear, "Pesticides in the Antarctic," *Journal of Applied Ecology* 3 (June 1, 1966): 155–167; Irston R. Barnes, chairman, Audubon Naturalist Society, "Antarctic Traces Show DDT Threat," *Washington Post, Times Herald*, October 23, 1966, p. L11; J. Tatton and J. H. A. Ruzicka, "Organochlorine Pesticides in Antarctica," *Nature* 215, no. 5099 (July 22, 1967): 346–348; John Davy, "Pesticides Even Contaminate Antarctic," *Jerusalem Post*, August 4, 1967, p. A18; John Davy, *Observer* (London), "DDT Traces in Antarctic Hint Global Danger: Heavier Elsewhere," *Washington Post*, August 8, 1967, p. A6.

92. OECD, "Activities of the Organisation for Economic Co-operation and Development Which Are Related to the UN Environment Programme," March 8, 1974, ENV (74) 13, OECD Archives, Paris, France, p. 6.

93. OECD, "Committee for Research Co-operation: First Policy Report of the Air Management Research Group," July 30, 1970, BAC 156/1990 No. 2388 Année(s) 1970–1971, European Commission Archives, p. 5.

94. On the complex environmental and geopolitical forces that helped set the stage for the UN Stockholm Conference, see Yannick Mahrane, Marianna Fenzi, Céline Pessis, and Christophe Bonneuil, "De la nature à la biosphère: L'invention politique de l'environnement global, 1945–1972," in *L'invention politique de l'environnement, Vingtième Siècle. Revue d'histoire* (Paris: Presses de Sciences Po, 2012), vol. 1, no. 113, pp. 127–141.

95. Organisation for Economic Co-operation and Development, *Ad Hoc Meeting on Acidity and Concentration of Sulphate in Rain, May 1969* (Paris: OECD, 1971); Svante Odén, "Acidification of Air and Precipitation and Its

Consequences on the Natural Environment," *Swedish State Natural Science Research Council*, Stockholm, Sweden (January 1, 1968).

96. Henning Rodhe, "Human Impact on the Atmospheric Sulfur Balance," *Tellus B* 51, no. 1 (February 1, 1999): 110–122.

97. C.-G. Rossby and H. Egnér, "On the Chemical Climate and Its Variation with the Atmospheric Circulation Pattern," *Tellus* 7, no. 1 (February 1955): 118–133.

98. On Rossby's work in the US, see Frederik Nebeker, *Calculating the Weather: Meteorology in the 20th Century* (San Diego: Academic Press, 1995), pp. 88–90; Kristine C. Harper, *Weather by the Numbers: The Genesis of Modern Meteorology* (Cambridge, MA: MIT Press, 2008), pp. 39–40, 69–73, 187–188.

99. Bert Bolin, "Carl-Gustaf Rossby," *Tellus A-B* 51, no. 1 (1999): 4–12, p. 10; Rodhe, "Human Impact on the Atmospheric Sulfur Balance," p. 111; Horace B. Byers, *Carl-Gustaf Arvid Rossby, 1898–1957: A Biographical Memoir* (Washington, DC: National Academy of Sciences, 1960), p. 261.

100. The Swedish government assisted Rossby in founding the Institute in the hopes of bringing the world-renowned meteorologist back to his home country to build up its own expertise in the field. See Bolin, "Carl Gustaf Rossby."

101. Rossby and Egnér, "On the Chemical Climate and Its Variation with the Atmospheric Circulation Pattern," p. 119.

102. Bert Bolin, "Angående försurningsproblemet," September 21, 1970, SE/RA/322619/~/8. Nationalkommittén för 1972 års FN-konferens om den mänskliga miljön. Kommittén för forskning och andra sakfrågor, protokoll jämte bilagor nr 91–141, Swedish National Archives, Marieberg, Sweden, p. 1.

103. C.-G. Rossby, "Current Problems in Meteorology," in Bert Bolin, ed., *The Atmosphere and the Sea in Motion: Scientific Contributions to the Rossby Memorial Volume* (New York: Rockefeller Institute Press, 1959), pp. 40–41.

104. Rossby died suddenly of a heart attack in his office at the age of 59. See "Preface," in Bolin, *The Atmosphere and the Sea in Motion*; Robert Angus Smith, *Air and Rain: The Beginnings of Chemical Climatology* (London: Longmans, Green, 1872).

105. Bjarne Sivertsen, *NILU; 40 år i lufta* (Kjeller: Skedsmo, Trykk & Grafisk, 2009), Norwegian Institute for Air Research, Archives and Library, pp. 27–28; Bolin, "Carl-Gustaf Rossby," p. 10; Arlid Holt Jensen, "Acid Rains in Scandinavia," in *Middle- and Long-Term Energy Policies and Alternatives, Part 4. Appendix to Hearings before Subcommittee on Energy and Power, Committee on Interstate and Foreign Commerce,* House of Representatives, Ninety-Fourth Congress, Second Session (Washington, DC: US G.P.O., 1976).

106. Ecologist Ellis Cowling believes this may have been because the research was published in a diverse array of specialized scientific journals without a wide readership. See Ellis B. Cowling, "Acid Precipitation in Historical Perspective," *Environmental Science & Technology* 16, no. 2 (February 1, 1982): 110A–123A, pp. 112A–113A.

107. World Meteorological Organization, "Interview with Erik Eriksson," *Bulletin,* 47 no. 4, p. 327.

108. Svante Odén, "Om bevätningsvärme hos ler och humus," *Geologiska föreningen i Stockholm förhandlingar* 78, no. 3 (1956): 536–552; Svante Odén,

"Genmäle till Prof. I. Th. Rosenqvist angående bevätningsvärme hos ler och humus," *Geologiska föreningen i Stockholm förhandlingar* 79, no. 1 (1957): 91–92; Svante Odén, "Differentiala och bidifferentiala sedimentationsvågar för kornstorleksbestämning samt vissa tillämpningar av denna metodik för studier av solers stabilitct och koagulation," *Geologiska föreningen i Stockholm förhandlingar* 81, no. 4 (1959): 620–640.

109. Bert Bolin, "The Need for Long Term Perspectives on the Acidification of the Environment," in *Ecological Effects of Acid Deposition* (National Swedish Environment Protection Board, 1983), Kommittén med uppgift att samordna aktiviteter i samband med uppföljningen av Stockholmskonferensen 1972 om den mänskliga miljön. SE/RA/323591/~/5, Swedish National Archives, Marieberg, Sweden, p. 35.

110. Svante Odén, "Nederbördens försurning," *Dagens nyheter*, October 24, 1967. Cited in Cowling, "Acid Precipitation in Historical Perspective." Odén was the first to use the phrase "acid rain" to describe transboundary air pollution, but the term had been utilized as early as the nineteenth century to describe precipitation with high levels of acidity. See R. A. Smith, *Air and Rain*; Peter Reed, *Acid Rain and the Rise of the Environmental Chemist in Nineteenth-Century Britain: The Life and Work of Robert Angus Smith* (Abingdon, UK: Routledge, 2016).

111. On the Scandinavian public's alarm concerning acid rain, see G. K., "Ren luft og sur nedbør," *Aftenpostens*, July 25, 1969, National Library of Norway; Dan Daniels, "The Long Fight to Shed Light on Acid Rain," *Globe and Mail*, January 7, 1985; Jensen, "Acid Rains in Scandinavia," p. 50; Anne La-Bastille, "Acid Rain: How Great a Menace?," *National Geographic* 160, July-December 1981, p. 661.

112. Though continuing to make media appearances and attend international symposia on acid rain issues, Odén did not undertake much further research on the problem following Sweden's case study. Instead, he became involved in work on other environmental pollutants, including DDT, heavy metals, and PCBs. See Andrew Jamison, "The Lesson of the Askö Lab," *New Scientist*, December 9, 1971, p. 79; Thorsten Ahl and Svante Odén, "River Discharges of Total Nitrogen, Total Phosphorus and Organic Matter into the Baltic Sea from Sweden / Общее Количество Нитрогена, Фосфора И Органического Вещества, Попадающего В Балтийское Море Со Стороны Швеции Из Рек," *Ambio Special Report* no. 1 (January 1, 1972): 51–56; Svante Odén and Jan Ekstedt, "PCB and DDT in Baltic Sediments / ПХБ И ДДТ В Осадках Балтийского Моря," *Ambio Special Report* no. 4 (January 1, 1976): 125–129; Svante Odén, "The Acidity Problem—An Outline of Concepts," *Water, Air, and Soil Pollution* 6 (March 26, 1976): 137–166; Erik Lotse, "NF Svante Odén," *Svenskt Biografiskt Lexikon* (Riksarkivet, 2013); Sivertsen, *NILU; 40 år i lufta*.

113. Harcourt, "Intervention," p. 9.

114. It was sponsored by the Council of Europe with assistance from the Netherlands government. See J. G. ten Houten, ed. *Air Pollution: Proceedings of the First European Congress on the Influence of Air Pollution on Plants and Animals, Wageningen, April 22 to 27, 1968* (Wageningen: Centre for Agricultural Publishing and Documentation, 1969), pp. 5, 7, 407–411.

115. Ibid., pp. 179–180. Experiments on plants exposed to sulfur dioxide concentrations by Swiss scientists suggested several possible explanations for these effects, although the exact mechanism of harm was still unknown. Sulfur dioxide was observed to affect the water permeability of cells, but the primary cellular points of damage were unclear. It was thought that photosynthesis might be affected, or that the production of amino acids was being disrupted. Similar retardation in plant growth had also been observed in the US from nitrogen oxides and hydrocarbons. See pp. 137–141, 180.

116. Ibid. As a point of comparison, 50 micrograms per cubic meter is the current maximum annual average deemed acceptable for human health by the WHO. Many areas throughout the world still exceed this, notably in China and India. See World Bank, *World Development Indicators* (World Bank Publications, 2005), p. 178.

117. ten Houten, *Air Pollution*, p. 379. Experiments of sulfur dioxide at .02 ppm had negative effects on some plant species. The conference attendees believed that their findings supported the inclusion of potential impacts on plants and animals in the formulation of pollution limits in addition to possible consequences for human health, as many of them were shown to be more susceptible to air pollution than human beings who, it was also said, could "protect themselves more effectively." Furthermore, since some plants demonstrated a vast range of specific sensitivity to different air pollutants, the scientists recommended using them as possible "biological pollution indicators" that could serve as warnings before pollution reached levels damaging to people. See pp. 50, 241.

118. Ibid., p. 7.

119. See, for instance, Eriksson's statements that the increase in acid rain was not as great as had been feared in a radio interview approximately one year after Odén's article came out. G. K., "Ren luft og sur nedbør," p. 22.

120. On the limited number of studies available concerning sulfur dioxide transport through the atmosphere, see Bert Bolin, "Angående försurningsproblemet," p. 2.

121. These discrepancies appeared to be the result of differences in sampling rates of the atmosphere. See William H. Schroeder and Paul Urone, "Sulfur Dioxide in the Atmosphere: A Wealth of Monitoring Data, but Few Reaction Rate Studies," *Environmental Science & Technology* 3, no. 5 (May 1969): 436–445, p. 436. For research regarding sulfur dioxide residence time, see C. E. Junge, "Sulfur in the Atmosphere," *Journal of Geophysical Research* 65, no. 1 (1960): 227–237; Erik Eriksson, "The Yearly Circulation of Sulfur in Nature," *Journal of Geophysical Research* 68, no. 13 (1963): 4001–4008; Erich Weber, "Contribution to the Residence Time of Sulfur Dioxide in a Polluted Atmosphere," *Journal of Geophysical Research* 75, no. 15 (1970): 2909–2914; Henning Rodhe, "On the Residence Time of Anthropogenic Sulfur in the Atmosphere," *Tellus* 22, no. 1 (1970): 137–139.

122. G. K., "Ren luft og sur nedbør," p. 22.

123. He applied for funding from the Swedish Environmental Protection Agency, which was created in 1967 as an organization within the Ministry of Agriculture. Svante Odén, "Försurningsproblemet som case study vid FN- konferensen 1972," May 20, 1970, SE/RA/322619/~/7. Nationalkommittén för 1972

års FN-konferens om den mänskliga miljön. Kommittén för forskning och andra sakfrågor, protokoll jämte bilagor nr 50–90, Swedish National Archives, Marieberg, Sweden, p. 3.

124. Maria Ivanova, "Designing the UN Environment Programme: A Story of Compromise and Confrontation," *International Environmental Agreements: Politics, Law and Economics* 7, no. 4 (December 1, 2007): 337–361, p. 341.

125. Ibid. Richard J. H. Johnston, "U.N. Health Study Urged by Sweden: Inquiry Sought on Effect of Scientific Gains on Man," *New York Times*, May 23, 1968, p. 95; Louis B. Fleming, "Sweden Seeks U. N. Aid in Fight on Pollution: Calls for Major Debate on Problem of Man-Made Perils to Human Environment," *Los Angeles Times*, May 29, 1968, p. 15; Kathleen Teltsch, "Thant Urges Concerted Action in Pollution Crisis: He Reports on Study Made for Global Conference on Man's Environment," *New York Times*, June 24, 1969, p. 4.

126. Roland Huntford, "The Factories May Be Poisoning All of Us," *Jerusalem Post*, June 28, 1968, p. 5; John M. Lee, "Pollution Drive Slows in Sweden: Joint Baltic Effort Entangled in East-West Politics," *New York Times*, October 26, 1969, p. 17; John M. Lee, "Sweden Vigorous in Pollution War: U.S. Group to Tour Projects That Draw World Attention," *New York Times*, May 31, 1970, p. 13.

127. Roland Huntford, "Sweden Bans DDT Following Reports of Harm to People: Sen. Nelson Introduces Bill against Compound," *Washington Post, Times Herald*, April 5, 1969, p. A12.

128. Cabinet Stockholm, "Sweden Proposes Pollution Pacts, Urges U.N. Parley to Draft International Accords," March 10, 1970, SE/RA/322619/~/7. Nationalkommittén för 1972 års FN-konferens om den mänskliga miljön. Kommittén för forskning och andra sakfrågor, protokoll jämte bilagor nr 50–90, Swedish National Archives, Marieberg, Sweden.

129. Hella Pick, "UN Help to Control Population Rise," *Guardian*, August 3, 1968, p. 9; Earl W. Foell, "Sweden to Ask U.N. for World Pollution Talks: Parley in 1972 Urged before Planetary Contamination Reaches Point of No Return," *Los Angeles Times*, October 28, 1968, p. 28.

130. "Pollution Worries Sweden," *Guardian*, May 26, 1969, p. 3.

131. Jan Mårtenson, "Nationalkommitténs informationsbroschyr, Bil. 6," October 22, 1971, SE-RA-322619-4-1971, Swedish National Archives, Marieberg, Sweden, p. 4.

132. Håkan Stenram, "Anteckningar från sammanträde den 16 april 1970," April 26, 1970, SE/RA/322619/~/7. Nationalkommittén för 1972 års FN-konferens om den mänskliga miljön. Kommittén för forskning och andra sakfrågor, protokoll jämte bilagor nr 50–90, Swedish National Archives, Marieberg, Sweden, p. 3.

133. Ibid., pp. 3–6.

134. Svante Odén, "Försurningsproblemet som case study vid FN- konferensen 1972."

135. Ibid. Odén added that these countries might be persuaded to adopt greener technologies in order to avoid causing acid rain.

136. Ibid., p. 2.

137. Ibid.

138. Stenram, "Anteckningar från sammanträde den 27 maj 1970," p. 3.

139. Bert Bolin was currently working as a professor of meteorology at Stockholm University and had been nominated to serve on the Committee on Research and Substantial Issues in the fall of 1969. He formally accepted at the end of November 1969. "Letter from Bert Bolin, Stockholms Universitet, Meteorologiska Institutionen, to Ingemund Bengtsson, Statsrådet, Jordbruksdepartementet," November 28, 1969, SE/RA/322619/~/7. Nationalkommittén för 1972 års FN-konferens om den mänskliga miljön. Kommittén för forskning och andra sakfrågor, protokoll jämte bilagor nr 50–90, Swedish National Archives, Marieberg, Sweden.

140. Håkan Stenram, "Rapport från möte med ledningsgruppen för nordiskt samarbete om orsakerna till nederbördens försurning (inom Nordforsk) den 11 juni 1970 i Oslo," June 22, 1970, SE/RA/322619/~/8. Nationalkommittén för 1972 års FN-konferens om den mänskliga miljön. Kommittén för forskning och andra sakfrågor, protokoll jämte bilagor nr 91–141, Swedish National Archives, Marieberg. H.S., "Svovelregn-Saken I Norden Prioriteres" (Aftenpostens, September 14, 1971), RA-PA-0641-E-Eg-Loo12, Norwegian National Archives.

141. Bert Bolin, "Angående försurningsproblemet," pp. 5–6.

142. "Letter from Bert Bolin, Stockholms Universitet, Meteorologiska Institutionen to Arne Engstrom, Forskningsberedningen, Utbildningsdepartementet," November 2, 1970, SE/RA/322619/~/8. Nationalkommittén för 1972 års FN-konferens om den mänskliga miljön. Kommittén för forskning och andra sakfrågor, protokoll jämte bilagor nr 91–141, Swedish National Archives, Marieberg, Sweden; Bert Bolin, "Case Study 'Nederbördens Försurning,'" November 2, 1970, SE/RA/322619/~/8. Nationalkommittén för 1972 års FN-konferens om den mänskliga miljön. Kommittén för forskning och andra sakfrågor, protokoll jämte bilagor nr 91–141, Swedish National Archives, Marieberg, Sweden; Stenram, "Anteckningar från sammanträde den september 24 1970," p. 2.

143. "Arbetsplan för presentation av nederbördens försurning som 'case study' till 1972 års konferens om den mänskliga miljön," November 2, 1970, SE/RA/322619/~/8. Nationalkommittén för 1972 års FN-konferens om den mänskliga miljön. Kommittén för forskning och andra sakfrågor, protokoll jämte bilagor nr 91–141, Swedish National Archives, Marieberg, Sweden.

144. Ibid. Royal Ministry for Foreign Affairs and Royal Ministry of Agriculture, *Air Pollution across National Boundaries: The Impact on the Environment of Sulfur in Air and Precipitation, Sweden's Case Study for the UN Conference on the Human Environment* (Stockholm: P. A. Norstedt, 1972), pp. 17–29.

145. Royal Ministry for Foreign Affairs and Royal Ministry of Agriculture, *Air Pollution across National Boundaries*, pp. 17–29. "Arbetsplan för presentation av nederbördens försurning som 'case study' till 1972 års konferens om den mänskliga miljön."

146. "Letter from Bert Bolin and Henning Rodhe to Kommittén För Forskning Och Andra Sakfrågor," August 5, 1971, SE/RA/322619/~/11. Nationalkommittén för 1972 års FN-konferens om den mänskliga miljön. Kommittén för forskning och andra sakfrågor, protokoll jämte bilagor nr 194–224, Swedish National Archives, Marieberg, Sweden.

147. Richard N. Gardner and Christian Herter Jr., "Can the U.N. Lead the Environmental Parade?," *American Journal of International Law* 64, no. 4 (September 1, 1970): 211–216.

148. Maurice F. Strong, "The Stockholm Conference: Where Science and Politics Meet," *Ambio* 1, no. 3 (June 1, 1972): 73–78, pp. 75, 76.

149. Ibid., p. 75.

150. Who exactly these scientists should be, however, was a matter of some debate in the US, which faced criticism for bringing scientific advisers only from the government to the UN conference; they were drawn from the Departments of State, Defense, and Commerce. See "The U.S. at Stockholm: Scientists All Federal," *Science News* 101, no. 22 (May 27, 1972): 342.

151. David A. Kay and Eugene B. Skolnikoff, "International Institutions and the Environmental Crisis: A Look Ahead," *International Organization* 26, no. 2 (April 1, 1972): 469–478, p. 472.

152. He dubbed this prospective organization the "International Environmental Agency." See George F. Kennan, "To Prevent a World Wasteland: A Proposal," *Foreign Affairs* 48, no. 3 (April 1, 1970): 401–413.

153. Claire Sterling, "The U.N. and World Pollution," *Washington Post, Times Herald*, July 28, 1970, sec. General, p. A14.

154. The Institute on Man and Science, "International Organization and the Human Environment: In Anticipation of the 1972 Stockholm Conference of the UN," 1971, GX 10/2/2/102. Economic Commission for Europe. Relations with International Organizations. UN Environment Programme (UNEP), United Nations Archive, Economic Commission of Europe, p. 24.

155. "Absences Said to Weaken U.N. Ecology Conference," *Sun*, June 8, 1972, p. A5. On the issue of developing nations, see Kay and Skolnikoff, "International Institutions and the Environmental Crisis," p. 474.

156. Kay and Skolnikoff, "International Institutions and the Environmental Crisis," p. 475.

157. Since it was not a member of the UN or any of its subsidiary organizations, East Germany was initially not invited to participate. Though a compromise proposal was put forward to allow it observer status, the US, alongside allies France and Britain, did not want East Germany to have voting rights at the Conference. The Soviets and the Communist bloc refused to attend unless the country was accorded equal opportunity to participate, including voting rights. See "The GDR and Stockholm: Politics before Ecology," *Science News* 101, no. 17 (April 22, 1972): 261; "Cold War Environment," *New York Times*, February 22, 1972, p. 36; "Soviets Standing By for World Meeting," *Washington Post, Times Herald*, June 8, 1972, sec. National News, p. A29; Thomas J. Hamilton, "Compromise on East Germany Reported to End Boycott Threat at Talks on the Environment," *New York Times*, February 13, 1972, p. 13; Philip H. Abelson, "After the Stockholm Conference," *Science* 175, no. 4022 (February 11, 1972): 585; Robert Alden, "Soviet Threatens Boycott of Environmental Talks," *New York Times*, March 31, 1972, p. 3; "U.N. Pollution Talks in Geneva Boycotted by Soviet and Allies," *New York Times*, February 8, 1972, p. 3.

158. The European Communities appealed to some foreign policy experts who felt environmental protection was best achieved through supranational law. See Brian Johnson, "The United Nation's Institutional Response to Stockholm: A Case Study in the International Politics of Institutional Change," *International Organization* 26, no. 2 (April 1, 1972): 255–301, p. 263–265.

159. Dominique Verguèse, "Europe and the Environment: Cooperation a Distant Prospect," *Science* 178, no. 4059 (October 27, 1972): 381–383.

160. Remarks by Konstantin Ananichev, director, International Organizations Department, State Committee of the USSR Council of Ministers for Science and Technology; The Institute on Man and Science, "International Organization and the Human Environment," p. 13.

161. Notably, it was the first international agreement to recognize environmental harm as a human rights issue, stipulating that the protection of the environment was essential to "the enjoyment of basic human rights." Declaration of the UN Conference on the Human Environment, Stockholm, June 16, 1972, UN Doc. A/CONF.48/14/Rev.1 (1972).

162. Bertil Hägerhäll, "Behandling av luftföroreningsproblemen i ett globalt perspektiv vid ministerkonferensen och expertmötena," December 21, 1981, SE/RA/323591/~/1. Kommittén med uppgift att samordna aktiviteter i samband med uppföljningen av Stockholmskonferensen 1972 om den mänskliga miljön, Swedish National Archives, Marieberg, Sweden, p. 1.

163. Utenriksdepartementet, Handelspolitiske kontor, "Norges deltakelse i det internasjonale arbeid vedrørende forurensningsproblemene," January 17, 1970, RA-S-1574-E-L0405, Norwegian National Archives, pp. 2–3. The Norwegians also felt this made it preferable to other groups like the Council of Europe, which focused on administrative tasks.

164. G. Bäckstrand, "Sammanträde med tjänstemän från de nordiska utrikesministerierna rörande samordning av miljöfrågans internationella handläggning," June 18, 1970, SE/RA/322619/~/8. Nationalkommittén för 1972 års FN-konferens om den mänskliga miljön. Kommittén för forskning och andra sakfrågor, protokoll jämte bilagor nr 91–141, Swedish National Archives, Marieberg, Sweden.

165. Ibid., pp. 5–6.

166. Observers of the conference cited acid rain as one of a few key issues likely to trigger international tension in the coming years. The others included loss of fish, polluted waters, modification of the weather, and tainted foodstuffs. See Kay and Skolnikoff, "International Institutions and the Environmental Crisis," p. 477.

CHAPTER TWO

1. J. E. Lovelock, "Air Pollution and Climatic Change," *Atmospheric Environment* 5, no. 6 (June 1971): 403–411, p. 410.

2. Clyde H. Farnsworth, "Norse Seek Curb on Acid Rainfall: Agency Is Set Up to Trace Source of Sulphur Fumes," *New York Times*, November 27, 1970, sec. Business & Finance, p. 66.

3. "Europeisk forskningsprosjekt over sur nedbør er i emning," *Stavanger Aftenblad*, August 7, 1970, National Library of Norway, p. 3. A number of scientists and public figures since the nineteenth century had hypothesized that increases in carbon dioxide from fossil fuel pollution could raise the earth's temperature and potentially cause dramatic environmental changes. However, research into climate change was limited to a relatively small number of scientists during the 1960s and did not become a topic of widespread scientific and public

concern until the mid-1970s. See James Rodger Fleming, *Historical Perspectives on Climate Change* (Oxford: Oxford University Press, 2005); Spencer R. Weart, *The Discovery of Global Warming*, revised and expanded edition (Cambridge, MA: Harvard University Press, 2008); James Rodger Fleming, *The Callendar Effect: The Life and Work of Guy Stewart Callendar (1898–1964), The Scientist Who Established the Carbon Dioxide Theory of Climate Change*, illustrated edition (American Meteorological Society, 2009); Paul N. Edwards, *A Vast Machine: Computer Models, Climate Data, and the Politics of Global Warming* (Cambridge, MA: MIT Press, 2010); Naomi Oreskes and Erik M. Conway, *Merchants of Doubt: How a Handful of Scientists Obscured the Truth on Issues from Tobacco Smoke to Global Warming* (New York: Bloomsbury Press, 2010).

4. Rachel Rothschild, "Environmental Awareness in the Atomic Age: Radioecologists and Nuclear Technology," *Historical Studies in the Natural Sciences* 43, no. 4 (September 1, 2013): 492–530. On European research into chemical contaminants beginning in the 1960s, see Michael Egan, "Communicating Knowledge: The Swedish Mercury Group and Vernacular Science, 1965–1972," in *New Natures: Joining Environmental History with Science and Technology Studies*, ed. Dolly Jørgensen, Finn Arne Jørgensen, and Sara B. Pritchard (Pittsburgh: University of Pittsburgh Press, 2013); Michael Egan, "Toxic Knowledge: A Mercurial Fugue in Three Parts," *Environmental History* 13, no. 4 (October 1, 2008): 636–642.

5. Utenriksdepartementet, Handelspolitiske kontor, "Norges deltakelse i det internasjonale arbeid vedrörende forurensningsproblemene," January 17, 1970, RA-S-1574-E-L0405, Norwegian National Archives, p. 1.

6. Allen Kent, Harold Lancour, and Jay Elwood Daily, *Encyclopedia of Library and Information Science* (CRC Press, 1977), p. 259.

7. Utenriksdepartementet, Handelspolitiske kontor, "Norges deltakelse i det internasjonale arbeid vedrörende forurensningsproblemene," p. 1.

8. For example, Denmark and Norway eventually joined NATO to take advantage of US aid, which marked a key split with its neighbors in foreign policy. For a discussion of NATO and Scandinavia during the Cold War, see Jurg M. Gabriel and Celina McEwen, *American Conception of Neutrality after 1941* (New York: Springer, 1988), pp. 94–108.

9. Ibid.

10. Harald Høydahl, Norges Naturvernforbund, "Forsuring av luft og vann," May 12, 1971, RA-PA-0641-E-Eg-L0012, Norwegian National Archives.

11. After founding the NTNF, the Norwegian government also established two other R & D bodies to sponsor non-military research: the Norwegian Council for Science and the Humanities and the Norwegian Agricultural Research Council. For a short overview of the NTNF and its history, see James Brian Quinn and Robert Major, "Norway: Small Country Plans Civil Science and Technology," *Science* 183, no. 4121 (January 18, 1974): 172–179. For scholarship that touches on the development of the NTNF, see Håkon With Andersen, "Technological Trajectories, Cultural Values and the Labour Process: The Development of NC Machinery in the Norwegian Shipbuilding Industry," *Social Studies of Science* 18, no. 3 (August 1, 1988): 465–482, p. 480; Jan Fagerberg, David Mowery, and Bart Verspagen, *Innovation, Path Dependency, and Policy:*

The Norwegian Case (Oxford: Oxford University Press, 2009), p. 74; John Peter Collett, *Making Sense of Space: The History of Norwegian Space Activities* (Oslo: Scandinavian University Press, 1995).

12. Bjarne Sivertsen, *NILU; 40 år i lufta* (Kjeller: Skedsmo, Trykk & Grafisk, 2009), Norwegian Institute for Air Research, Archives and Library. For the dramatic increase in pollution funding, which rose by a greater percentage than that of any other sector except "engineering" between 1967 and 1971, see Quinn and Major, "Norway," p. 175. Funding for research on the continental shelf was a close third.

13. "NILU gjennom 25 år," Norwegian Institute for Air Research, Archives and Library, p. 4; Sivertsen, *NILU; 40 år i lufta*, p. 9.

14. Sivertsen, *NILU; 40 år i lufta*, p. 26; Interview with Øystein Hov, atmospheric scientist and research director at the Norwegian Institute of Meteorology, by the author, May 19, 2011, Oslo, Norway.

15. Ottar, along with several other prominent Norwegian scientists working in and around the University of Oslo, was involved in the Norwegian Intelligence agency "XU." Ottar helped to establish underground headquarters in Stockholm and Northern Norway and would later receive several commendations from the Norwegian and Danish governments for his wartime service. See Kenneth A. Rahn, "Biographical Sketch of Dr B. Ottar," *Atmospheric Environment*, Arctic Air Chemistry, 23, no. 11 (1989): 2347; Interview with Øystein Hov.

16. Interview with Øystein Hov.

17. After 1972, the growth in staff was much more gradual, reaching 80 by 1990. There was then another period of significant hiring, with the number of employees totaling over 120 by 1994. "NILU gjennom 25 år," p. 6.

18. Sivertsen, *NILU; 40 år i lufta*, pp. 26-27.

19. Sweden had approached both the Council of Europe and OECD about the problem of acid rain in addition to bringing the issue before the United Nations as part of the 1972 Conference on the Human Environment. However, given the OECD's greater experience in air pollution research, the Council of Europe agreed that scientific examination of the problem should be managed by the OECD. The OECD's Air Management group subsequently asked that member countries nominate experts for an ad hoc meeting on the Swedish report. Sweden's Erik Eriksson, who had worked closely with Odén and served as Sweden's delegate to the OECD, first presented a report in 1969 at the OECD on the current evidence for increasing amounts of sulfur compounds in the atmosphere. See OECD Directorate for Scientific Affairs, "Air Management Research Group: Decisions of the Second Meeting. 5th-7th February 1969. Note by the Secretariat," March 14, 1969, BAC 156/1990 No. 2386 Année(s) 1967-1969, European Commission Archives, p. 3; Organisation for Economic Co-operation and Development, *Ad Hoc Meeting on Acidity and Concentration of Sulphate in Rain, May 1969*, (Paris: OECD, 1971); Organisation for Economic Co-operation and Development, *Air Management Problems and Related Technical Studies; Policy Report of the Air Management Research Group* (Paris: Organisation for Economic Co-operation and Development, 1972), p. 145.

20. Organisation for Economic Co-operation and Development, Environment Directorate, *Plan for a Project to Study the Long Range Transport of Air Pollut-*

ants, June 25, 1971, OECD Archives, NR/ENV/71.27, p. 2. Norwegian officials had already begun lobbying heavily for the Scandinavian Council for Applied Research to finance further work on acid rain as part of its new focus on pollution problems. They succeeded in gaining the support of the Nordic Council of Ministers for this issue, and in 1969 it issued a recommendation that the Scandinavian Council for Applied Research begin supporting cooperative research projects on pollution problems, citing Norway's attempts to "sound the alarm" about increasing amounts of sulfur dioxide in rainfall. See Sven Mellqvist, Gerda Møller, and Niels Mørk, "Medlemsforslag om åtgärder mot vissa luftföroreringar" (Nordiska Rådet, May 24, 1971), RA-PA-0641-E-Eg-Loo12, Norwegian National Archives, p. 1.

21. Organisation for Economic Co-operation and Development, *Air Management Problems and Related Technical Studies*, pp. 145, 175–190; Organisation for Economic Co-operation and Development, Environment Directorate, *Plan for a Project to Study the Long Range Transport of Air Pollutants*, p. 3.

22. "En 'kake' av luftforurensninger over det europeiske kontinent: bygges opp under spesielle meteorologiske forhold, sier instituttsjef Ottar," *Stavanger Aftenblad*, April 23, 1970, National Library of Norway, pp. 1, 19; "Europeisk forskningsprosjekt over sur nedbør er i emning."

23. The International Geophysical Year from 1957 to 1958 incorporated cooperative measurements of atmospheric processes but did not directly examine air pollution processes. However, the measurements of carbon dioxide begun by Charles David Keeling would become central to later research on climate change. See Roger D. Launius, David H. DeVorkin, and James Rodger Fleming, *Globalizing Polar Science: Reconsidering the International Polar and Geophysical Years* (New York: Palgrave Macmillan, 2010), p. 367.

24. "Environment Committee: Proposal for a Co-operative Technical Project to Measure the Long Range Transport of Air Pollutants," September 7, 1971, OECD Archives, ENV(71)28. 82.075, p. 3.

25. For the most recent historical analysis of the American environmental movement, see Robert Gottlieb, *Forcing the Spring: The Transformation of the American Environmental Movement* (Washington, DC: Island Press, 2005). For a discussion of the emergence of environmental movements in Western Europe, see Russell J. Dalton, *The Green Rainbow: Environmental Groups in Western Europe* (New Haven, CT: Yale University Press, 1994).

26. He later became an outspoken critic of US policies on acid rain upon his departure from the OECD in 1978. Hilliard Roderick, "The Future Natural Sciences Programme of Unesco," *Nature* 195, no. 4838 (7, 1962): 215–222; Jurgen Schmandt and Hilliard Roderick, *Acid Rain and Friendly Neighbors: The Policy Dispute Between Canada and the United States*, Duke Press Policy Studies (Durham: Duke University Press, 1985).

27. Erik Lykke, "Europe versus Itself," *Nature* 269, no. 5627 (1977): 372; Erik Lykke, "Economic Development and Environment," *Environmentalist* 6, no. 4 (12, 1986): 245–246.

28. "Environment Committee: Meeting at Ministerial Level: Minutes of the 13th Session," March 6, 1975, OECD Archives, ENV/M(74) 4 12.484, pp. 6–8.

29. "Co-Operative Technical Programme to Measure the Long-Range Transport of Air Pollutants: Note by the Secretary General," February 24, 1972, OECD Archives, CE/M(72)7(Prov.) Part I, p. 4.

30. "Preliminary Outline of the Technical Work in Relation to a Proposed Project to Study the Long Range Transport of Air Pollutants," October 25, 1971, OECD Archives, NR/JAH/A.71.163, Annex to Letter, pp. 2–3.

31. S. C., Technical Intelligence Branch, "OECD Study of the Long-range Transport of Pollutants," August 26, 1971, COAL 97/541, UK National Archives, p. 2.

32. Ibid. Also see Organisation for Economic Co-operation and Development, Environment Directorate, *Planning Group—Long Range Transport of Air Pollutants, Third Meeting, 9th–11th June, 1971,* June 25, 1971, OECD Archives, NR/ENV/71.26, p. 3.

33. Organisation for Economic Co-operation and Development, Environment Directorate, *Plan for a Project to Study the Long Range Transport of Air Pollutants*, p. 4.

34. This is in contrast to support for weather forecasting, particularly through the World Meteorological Organization, as well as work in dynamical meteorology that developed out of the University of Bergen in Norway under Vilhelm Bjerknes. See Robert Marc Friedman, *Appropriating the Weather: Vilhelm Bjerknes and Construction of a Modern Meteorology* (Ithaca, NY: Cornell University Press, 1993); Frederik Nebeker, *Calculating the Weather: Meteorology in the 20th Century* (San Diego: Academic Press, 1995); Kristine C. Harper, *Weather by the Numbers: The Genesis of Modern Meteorology* (Cambridge, MA: MIT Press, 2008); H. H. Lamb, "The New Look of Climatology," *Nature* 223 (September 20, 1969): 1209–1215; and Lovelock, "Air Pollution and Climatic Change," p. 410. Lovelock specifically uses sulfate as an example of the lack of appropriate monitoring stations for air pollution.

35. Brynjulf Ottar, "Notat: Planlegging, forberedelse og igangsetting av det Europeiske Monitoring Program for Luftforurensninger (EMP)," March 2, 1976. RA-S-2532-2-Dca-L0292, FN (med særorganisasjoner) UN ECE—Task force—European Monitoring Programme EMP. Norwegian National Archives, p. 2.

36. Organisation for Economic Co-operation and Development, *Ad Hoc Meeting on Acidity and Concentration of Sulphate in Rain, May 1969*, p. 3.

37. Air Management Sector Group, "Air Management in Relation to Emissions of Sulphur Oxides. Committee Room Document No. 1," October 25, 1971, OECD Archives, E.48.414, p. 2.

38. Ibid.

39. Nils Brandt, "Referat fra kontaktmøte med emnet sur nedbør og dens virkninger på det norske skogforsøksvesen, den 27 April 1971," June 21, 1971, RA-PA-0641-E-Eg-L0012, Norwegian National Archives.

40. "En 'kake' av luftforurensninger over det europeiske kontinent," pp. 1, 19; "Europeisk forskningsprosjekt over sur nedbør er i emning."

41. H. C. Christensen, "Møte mellom NILU's styre og komite for forurensningsspørsmål, 25.11.71," November 30, 1971, RA-S-1574-E-L0065, Norwegian National Archives, p. 2. For Brandt's expertise, see Nils Roll-Hansen, *Sur nedbørs virkning på skog og fisk: 1972–1976* (Oslo: NAVF, 1986), p. 28.

42. Christensen, "Møte mellom NILU's styre og komite for forurensningsspørsmål, 25.11.71," p. 2.

43. Ibid. H. S., "Luftens giftvirkning i skog må klarlegges," *Aftenpostens*, September 16, 1971, RA-PA-0641-E-Eg-Loo12, Norwegian National Archives.

44. Roll-Hansen, *Sur nedbørs virkning på skog og fisk: 1972–1976*, pp. 20–21.

45. Ibid., p. 25.

46. Several scientists from Norway's Agricultural University also participated in these initial meetings, which specialized in training practitioners in land use management, forestry, and agricultural sciences. Brandt, "Referat fra kontaktmøte med emnet sur nedbør og dens virkninger på det norske skogforsøksvesen, den 27 april 1971," p. 1.

47. Christensen, "Møte mellom NILU's styre og komite for forurensningsspørsmål, 25.11.71," p. 2.

48. Brandt, "Referat fra kontaktmøte med emnet sur nedbør og dens virkninger på det norske skogforsøksvesen, den 27 april 1971," p. 5.

49. Lars Overrein, "NLVF-NTNF's forskningsprosjekt 'luftforurensning—virkning på jord, vegetasjon og vann' nytt forslag til arbeidsprogram for perioden 1/7–31/12 1972," June 1972, RA-S-2939-D-Db-Dbk-Dbkj-Lo734, Norwegian National Archives, p. 2.

50. Ibid., p. 6.

51. After the first year of investigations in the SNSF project, the extent of the impact on aquatic ecosystems appeared to be worse than expected and underscored the importance of intensifying research efforts. These results also caused members of the NTNF pollution committee to become increasingly fearful about the potential impact on Norwegian forests, which had far greater economic importance. Historian Nils Roll-Hansen, who was asked to compile an evaluation of the SNSF project's successes and failures during the 1980s, stated in his work that it was the fear of reduced growth in the forest that triggered the project and that studies into fish death were incorporated later into the project. However, this does not quite conform to the archival material now available for public viewing from the NTNF and SNSF project, as explained in greater detail in the text. See H. C. Christensen, "Fellesprosjekt 'Sur Nedbørs Virkning på Skog og Fisk' søknad om ekstrabevilgning for 1973," February 8, 1973, RA-S-1574-E-Loo65, National Library of Norway, p. 2; Nils Brandt, "Referat fra kontaktmøte med emnet sur nedbør og dens virkninger på det norske skogforsøksvesen, den 27 april 1971." On the importance of the SNSF project's completion with the OECD study, see "NLVF-NTNF's felles forskningsprosjekt 'Sur Nedbørs Virkning På Skog Og Fisk,' " 1973, RA-S-2532-2-Daa-Lo506, Norwegian National Archives, p. 2.

52. Lars N. Overrein, "A Presentation of the Norwegian Project 'Acid Precipitation—Effects on Forest and Fish,' " *Water, Air, & Soil Pollution* 6, no. 2 (June 1, 1976): 167–172.

53. Magne Stubsjøen and Rolf Marstrander, "Vedlegg 2 (Referat: Fra møte 4/76 i styringsutvalget for SNSF-prosjektet l. del mandag 8. mars 1976 Kl. 1300—1630 på NISK, ås 2. del onsdag 10. mars 1976 kl. 0830—1030 i Stortinget)," January 15, 1976, RA-S-2532-2-Dca-Loo14, Norwegian National Archives, p. 4.

54. Ibid. Also see P. J. W. Saunders, "Impact of Acid Precipitation on Forest and Freshwater Ecosystems in Norway by H. Braekke Finn," *Journal of Applied Ecology* 15, no. 1 (April 1978): 338–339; "Letter from Norman Glass, Director, National Ecological Research Laboratory, Environmental Protection Agency, to Dr. Endsjo, Head of Division, Ministry of Environment," October 31, 1973, RA-S-2532-2-Dca-L0437, Norwegian National Archives; "Letter from Ellis B. Cowling, Professor, Plant Pathology and Forest Resources, to Dr. Donald King, Deputy Director for Science, Office of Environmental Affairs, Department of State," October 13, 1976, RA-S-2532–2-Dca-L0292, FN (med særorganisasjoner) UN ECE—Task force—European Monitoring Programme EMP, Norwegian National Archives.

55. Quote translated from Norwegian. H. C. Christensen, "Referat fra møte 10/72 i styringsutvalget for prosjektet 'Sur Nedbørs Virkning På Skog Og Fisk,' " January 4, 1973, RA-S-2532-2-Dca-L0437, Norwegian National Archives.

56. "Executive Committee: Summary Record of the 306th Meeting held at the Chateau de la Muette, Paris," April 18, 1972, OECD Archives, C/M(72)11(Prov.), p. 7.

57. "Sulphur in Oil," 1972, UK National Archives, HLG 120/1571.

58. "Executive Committee: Summary Record of the 306th Meeting held at the Chateau de la Muette, Paris," pp. 8–9.

59. Ibid.

60. NILU, "NILU gjennom 25 år," pp. 15–16.

61. "Letter from Leslie Reed to John Reay," February 15, 1973, AIR 20-12566, UK National Archives.

62. "Cooperative Technical Programme to Measure the Long-range Transport of Air Pollutants. Steering Committee: Summary Record of the Third Meeting," December 14, 1973, OECD Archives, NR/ENV/73.62, pp. 12–19.

63. Ibid.

64. Ibid., pp. 12–19; J. S. S. Reay, "Pollution by Smoke and Sulphur Dioxide and the National Survey," January 1976, AT 34–12, UK National Archives, p. 3.

65. "Cooperative Technical Programme to Measure the Long-range Transport of Air Pollutants. Steering Committee: Summary Record of the Third Meeting," pp. 12–19.

66. Interdepartmental Committee on Air Pollution Research, "Progress in Research on Air Pollution in the United Kingdom: Report by the Interdepartmental Committee on Air Pollution Research to the Clean Air Council," September 1975, AT 34/89, UK National Archives, p. 31.

67. N. G. Stewart, "Atmospheric Pollution Project at A.E.R.E. Harwell. 2nd Revision," February 1972, AB 88–113, UK National Archives.

68. "Atmospheric Pollution Project at A.E.R.E. Harwell," January 27, 1972, AB 88/113, UK National Archives, p. 2. Despite Harwell lab's enthusiasm for this work, the director of the Warren Spring Laboratory was determined to keep his research teams at the forefront of the project, and there appears to have been some internal competition for participating in research on acid rain. See "Letter from AJ Robinson, Director, Warren Spring Laboratory, to LEJ Roberts, Atomic Energy Research Establishment," January 11, 1972, AB 88/113, UK National Archives.

69. In an attempt to change Britain's critical attitude, Ottar began sending periodic updates on the project's status to Leslie Reed, who represented Britain on the project's steering committee. Leslie Reed, "Long Range Transport of Pollutants," June 4, 1973, AIR 20–12566, UK National Archives, p. 2; Central Unit on Environmental Pollution, "Project Committee for Long-range Transport of Pollutants: Minutes of a Meeting Held in Room N19/06, DOE, 2 Marsham Street, Sw1 on Wednesday 12 September 1973," September 17, 1973, AIR 20–12566, UK National Archives.

70. "Letter from D. H. Lucas, CEGB, to Leslie Reed, CUEP," April 11, 1973, AIR 20–12566, UK National Archives; "Letter from G. G. Thurlow, National Coal Board, to Leslie Reed, CUEP," March 11, 1974, AIR 20–12566, UK National Archives; "Letter from D. H. Lucas, CEGB, to Leslie Reed, CUEP," February 4, 1974, AIR 20–12566, UK National Archives.

71. Brynjulf Ottar, "Notat: Planlegging, forberedelse og igangsetting av det Europeiske Monitoring Program for Luftforurensninger (EMP)," p. 2.

72. "Cooperative Technical Programme to Measure the Long-range Transport of Air Pollutants. Steering Committee: Summary Record of the Third Meeting," pp. 12–19.

73. Brynjulf Ottar, "Monitoring Long-Range Transport of Air Pollutants: The OECD Study," *Ambio* 5, no. 5/6 (1976): 203–206.

74. Ibid. Also see chapter 9 and 11 in *The OECD Programme on Long Range Transport of Air Pollutants: Measurements and Findings* (Paris: OECD, 1977), Norwegian Institute of Air Research (NILU) Archives.

75. Central Unit on Environmental Pollution, "Project Committee for Long Range Transport of Pollutants: Minutes of a Meeting Held in Room N19/09, Department of the Environment, London SW1, on Tuesday 6 November 1973," November 13, 1973, AIR 20–12566, UK National Archives.

76. Richard G. Fort, "OECD. Miljøkomiteen. LRTAP-prosjektet. Møte i styringsgruppen 13.-14. november 1973. Kort referat fra forberedende møte i Miljøverndepartementet 9. november 1973," November 20, 1973, RA-S-2532-2-Dca-L0437, Norwegian National Archives.

77. "Environment Committee: Summary Record of the 3rd Session," November 2, 1971, OECD Archives, ENV/M(71)2, p. 6.

78. The idea of the polluter-pays principle was first articulated in the Trail Smelter case in 1941 concerning pollution from Canada crossing the US border. Rebecca M. Bratspies and Russell Miller, eds., *Transboundary Harm in International Law: Lessons from the Trail Smelter Arbitration* (Cambridge: Cambridge University Press, 2006).

79. "Environment Committee: Issues Identified by the Sub-committee of Economic Experts for Consideration by the Committee," September 21, 1971, OECD Archives, ENV(71)30.

80. "Environment Committee: Summary Record of the 9th Session," September 24, 1973, OECD Archives, ENV/M(73)3, p. 8.

81. Organisation for Economic Co-operation and Development, *International Scientific Cooperation* (Paris: OECD, 1965), p. 64.

82. "Environment Committee: Meeting at Ministerial Level: Minutes of the 13th Session," pp. 6–8; United Nations, *Environmental Conventions: Elaborated*

under the Auspices of the United Nations Economic Commission for Europe (New York: United Nations, 1994).

83. Daniel Yergin, *The Prize: The Epic Quest for Oil, Money & Power* (Simon and Schuster, 2011), p. 598.

84. Organisation for Economic Co-operation and Development, "Meeting of the Environment Committee at Ministerial Level: Draft Communiqué," November 5, 1974, OECD Archives, CES/74.101 (1st Revision), p. 3.

85. Before the November gathering, the transboundary air pollution recommendations were the only topic of the ten environmental issues that was still unresolved. See "Strictly Confidential: Environment Committee 13th Session. Meeting at Ministerial Level: Briefing Notes," 1974, RA-S-2532-2-Dca-L0409, Norwegian National Archives, p. 8.

86. "Environmental Committee: Draft Action Proposals for Submission to the Environment Committee at Ministerial Level," June 19, 1984, OECD Archives, Annex VIII to ENV/MIN(74)7. 6.882, p. 2. Also see "Preparation of the Environment Committee Meeting at Ministerial Level," July 1, 1974, OECD Archives, ENV/M(74)2, p. 7.

87. Erik Lykke, "Sur nedbør—vedtak i OECD's miljøkomite om tiltak mot utslipp av svoveldioksyd fra stasjonære kilder," March 27, 1974, RA-S-2532-2-Dca-L0437, Norwegian National Archives; "Air Management Sector Group: Note by the Secretariat," March 12, 1974, OECD Archives, Addendum 1 to NR/ENV/74.9. E.64293, p. 3.

88. "Council: Draft Declaration on Environmental Policy Submitted to the Environment Committee at Ministerial Level," September 23, 1974, OECD Archives, C/M(74)21 (Prov.), pp. 12–13.

89. "Executive Committee: Summary Record of the 380th Meeting," October 3, 1974, OECD Archives, CE/M(74)16(Prov.), p. 4.

90. Ibid., pp. 4–5.

91. "Executive Committee: Summary Record of the 381st Meeting," October 1, 1974, OECD Archives, CE/M(74)17(Prov.) Part I, p.3.

92. "Council: Action Proposal on Principles Concerning Transfrontier Pollution /C(74)158 Annex X (1st Revision) dated 4th October, 1974/ Statement by the Delegate for Canada at the Meeting of the Council on 8th October, 1974," October 9, 1974, OECD Archives, CES/74.95, pp. 1–4.

93. Declaration of the United Nations Conference on the Human Environment, Stockholm, June 16, 1972, UN Doc. A/CONF.48/14/Rev.1 (1972).

94. Louis B. Sohn, "The Stockholm Declaration on the Human Environment," *Harvard International Law Journal* 14 (1973): 423–515, p. 427.

95. Environment Committee Transfrontier Pollution Group, "Mandate of the Group," September 9, 1976, OECD Archives, ENV/TFP/76.18, p. 29.

96. "Annex VI: Proposals for a Draft Declaration by the Delegation for Norway," April 16, 1974, OECD Archives, ENV/MIN(74)6, p. 19.

97. OECD-delegasjonen, Paris, "Inkommet melding. Sak: Luftforurening," February 7, 1974, RA-S-2532-2-Dca-L0437, Norwegian National Archives. "Telegram from Kristiansen, Norwegian Delegation at OECD to Ministry of the Environment," April 19, 1974, RA-S-2532-2-Dca-L0437, Norwegian National Archives.

98. "Telegram from Norwegian Delegation at the OECD to Erik Lykke," April 30, 1974, RA-S-2532-2-Dca-L0437, Norwegian National Archives. The 1973 oil crisis occurred in response to the outbreak of the Yom Kippur War between Israel, Egypt, Syria, Jordan, and Iraq. In an effort to penalize Western nations supporting Israel, Arab members of OPEC raised the price of oil and placed an embargo on exports to Israel's allies, notably the US and the Netherlands. Britain supported the Arab states during the Yom Kippur war, refusing to allow US troops to use British bases to assist the Israelis as well as refusing to share oil with European nations affected by the embargo. For an account of British foreign policy during the 1973 oil crisis, see Roy Licklider, "The Power of Oil: The Arab Oil Weapon and the Netherlands, the United Kingdom, Canada, Japan, and the United States," *International Studies Quarterly* 32, no. 2 (June 1988): 205–226.

99. "Telegram from Norwegian Delegation at the OECD to Erik Lykke," April 30, 1974. The exact quote from the telegram reads as follows, translated from Norwegian by the author: "The British put the matter in the context of relations with the oil-producing countries . . . The oil-producing countries could conceivably exploit the matter during discussions of price issues, claiming that if one is willing to assume the financial burden for this decision one can also afford to pay higher prices for oil."

100. Ibid. Norway's access to oil from the North Sea may have muted its concerns about OPEC in comparison to the British, although Britain had access to North Sea oil as well though it was not producing extensively before the OPEC crisis.

101. Whether the Netherlands made this suggestion in response to Britain's refusal to supply the country with oil while it was subject to OPEC's embargo is not clear, and merits further historical investigation. "Letter from Richard Fort to Erik Lykke," May 2, 1974, RA-S-2532-2-Dca-L0437, Norwegian National Archives.

102. Ibid. See also Richard G. Fort, "Kort referat fra SO$_2$-møte i Miljøvernde-partementet 25. April 1974," May 8, 1974, RA-S-2532-2-Dca-L0437, Norwegian National Archives.

103. "Environment Committee: Meeting at the Ministerial Level. Minutes of the 13th Session held at the OECD Headquarters in Paris on the 13th and 14th November 1974," March 6, 1975, OECD Archives, ENV/M(74)4 & Corrigendum, p. 46.

104. Ibid.

105. The US took a milder stance than Canada, however, stating it would only issue a reservation on particular paragraphs of the text. See "Strictly Confidential: Environment Committee 13th Session. Meeting at Ministerial Level: Briefing Notes," p. 11.

106. Ibid., pp. 7–9.

107. "Council: Action Proposal on Principles Concerning Transfrontier Pollution /C(74)158 Annex X (1st Revision) dated 4th October, 1974/ Statement by the Delegate for Canada at the Meeting of the Council on 8th October, 1974," pp. 1–4.

108. "Procedure to Be Followed for the Adoption by the Council of Certain Texts Approved by the Environment Committee at Ministerial Level," October 24,

1974, OECD Archives, C/M(74)24(Prov.) Part I, pp. 14–15; "Environment Committee: Summary Record of the 12th Session," September 11–12, 1974, OECD Archives, ENV/M(74)3 & Corrigendum, pp. 5–6.

109. Organisation for Economic Co-operation and Development, *Legal Aspects of Transfrontier Pollution* (Paris: OECD, 1977). pp. 13–18.

110. "Environment Committee: Draft Action Proposals for Submission to the Environment Committee at Ministerial Level," May 8, 1974, OECD Archives, Annex VIII to ENV/MIN/74.7, pp. 1–9.

111. Ibid., p. 7.

112. Organisation for Economic Co-operation and Development, *Legal Aspects of Transfrontier Pollution*, p. 125.

113. Environment Committee Transfrontier Pollution Group, "The Role of Domestic Procedures in Transnational Environmental Disputes, by Peter Sand, United Nations FAO Legal Office," May 29, 1976, OECD Archives, ENV/TFP/ "Divers," p. 14.

114. "Environment Committee: Draft Action Proposals for Submission to the Environment Committee at Ministerial Level," May 8, 1974, p. 6. For the final version of the document, see Organisation for Economic Co-operation and Development, *Legal Aspects of Transfrontier Pollution*, pp. 7–16.

115. "Environment Committee: Draft Action Proposals for Submission to the Environment Committee at Ministerial Level," May 8, 1974, p. 5. "Council: Minutes of the 368th Meeting," October 17, 1974, OECD Archives, C/M(74)22(Prov.), p. 6.

116. Emile van Lennep, "Opening Address by the Secretary-General of the O.E.C.D. Mr. Emile van Lennep, to the Environment Committee Meeting at Ministerial Level on 13th-14th November, 1974," n.d. RA-S-2532-2-Dca-Lo409, Norwegian National Archives, pp. 2–3.

117. In the aftermath of the meeting, the OECD Secretariat decided to form a new Transfrontier Pollution Group as a subdivision of the Environment Committee with the goal of intensifying cooperation on transboundary pollution. See Organisation for Economic Co-operation and Development, *Legal Aspects of Transfrontier Pollution*, p. 7.

118. Peder Anker, "Økologisk ukorrekt," in *Universitetet i Oslo: Samtidshistoriske perspektiver*, ed. Peder Anker, Magnus Gulbrandsen, Eirinn Larsen, Johannes W. Løvhaug, and Bent Sofus Tranøy (Auskog, Norway: Unipub, 2011), p. 138.

119. "Environment Committee: Meeting at the Ministerial Level. Minutes of the 13th Session held at the OECD Headquarters in Paris on the 13th and 14th November 1974," p. 23.

CHAPTER THREE

1. Gene Likens, quoted in Anson Smith, "Acid Rain Falling in Northeast May Stunt Plants and Kill Fish," *Boston Globe* (June 22, 1975), p. 28.

2. George Hidy, interview transcript, March 2, 2013, Niels Bohr Library & Archives, American Institute of Physics, College Park, Maryland, p. 35.

3. G. E. Likens and F. H. Bormann, "Acid Rain: A Serious Regional Envi-

ronmental Problem," *Science* 184, no. 4142 (June 14, 1974): 1176–1179. Likens and Bormann had published an earlier article on acid rain in the White Mountains in 1972, but it received little attention outside the immediate circle of scientists working on the problem. See Gene E. Likens, F. Herbert Bormann, and Noye M. Johnson, "Acid Rain," *Environment: Science and Policy for Sustainable Development* 14, no. 2 (March 1, 1972): 33–40.

4. Boyce Rensenberger, "Acid in Rain Found Up Sharply in East; Smoke Curb Cited," *New York Times*, June 13, 1974, p. 89; "The Northeastern U.S.'s Acid Rain," *Science News* 105, no. 24 (June 15, 1974): 383; "Smokestack Filters Tied to Rain Acidity," *Los Angeles Times*, June 14, 1974, p. A5; "Scientist Warns of 'Acid Rains,'" *Hartford Courant*, June 22, 1974, p. 36.

5. Author interview with Gene Likens, Cary Institute, March 9, 2012, and August 18, 2016.

6. Allan M. Brandt, *The Cigarette Century: The Rise, Fall, and Deadly Persistence of the Product That Defined America* (New York: Basic Books, 2007); Gerald Markowitz and David Rosner, *Lead Wars: The Politics of Science and the Fate of America's Children* (Berkeley: University of California Press; Milbank Memorial Fund, 2013); Naomi Oreskes and Erik M. Conway, *Merchants of Doubt: How a Handful of Scientists Obscured the Truth on Issues from Tobacco Smoke to Global Warming* (New York: Bloomsbury Press, 2010); Jacob Darwin Hamblin, *Poison in the Well: Radioactive Waste in the Oceans at the Dawn of the Nuclear Age* (New Brunswick, NJ: Rutgers University Press, 2009).

7. Naomi Oreskes and Erik Conway, for instance, place the blame of acid rain and climate change doubt-mongering on several scientists like William Nierenberg and Fred Seitz who were largely outside the coal industry, though they had some ties to EPRI. They also treat Europe as operating separately from the US and achieving political progress on acid rain far in advance of the Americans, which does not accurately reflect developments in the 1970s and 1980s. See *Merchants of Doubt*, pp. 66–106.

8. Committee on Environmental Pollution, "Fuels, Sulphur Oxides, and Air Pollution," October 1972, HLG 120/1571, UK National Archives.

9. US spending on control technologies and acid rain research calculated by the author based on R&D Status Reports, *EPRI Journal* 2, nos. 1–10 (1977).

10. George Hidy, interview transcript, March 2, 2013, Niels Bohr Library & Archives, American Institute of Physics, College Park, Maryland, p. 35.

11. Author interview with Robert Goldstein, Electric Power Research Institute, November 3 and November 19, 2015; Author interview with Anthony Kallend, retired, head of environmental chemistry at CERL, March 25, 2013; Author interview with Richard Skeffington, professor, University of Reading, former biologist at the CERL, March 15, 2013.

12. "Keeping Tabs on an Ill Wind," *New Scientist* (May 25, 1972), p. 438.

13. Committee on Environmental Pollution, "Fuels, Sulphur Oxides, and Air Pollution," October 1972, HLG 120/1571, UK National Archives; Martin W. Holdgate, "Sulphur," October 6, 1972, HLG 120/1571, UK National Archives.

14. The impetus for research into flue gas desulfurization technologies was a 1929 case that came before the House of Lords claiming damages against the Barton Electricity Works of the Manchester Corporation. The House of Lords

ruled in favor of the complainant, awarding him compensation for damages to his land from sulfur dioxide emissions downwind from a Barton Electricity Works power plant. See S. J. Biondo and J. C. Marten, "History of Flue-Gas Desulfurization Systems since 1850: Research, Development, and Demonstration," *Journal of the Air Pollution Control Association (United States)* 27, no.10 (October 1, 1977), p. 955; R. Lessing, "Development of a Process of Flue-Gas Washing Without Effluent. [Washing with CaO or Chalk Suspension]," *Journal of the Society of Chemical Industry, London* 57 (January 1, 1938).

15. The first prototypes of this technology were installed at the Bankside, Battersea, and Fulham Power stations. See Central Electricity Generating Board, "The Removal of Sulphur Dioxide from Flue Gases," July 1965, BT 328/105, UK National Archives, p. 1.

16. Ibid., p. 2. Also see Central Electricity Generating Board, "Chimney Emissions and Power Stations," July 1965, BT 328/105, UK National Archives.

17. "Letter from D. L. Nicol to Mr. Jenkyns, Mr. Ireland, Dr. Holdgate," August 3, 1972, HLG 120/1571, UK National Archives; H. B. Locke, "Fluidized Combustion: The Background and Problems of the Proposed EPA/NRDC Contract Tor Development Work at BCURA," June 30, 1972.

18. "Letter from Dr. Reed to Martin Holdgate," August 16, 1972, HLG 120/1571, UK National Archives.

19. Daniel Yergin, *The Quest: Energy, Security, and the Remaking of the Modern World* (New York: Penguin, 2011).

20. Fraser Ross, "Sulfur Dioxide over Britain and Beyond" (*New Scientist and Science Journal,* May 13, 1971), COAL 97/541, UK National Archives; "Letter from F. F. Ross to Dr. L. C. F. Blackman," November 13, 1970, COAL 97/541, UK National Archives.

21. National Society for Clean Air: Technical Committee, "SO_2: Sulphur Dioxide, an Air Pollutant," 1971, HLG 120/1571, UK National Archives, pp. 1, 14, 22.

22. Ibid., p. 21. However, the report did acknowledge that little research had been done on the effects of sulfate on vegetation. See ibid., pp. 5–6.

23. Ibid.

24. "Letter from RCC [R. C. Chilver] to Mr. Jukes. Sulphur in the Atmosphere," March 24, 1972, HLG 120/1571, UK National Archives, p. 2; "Letter from J. E. Beddoe to Mr. Chilver," November 24, 1971, HLG 120/1571, UK National Archives; "Letter from R. C. Chilver to Mr. Ireland, Mr. Beddoe, Dr. Holdgate," July 22, 1971, HLG 120/1571, UK National Archives; "Letter from J. E. Beddoe to Mr. Chilver," July 15, 1971, BT 328/152, UK National Archives.

25. See marginalia notes, written by H. L. Jenkins, in Committee on Environmental Pollution, "Fuels, Sulphur Oxides, and Air Pollution," October 1972, HLG 120/1571, UK National Archive, pp. 14, 19.

26. "Letter from J. E. Beddoe to Dr. Chilver," July 15, 1971, BT 328/152, UK National Archives.

27. "Letter from Richard Bullock, Deputy Secretary, Department of Trade and Industry, to R. C. Chilver, Department of the Environment," July 2, 1971, HLG 120/1571, UK National Archives.

28. Ibid.

29. Ibid. The Department of the Environment initially reached a different conclusion about the effect of foreign restrictions on emissions. It believed that such regulations would make low sulfur fuel incredibly expensive and potentially profitable for oil companies to sell abroad. See "Letter from RC Chilver to RHW Bullock," March 10, 1971, HLG 120/1571, UK National Archives.

30. Victor Keegan, "From the Archive, 20 October 1970: BP Finds Big Oilfield in the North Sea," *Guardian*, October 20, 2010, http://www.theguardian.com /theguardian/2010/oct/20/archive-bp-finds-big-oilfield-in-the-north-sea-1970.

31. "Letter from D. G. Brandrick, National Coal Board, to L. H. Leighton, Department of Trade and Industry," November 11, 1970, COAL 97/541, UK National Archives; D. H. Broadbent, "Notes on Paper by F. F. Ross of 29.10.70. for E.C.E. Seminar," November 9, 1970, COAL 97/541, UK National Archives; "Letter from L. H. Leighton to D. G. Brandrick," November 4, 1970, COAL 97/541, UK National Archives; "Letter from D. G. Brandrick to B. C. Wiliams," October 23, 1970, COAL 97/541, UK National Archives.

32. Committee on Environmental Pollution, "Fuels, Sulphur Oxides, and Air Pollution," October 1972, HLG 120/1571, UK National Archives, p. 11; "Letter from MW Holdgate to Mr. Davis and Dr. Reed," September 9, 1971, HLG 120/1571, UK National Archives; "Letter from JE Beddoe to Mr. Chilver," July 15, 1971, HLG 120/1571, UK National Archives; "Letter from R. C. Chilver to Mr. Ireland, Mr. Beddoe, Dr. Holdgate," July 22, 1971, HLG 120/1571, UK National Archives; "Letter from FE Ireland to Mr. Chilver," September 3, 1971, HLG 120/1571, UK National Archives.

33. D. McHugh, "Good News for World Conservation: Dr Martin W. Holdgate Accepts Administrative Headship of IUCN," *Environmental Conservation* 14, no. 4 (Winter 1987): 365; M. W. Holdgate, *A Perspective of Environmental Pollution* (Cambridge: Cambridge University Press, 1980).

34. "Letter from MW Holdgate to Mr. Chilver," December 8, 1971, HLG 120/1571, UK National Archives.

35. Holdgate wrote to colleagues that this chance, coupled with the prospect that low levels of sulfur dioxide could potentially pose a risk to vulnerable populations, such as children and the elderly, or damage British agricultural crops, seemed ample justification to pursue work on abatement technologies. See "Letter from MW Holdgate to Mr. Chilver," December 8, 1971, HLG 120/1571, UK National Archives; "Letter from Martin Holdgate to Mr. Ireland," September 7, 1972, HLG 120/1571, UK National Archives.

36. "Letter from Dr. Reed to Martin Holdgate," August 16, 1972, HLG 120/1571, UK National Archives.

37. Leslie E. Reed, "The Long-Range Transport of Air Pollutants," *Ambio* 5, no. 5/6 (January 1, 1976): 202.

38. Armin Rosenkranz, "The Stockholm Conference," *Acid News: The Swedish NGO Secretariat on Acid Rain* 1/83 (1983), p. 7.

39. "Letter from M. W. Holdgate to Mr. Chilver," July 29, 1971, HLG 120/ 1571, UK National Archives.

40. "Sulphur Dioxide: Fluidised Bed Combustion," November 18, 1971, HLG 120/1571, UK National Archives.

41. When Britain's coal industry was nationalized after the Second World War, the National Coal Board was created to manage mining and extraction of the country's coal resources. It reported to the Ministry of Power. For an overview of the nationalization of the coal industry, see John Hatcher, *The History of the British Coal Industry* (Oxford: Clarendon Press, 1984). On the National Coal Board's support of desulfurization investment, see "Letter from D. H. Broadbent to Mr. D. G. Brandrick. Comments on E.C.E. Seminar on Desulphurisation of Fuels and Combustion Gases—Paper by Mr. Ross," October 21, 1970, COAL 97/541, UK National Archives.

42. Many members of the Department of the Environment believed it was not their responsibility to fund this work, though they welcomed funding from other departments or private industry. See "Letter from H. L. Jenkyns to Mr. Adams," October 16, 1972, HLG 120/1571, UK National Archives; "Letter from R. G. Adams to Mr. Jenkyns," September 26, 1972, HLG 120/1571, UK National Archives.

43. "CONFIDENTIAL: Sulphur in Oil," n.d., very likely sometime in 1972, HLG 120/1571, UK National Archives.

44. Although on a global basis about half of sulfur dioxide is produced by natural sources such as volcanoes and sea spray, in industrial areas like Europe and the US most sulfur dioxide is produced from burning fossil fuels. Frank R. Spellman and Nancy E. Whiting, *Environmental Science and Technology: Concepts and Applications* (Lanham, MD: Government Institutes, 2006), p. 291.

45. *Clean Air Act Amendments–1975, Part II: Hearings before the Subcommittee on Health and the Environment, Committee on Interstate and Foreign Commerce*, House of Representatives, 94th Congress, First Session (Washington, DC: US G.P.O., 1975), pp. 1223–1232; E. W. Kenworthy, "Donald Cook vs. E.P.A.: Wide Open Clash Over Coal and Clean Air; Donald Cook Battles Clean Air Standards—At a Glance How Scrubbers Work," *New York Times*, November 24, 1974.

46. *Implementation of the Clean Air Act–1975, Part I, Hearings before the Subcommittee on Environmental Pollution, Committee on Public Works*, Senate, 94th Congress, First Session (Washington, DC: US G.P.O., 1975).

47. National Academy of Sciences, *Air Quality and Stationary Source Emission Control* (Washington, DC, 1975); *Research and Development Related to Sulfates in the Atmosphere, Hearings before the Subcommittee on the Environment and the Atmosphere, Committee on Science and Technology*, House of Representatives, 94th Congress, First Session (Washington, DC: U.S. G.P.O., 1975); *Review of Research Related to Sulfates in the Atmosphere*, Report Prepared for the Subcommittee on the Environment and the Atmosphere, Committee on Science and Technology, House of Representatives, Ninety-Fourth Congress, Second Session (Washington, DC: US G.P.O., 1976).

48. Edward Cowan, "Limited Coal Use Is Seen for Utilities: U.S. Aide Says Few Plants Meet Rules; Few Utilities Expected to Meet Coal Standards," *New York Times*, July 11, 1974.

49. Rachel Rothschild, "Environmental Awareness in the Atomic Age: Radioecologists and Nuclear Technology," *Historical Studies in the Natural Sciences* 43, no. 4 (September 1, 2013): 492–530.

50. Jongming Lee, "CHESS Lessons: Controversy and Compromise in the Making of the EPA," in *Toxic Airs: Body, Place, Planet in Historical Perspective*, ed. James Rodger Fleming and Ann Johnson (Pittsburgh: University of Pittsburgh Press, 2014).

51. *Fiscal Year 1976 EPA R & D Authorization, Hearings before the Subcommittee on the Environment and the Atmosphere, Committee on Science and Technology*, House of Representatives, 94th Congress, First Session (Washington, DC, 1975), pp. 110–121, 260–262.

52. A. J. Robinson, "II. Air Pollution," *Journal of the Royal Society of Arts* 119, no. 5180 (July 1, 1971): 505–519, p. 506; Martin W. Holdgate, "Fuels, Sulphur Oxides, and Air Pollution," October 1972, HLG 120/1571, UK National Archives, pp. 5–8.

53. David Leslie Hawksworth, "Lichens as Litmus for Air Pollution: A Historical Review," *International Journal of Environmental Studies* 1, no. 1–4 (1970): 281–296.

54. CUEP, DOE, "Project Committee for Long-Range Transport of Pollutants: Minutes of a Meeting Held in Room N19/06, DOE, 2 Marsham Street, Sw1 on Monday 2 July 1973," July 6, 1973, AIR 20–12566, UK National Archives.

55. Ibid., p. 3.

56. Ibid. Holden had already developed extensive experience studying the effects of pesticides on the environment during the 1960s. See A. V. Holden, "The Effects of Pesticides on Life in Fresh Waters," *Proceedings of the Royal Society of London B: Biological Sciences* 180, no. 1061 (March 21, 1972): 383–394.

57. A. Tollan, "Effects of Sulphur Compounds on Aquatic Ecosystems," June 30, 1980, RA-S-2532-2-Dca-L0295, Norwegian National Archives, p. 2.

58. Central Electricity Generating Board, "30 Years of Achievements: CEGB Research" (Chichester Press, 1990), Personal Papers, Richard Skeffington, obtained during interview, pp. 40, 114.

59. Ian Mogford, "CERL: From Shed to Watershed" (National Power, 1993), 7802/1/7–13, Surrey History Centre Archives, p. 15; Author interview with Anthony Kallend, retired, head of environmental chemistry at CERL, March 25, 2013; Author interview with Alan Webb, retired, environmental chemist at CERL, March 16, 2013; Author interview with Richard Skeffington, professor, University of Reading, former biologist at the CERL, March 15, 2013.

60. The physics section oversaw atmospheric research, including the CEGB's contributions to the OECD project. Author interview with Alan Webb, retired, environmental chemist at CERL, March 16, 2013; Author interview with Richard Skeffington, professor, University of Reading, former biologist at the CERL, March 15, 2013; Author interview with Anthony Kallend, retired, head of environmental chemistry at CERL, March 25, 2013.

61. Author interview with Alan Webb, retired, environmental chemist at CERL, March 16, 2013; Author interview with Richard Skeffington, professor, University of Reading, former biologist at the CERL, March 15, 2013.

62. Author interview with Alan Webb, retired, environmental chemist at CERL, March 16, 2013.

63. Howells was the only female manager of a scientific research division in CEGB history. She reported general hostility from other lab managers and rare

invites to social events while working there. See Mogford, "CERL: From Shed to Watershed," p. 26; Author interview with Anthony Kallend, retired, head of environmental chemistry at CERL, March 25, 2013; Author interview with Alan Webb, retired, environmental chemist at CERL, March 16, 2013; Author interview with Richard Skeffington, professor, University of Reading, former biologist at the CERL, March 15, 2013.

64. Nilo Lindgren, "The First Five Years: Chauncey Starr and the Building of EPRI," *EPRI Journal* 3, no. 1 (February 1978).

65. Ibid., p. 6.

66. Chauncey Starr, "Energy and Power," *Scientific American* 225, no. 3 (September 1971): 36–49.

67. Paul M. Grant, "Chauncey Starr," *Physics Today* 60, no. 7 (July 1, 2007): 79.

68. Lindgren, "The First Five Years," p. 8.

69. Ibid., p. 9

70. For Starr's connection to the George Marshall Institute and physicists like Fred Seitz and William Nierenberg, see Oreskes and Conway, *Merchants of Doubt*, pp. 161, 193–194.

71. "Inhaling the Invisible," *EPRI Journal* 3, no. 8 (October 1978).

72. "A Round of Response," *EPRI Journal* 3, no. 1 (February 1978), p. 31.

73. "Statement of Ralph Perhac," *Acid Rain: Hearings before the Subcommittee on Oversight and Investigations of the Committee on Interstate and Foreign Commerce. Ninety-Sixth Congress, Second Session, February 26 and 27 1980.* Serial No. 96–150, 1980, pp. 457–458.

74. "Letter from Walter Marshall to Chauncey Starr," December 12, 1974, AB 48/1469, UK National Archives; Central Electricity Generating Board, "30 Years of Achievements: CEGB Research" (Chichester Press, 1990), Personal Papers, Richard Skeffington, obtained during interview, p. 38.

75. "Letter from Dr. W. Marshall, C.B.E., F.R.S., to Dr. Chauncey Starr, President, Electric Power Research Institute," December 12, 1974, AB 48/1469, UK National Archives; "Letter from Chauncey Starr, President, Electric Power Research Institute, to Dr. Walter Marshall," January 10, 1975, AB 48/1469, UK National Archives; "Letter from Roberts L. Loftness, Director, Washington Office, Electric Power Research Institute, to R.L.R. Nicholson, Principle Economics and Finance Officer, UK Atomic Energy Authority," February 11, 1975, AB 48/1469, UK National Archives.

76. On discussions concerning the potential for US and British control of acid rain and its implications for the energy industry, see W. D. Halstead, "CEGB Confidential Report on a Visit to the USA May 9th to 23rd 1985," January 8, 1985, COAL 29/655, UK National Archives, p. 11.

77. Joseph A. O'Brien, "'Acid Rain' Stirs Concern Among Environmentalists," *Hartford Courant*, September 25, 1977, p. 1A; Interview with Richard Goldstein, EPRI, November 19, 2015.

78. Many of these grants went to American universities as well as large US research institutes such as Westinghouse and General Electric, in addition to the CERL. Mogford, "CERL: From Shed to Watershed," p. 32.

79. Joanne Omang, "Acid Rain: Push toward Coal Makes Global Pollution Worse: Push to Coal Raises Fear of Acid Rain," *Washington Post*, December 30, 1979, p. A1.

80. Lars N. Overrein, "A Presentation of the Norwegian Project 'Acid Precipitation—Effects on Forest and Fish,' " *Water, Air, and Soil Pollution* 6, no. 2 (June 1, 1976): 167–172, p. 171. For discussion in the US on the relationship between acid rain and *Silent Spring*, see George Hendry and Frederick Lipfert, Acid Precipitation and the Aquatic Environment, Effects of Acid Rain, Committee on Energy and Natural Resources, United States Senate, Ninety-Sixth Congress, Second Session on the Phenomenon of Acid Rain and its Implications for a National Energy Policy, May 28, 1980.

81. On the importance of public health and toxicology research to the DDT ban after *Silent Spring*, see Frederick Rowe Davis, *Banned: A History of Pesticides and the Science of Toxicology* (New Haven, CT: Yale University Press, 2014).

82. Author interview with Alan Webb, retired, environmental chemist at CERL, March 16, 2013.

83. Chester had first approached a Norwegian representative about the SNSF project at a meeting of the International Council on Large Electric Systems (C.I.G.R.E.) in Paris in August 1974. "Letter from Asbjørn Vinjar to Erik Lykke, Fjerntransport Av Luftforurensninger," September 6, 1974, RA-S-2532–2-Dca-L0437, Norwegian National Archives.

84. "Letter from H. C. Christensen Secretary to the Steering Committee of the SNSF-Project, to Dr. P. F. Chester, Director Central Electricity Research Laboratories," August 28, 1975, RA-S-2532–2-Dca-L0014, Norwegian National Archives.

85. Author interview with Alan Webb, retired, environmental chemist at CERL, March 16, 2013.

86. H. Leivestad and I. P. Muniz, "Fish Kill at Low pH in a Norwegian River," *Nature* 259 (February 5, 1976): 391–392.

87. Ibid., p. 391.

88. Author interview with Alan Webb, retired, environmental chemist at CERL, March 16, 2013. On the widespread distribution of this research among biologists working on acid rain, see Carl L. Schofield, "Acid Precipitation: Effects on Fish," *Ambio* 5, no. 5/6 (January 1, 1976): 228–230; George R. Hendrey, Kjell Baalsrud, Tor S. Traaen, et al., "Acid Precipitation: Some Hydrobiological Changes," *Ambio* 5, no. 5/6 (January 1, 1976): 224–227; Gene Likens, "Acid Precipitation," *Chemical and Engineering News* (November 22, 1976): 29–44. Cited in Oreskes and Conway, *Merchants of Doubt*, p. 76.

89. Author interview with Alan Webb, retired, environmental chemist at CERL, March 16, 2013.

90. Ivar Muniz et al., *Fiskedød i forbindelse med snøsmelting i tovdalvassdraget våren 1975* (SNSF-prosjektet, November 1975), IR 13/75, National Library of Norway, p. 39.

91. Leivestad and Muniz, "Fish Kill at Low pH in a Norwegian River."

92. Author interview with Richard Skeffington, professor, University of Reading, former biologist at the CERL, March 15, 2013; Author interview with

Alan Webb, retired, environmental chemist at CERL, March 16, 2013; Author interview with Anthony Kallend, retired, head of environmental chemistry at CERL, March 25, 2013.

93. "Letter from I. Th. Rosenqvist to Miljøvernminister, Statsråd Gro Harlem Brundtland," December 20, 1976, RA-S-2532-2-Dca-Loo15, Norwegian National Archives.

94. Historians Nils Roll-Hansen and Geir Hestmark did an evaluation of the SNSF project at the request of the NAVF; it is critical of the decision to conduct the project at applied research institutes, concluding that "for politically sensitive environmental issues the academic research system may in general be more able to sufficiently differentiate between science and politics than special institutions for applied research." They base this conclusion on the fact that "basic" research questions, notably those pertaining to soil chemistry, did not receive sustained focus until the second phase of the project. Roll-Hansen and Hestmark are clear that it was the scientific leadership of the project that organized the work in this way, not the Ministry of the Environment. While this chapter is not intended to evaluate whether government or university researchers are better suited to environmental research, it is worth noting that it does not support Roll-Hansen and Hestmark's point on this matter. The neglect of "basic" research questions in the SNSF project does not appear to have been simply a function of bias against such work by applied research institutes. It was also because the process of how acid rain damages the environment turned out to be much more complicated than ecologists initially thought, thus leading to greater investigations into soil chemistry in the second phase. On this question and for an in-depth examination of the importance of mass media in fueling the controversy between Rosenqvist and the SNSF, see Nils Roll-Hansen and Geir Hestmark, *Miljøforskning mellom vitenskap og politikk: En studie av forskningspolitikken omkring andre fase av storprosjektet "sur nedbørs virkning på skog og fisk" (SNSF), 1976–1980* (Oslo: NAVF, 1990).

95. Ivan Rosenqvist, "Et bidrag til analyse av geologiske materialers bufferegenskaper mot sterke syrer i nedbørsvann: Rapport utarbeidet for rådet for naturvitenskapelig forskning," November 16, 1976, RA-S-2939-D-Db-Dbk-Dbkj-Lo738, Norwegian National Archives; I. Th. Rosenqvist, "Acid Rain and Freshwater Ecosystems in Norway," Enclosure in Letter from I. Th. Rosenqvist to the Editor of Chemical and Engineering News, December 16, 1976, RA-S-2532-2-Dca-Loo15, Norwegian National Archives.

96. H. C. Christensen, "Møte Mellom NILU's Styre Og Komite for Forurensningsspørsmål, 25.11.71," November 30, 1971, RA-S-1574-E-Loo65, Norwegian National Archives, p. 2.

97. Rosenqvist also argued that the SNSF research could easily be picked apart by scientists from other countries involved in the acid rain negotiations. See I. Th. Rosenqvist, "Ad. forskerseminar: Sur nedbørs virkning på skog og fisk 26. og 27. februar, hurdalsjøens hotel," March 1, 1976, RA-S-2532-2-Dca-Loo14, Norwegian National Archives.

98. I. Th. Rosenqvist, "Ad. forskerseminar: Sur nedbørs virkning på skog og fisk 26. og 27. februar, hurdalsjøens hotel," March 1, 1976, RA-S-2532-2-Dca-Loo14, Norwegian National Archives. Historians Nils Roll-Hansen and

Geir Hestmark believe that Rosenqvist's critiques stemmed from genuine concern for the scientific integrity of the research, and that he did not actively seek to foment public controversy. However, Rosenqvist's numerous public attacks on the use of government rather than university researchers, his frequent mentions of the finances involved, and his attempts to widely circulate his criticisms in the English press suggest that other motivations may have undergirded his behavior. These various circumstances indicate that the polemical nature of his work was perhaps rooted in feeling slighted and overlooked by the organizers of the SNSF project, which would also help explain why other university scientists came to his defense. Historian Peder Anker has also suggested that Rosenqvist was driven by concern for the importance of the Norwegian power industry to the preservation of a socialist society. For the perspectives of Roll-Hansen and Hestmark on Rosenqvist, see Roll-Hansen and Hestmark, *Miljøforskning mellom vitenskap og politikk*, particularly pp. 43–47 and 135–137.

99. Arne Løvlie and Axel Anderson, "Forord," in Ivan Rosenqvist, "Et bidrag til analyse av geologiske materialers bufferegenskaper mot sterke syrer i nedbørsvann: rapport utarbeidet for rådet for naturvitenskapelig forskning," November 16, 1976, RA-S-2939-D-Db-Dbk-Dbkj-L0738, Norwegian National Archives.

100. Erik Lykke, "SNSF-prosjektet. Kritikk fra professorene Løvlie, Kjensmo og Rosenqvist," April 22, 1976, RA-S-2532–2-Dca-L0014, Norwegian National Archives, p. 3.

101. Christensen added that the project would need to be careful to avoid further polarization between the SNSF and university researchers down the road. See H. C. Christensen, "Notat: Orientering om Rosenqvists rapport," December 10, 1976, RA-S-2532–2-Dca-L0441, Norwegian National Archives. For further comments Rosenqvist made to the Ministry of Environment expressing his frustration that he and other university scientists were not asked to participate more in the work, see Erik Lykke, "Professor Rosenqvists kommentarer vedrørende sur nedbør-problemene," January 26, 1977, RA-S-2532–2-Dca-L0015, Norwegian National Archives.

102. Author interview with Anton Eliassen, Director of the Norwegian Meteorological Institute, May 16, 2011.

103. Brynjulf Ottar, "Notat: Forsurningsproblemene—kommentarer fra Professor Rosenqvist," November 8, 1976, RA-S-2532–2-Dca-L0015, UK National Archives, p. 2.

104. Author interview with Alan Webb, retired, environmental chemist at CERL, March 16, 2013.

105. Ibid.

106. G. M. Glover and A. H. Webb, "Weak and Strong Acids in the Surface Waters of the Tovdal Region in S. Norway," *Water Research* 13, no. 8 (1979): 781–783, p. 781.

107. Ibid.

108. Author interview with Alan Webb, retired, environmental chemist at CERL, March 16, 2013.

109. Alan Webb, "On the Contribution of Weak Acids to the pH of Surface Waters. Laboratory Note No. Rd/L/N 207/76" (Central Electricity Research Laboratories, October 29, 1976). Copy of the original report obtained from

Alan Webb. Author interview with Alan Webb, retired, environmental chemist at CERL, March 16, 2013.

110. Ibid. Hart began leading this division in 1975. See Mogford, "CERL: From Shed to Watershed," p. 27.

111. Author interview with Alan Webb, retired, environmental chemist at CERL, March 16, 2013.

112. Ibid. While deeply regretting his actions and having his arm twisted, Webb felt at the time that these manipulations were no worse than what Norwegian SNSF scientists were doing under political pressure from the Norwegian Ministry of the Environment. While such accusations were made by CERL scientists against the SNSF scientists, as well as by a few university researchers in Norway who sided with Rosenqvist, there is no evidence to suggest that there was any manipulation or influence on research results by Norway's Ministry of the Environment. For a discussion of this debate domestically in Norway between Rosenqvist and one of his colleagues, Professor N. A. Sørensen, see Roll-Hansen and Hestmark, *Miljøforskning mellom vitenskap og politikk,"* pp. 109–124. For Webb's article, see Glover and Webb, "Weak and Strong Acids in the Surface Waters of the Tovdal Region in S. Norway."

113. Author interview with Ronald Barnes, Retired, Former Principal Environmental Adviser, Esso, March 4, 2013, Wantage, England; Author interview with Alan Webb, retired, environmental chemist at CERL, March 16, 2013; Author interview with Anthony Kallend, retired, head of environmental chemistry at CERL, March 25, 2013. It should be noted that at least one CERL scientist, Richard Skeffington, claims there was never any pressure to modify or alter his research to be in line with the CEGB's position on acid rain. Author interview with Richard Skeffington, professor, University of Reading, former biologist at the CERL, March 15, 2013.

114. Chauncey Starr, "The Electric Power Research Institute," *Science* 219, no. 4589 (March 11, 1983): 1190–1194.

115. Author interview with Robert Goldstein, Electric Power Research Institute, November 3 and November 19, 2015.

116. R. A. Goldstein et al., "Integrated Acidification Study (ILWAS): A Mechanistic Ecosystem Analysis [and Discussion]," *Philosophical Transactions of the Royal Society of London B: Biological Sciences* 305, no. 1124 (May 1, 1984): 409–425.

117. John O. Reuss, Nils Christophersen, and Hans M. Seip, "A Critique of Models for Freshwater and Soil Acidification," *Water, Air, and Soil Pollution* 30, no. 3–4 (October 1986): 909–930.

118. Susan West, "Acid Solutions," *Science News* 117, no. 7 (February 16, 1980): 106.

119. *Integrated Lake-Watershed Acidification Study*, Special Issue: *Water, Air and Soil Pollution*, vol. 26, no. 4 (Dordrecht: Springer Netherlands, 1985).

120. *Acid Rain*, Subcommittee on Oversight and Investigations, Committee on Interstate and Foreign Commerce, House of Representatives, Ninety-Sixth Congress (February 26–27, 1980) (Testimony of Ralph Perhac, Electric Power Research Institute), p. 469; Ted Williams, "Clearing the Air on Acid Rain," *Gray's Current*, Summer 1981, p. 95.

121. "Statement of Dr. Robert Brocksen, Program Manager for Ecological Effects, Electric Power Research Institute," Effects of Acid Rain, Committee on Energy and Natural Resources, United States Senate, Ninety-Sixth Congress, Second Session on the Phenomenon of Acid Rain and Its Implications for a National Energy Policy, May 28, 1980.

122. Effects of Acid Rain, Committee on Energy and Natural Resources, United States Senate, Ninety-Sixth Congress, Second Session on the Phenomenon of Acid Rain and Its Implications for a National Energy Policy, May 28, 1980 (Testimony of George Hendrey, Brookhaven National Laboratory, p. 98).

123. Edison Electric Institute, "Acid Precipitation: The Issue in Perspective," Effects of Acid Rain, Committee on Energy and Natural Resources, United States Senate, Ninety-Sixth Congress, Second Session on the Phenomenon of Acid Rain and Its Implications for a National Energy Policy, May 28, 1980, pp. 29–33.

124. Ralph Perhac, interview transcript, April 5, 2013, Niels Bohr Library & Archives, American Institute of Physics, College Park, Maryland, p. 18.

125. Raymond G. Wilhour, EPA, Chief, Air Pollution Effects Branch, "Minutes, Acid Precipitation Staff Meeting," March 17, 1982, RG 412—Studies and Other Related Records, Acid Rain, U.S. National Archives and Record Administration II, Box 8.

126. "Letter from Ellis B. Cowling to Raymound G. Wilhour, EPA," April 27, 1982, RG 412—Studies and Other Related Records, Acid Rain, U.S. National Archives and Record Administration II, Box 7.

127. The report found that uncertainties surrounding acid rain were too high for policymakers to enact regulations. James Galloway, a University of Virginia scientist, also had received funding from EPRI and participated in the panel without disclosing this. It is important to note, however, that Galloway had been an outspoken advocate of controls on acid rain. See Committee on Atmospheric Transport and Chemical Transformation in Acid Precipitation et al., *Acid Deposition: Atmospheric Processes in Eastern North America* (Washington, DC: National Academies Press, 1983).

128. White House Office of the Press Secretary, "Acid Precipitation Task Force: Press Release," December 10, 1982, RG 412—Studies and Other Related Records, Acid Rain, U.S. National Archives and Record Administration II, Box 14.

CHAPTER FOUR

1. Original quote in Norwegian, translation by the author. H. C. Christensen, "Rapport fra deltakelse i underkomité for miljøvern, annen fase av Konferansen om Sikkert og Samarbeid i Europe, Geneve 22. oktober–6. november 1973," November 12, 1973, RA-S-1574-E-L0065, Norwegian National Archives, p. 5.

2. Original quote in Norwegian, translation by the author. "En del momenter fra møtet i NILU's regi 1.12.75 om OECD-prosjektet for langtransport av luftforurensninger," January 28, 1976, RA-S-2532-2-Dca-L0405, Norwegian National Archives, pp. 1–2.

3. Tony Judt, *Postwar: A History of Europe since 1945* (New York: Penguin, 2006), p. 32. On the history of a divided Europe, see Sari Autio-Sarasmo and Katalin Miklóssy, *Reassessing Cold War Europe* (Abington, UK: Routledge,

244

2010); Timothy Garton Ash, *In Europe's Name: Germany and the Divided Continent* (New York: Vintage, 1994); Charles S. Maier, *The Cold War in Europe: Era of a Divided Continent* (Princeton: Markus Wiener Publishers, 1996); William I. Hitchcock, *The Struggle for Europe: The Turbulent History of a Divided Continent 1945 to the Present* (New York: Anchor, 2004).

4. John Krige, *American Hegemony and the Postwar Reconstruction of Science in Europe* (Cambridge, MA: MIT Press, 2008); Erik van der Vleuten and Arne Kaijser, *Networking Europe: Transnational Infrastructures and the Shaping of Europe, 1850–2000* (Sagamore Beach, MA: Science History Publications, 2006); and David A. Hounshell, "Epilogue: Rethinking the Cold War; Rethinking Science and Technology in the Cold War; Rethinking the Social Study of Science and Technology," *Social Studies of Science* 31, no. 2 (April 1, 2001): 289–297.

5. On the earth sciences and the military, see S. Turchetti and P. Roberts, *The Surveillance Imperative: Geosciences during the Cold War and Beyond* (New York: Palgrave MacMillan, 2014). For some exceptions, see Bruno J. Strasser, "The Coproduction of Neutral Science and Neutral State in Cold War Europe: Switzerland and International Scientific Cooperation, 1951–69," *Osiris* 24, no. 1 (January 1, 2009): 165–187; Kai-Henrik Barth, "Catalysts of Change: Scientists as Transnational Arms Control Advocates in the 1980s," *Osiris*, 2nd Series, 21 (January 1, 2006): 182–206.

6. This statement was made by the UN Secretary General, Kurt Waldheim. See "34 Nations Act to Reduce Pollution Crossing Borders," *Los Angeles Times,* November 13, 1979, sec. Part I. On the importance of transnational pollution to understanding European integration, see Arne Kaijser, "The Trail from Trail: New Challenges for Historians of Technology," *Technology and Culture* 52, no. 1 (2011): 131–142.

7. Komité for Forurensningsspørsmål, "5-årsplan 1974–78 for aktivitetsområdet forurensningsspørsmål," May 21, 1973, RA-S-1574-E-L0025, Norwegian National Archives, pp. 1, 7.

8. Norwegian Institute for Air Research, "The OECD-Project 'LRTAP' 1st Quartely [*sic*] Report, July–Sept. 1972," November 7, 1972, RA-S-2532-2-Dca-L0403, Norwegian National Archives, p. 6.

9. Brynjulf Ottar, "Internt notat: Vurderinger i forbindelse med en koordinert begrensning av utslippene av luftforurensninger i Europa," August 27, 1973, RA-S-2532-2-Daa-L0506, Norwegian National Archives.

10. Ibid., p. 2. This was highly dependent on the level of acidity of the soil in question. Scandinavian soils were somewhat acidic due to the presence of siliceous rock formations, making them much more susceptible to acid rain pollution. While ecologists had hypothesized that additional acidity might not harm very basic soils in Central Europe, soils in Canada, in the Eastern US, and in tropical rainforests were viewed as similarly vulnerable to acid rain. See "Sulphur Pollution across National Boundaries," *Ambio* 1, no. 1 (February 1, 1972): 15–20. On studies concerning acid rain's effects on soil at the time, see L. N. Overrein, "Sulphur Pollution Patterns Observed; Leaching of Calcium in Forest Soil Determined," *Ambio* 1, no. 4 (September 1, 1972): 145–147.

11. "Letter from Brynjulf Ottar to Erik Lykke," February 16, 1973, RA-S-2532-2-Dca-L0403, Norwegian National Archives.

12. There is no evidence that Norwegian embassies in the USSR, Poland, or East Germany were of any assistance in this first outreach attempt, and there is no documentation of any contact between the Norwegian Ministry of the Environment and these governments in the spring of 1973. It appears the Norwegian embassy in Czechoslovakia did manage to convey Ottar's request to the Czech Research and Development Centre for Control of Environmental Pollution. It had been established just a year prior and told the Norwegian embassy that while it was very interested in NILU's work on acid rain, it did not have the data Ottar sought and would not be able to easily obtain it. The Embassy also cautioned the Ministry of the Environment that there was no institutional framework for scientific cooperation between Norway and Czechoslovakia. See "Letter from Thor Brodtkorb to Utenriksdepartementet," March 8, 1973, RA-S-2532-2-Dca-L0403, Norwegian National Archives.

13. The NTNF was formed after the Second World War to provide government support for scientific projects, notably in atomic research. An NTNF committee on pollution issues (Komité for Forurensningsspørsmål) was first established in 1972, the same year Norway established its Ministry of the Environment and the OECD project began. The head of the NTNF was also heavily involved with the OECD Science Committee in the decades after its founding. See Robert Leon Stern, *Technology and World Trade: Proceedings* (US Department of Commerce, National Bureau of Standards, 1967). pp. 76–77.

14. Richard G. Fort and Erik Lykke, "Referat fra SO$_2$-møtet i Miljøverndepartementet 6. juni 1973," June 14, 1973, RA-S-2532-2-Dca-L0403, Norwegian National Archives, p. 2.

15. Ottar, "Internt notat: Vurderinger i forbindelse med en koordinert begrensning av utslippene av luftforurensninger i Europa."

16. Ibid., pp. 2–3.

17. Fort and Lykke, "Referat fra SO$_2$-møtet i Miljøverndepartementet 6. juni 1973."

18. Environment Committee, "Meeting at Ministerial Level: Minutes of the 13th Session held at the OECD Headquarters in Paris on 14th and 14th November 1974," March 6, 1975, ENV/M(74)4, OECD Archives, p. 36; Brynjulf Ottar, "Organization of Long Range Transport of Air Pollution Monitoring in Europe," *Water, Air, and Soil Pollution* 6, no. 2 (1976): 219–229, p. 220.

19. H. C. Christensen, "Rapport fra deltakelse i underkomité for miljøvern, annen fase av Konferansen om Sikkert og Samarbeid i Europe, Geneve 22. oktober–6. november 1973."

20. Angela Romano, *From Détente in Europe to European Détente: How the West Shaped the Helsinki CSCE* (Brussels: Peter Lang, 2009), pp. 55–56.

21. Andreas Wenger, Vojtech Mastny, and Christian Nuenlist, *Origins of the European Security System: The Helsinki Process Revisited, 1965–75* (Abingdon, UK: Routledge, 2008), pp. 5–6. For an overview of the inclusion of environmental issues in the Helsinki process see Thomas Gehring, *Dynamic International Regimes: Institutions for International Environmental Governance* (Bern: Peter Lang, 1994), pp. 67–78. Gehring's work is primarily concerned with elaborating political theories of environmental conventions, with acid rain as a case study alongside ozone depletion. It is helpful in providing a synopsis of the events

leading up to Norway's proposal during the Helsinki negotiations, but the work has some historical inaccuracies with regard to these negotiations, as well as the OECD project, the negotiation of the 1979 Convention on Long-Range Transboundary Air Pollution, and the 1985 Sulfur Protocol.

22. H. C. Christensen, "Rapport fra deltakelse i underkomité for miljøvern, annen fase av Konferansen om Sikkert og Samarbeid i Europe, Geneve 22. oktober–6. november 1973," p. 2.

23. Ibid., p. 3.

24. Finland had already contacted Norway to inquire about its reasons for cooperating with Eastern Europe. H. C. Christensen, "Rapport fra møte i styringskomiteen for OECD-prosjektet 'Long Range Transport of Air Pollutants', Paris 13.–14 november 1973," November 26, 1973, RA-S-2532-2-Dca-L0403, Norwegian National Archives.

25. "Cooperative Technical Programme to Measure the Long-range Transport of Air Pollutants. Steering Committee: Summary Record of the Third Meeting," December 14, 1973, NR/ENV/73.62, OECD Archives, p. 5.

26. Ibid. Also see Christensen, "Rapport fra møte i styringskomiteen for OECD-prosjektet 'Long Range Transport of Air Pollutants', Paris 13.–14 november 1973," p. 3.

27. Richard G. Fort, "OECD. Miljøkomiteen. LRTAP-prosjektet. Møte i styringsgruppen 13.–14. november 1973. Kort referat fra forberedende møte i Miljøverndepartementet 9. november 1973," November 20, 1973, RA-S-2532-2-Dca-L0403, Norwegian National Archives, p. 1.

28. Ibid., p. 2.

29. Richard G. Fort, "Kort referat fra nordisk SO$_2$- møte i Miljøverndepartementet, Oslo, 6. desember 1973," December 19, 1973, RA-S-2532-2-Dca-L0403, Norwegian National Archives.

30. Ibid., p. 4. Each country was given responsibility for preparing a different potential topic for the expert meeting. Norway assumed responsibility for atmospheric and ecological effects of acid rain, Sweden was tasked with covering policy instruments like taxes on emissions, Denmark was assigned health effects, and Finland was asked to investigate corrosive effects on buildings.

31. It was formally submitted for consideration in the Helsinki Final Act in May 1974. See "Informasjon fra Miljøverndepartementet. Nr 12," December 1974, RA-PA-0641-E-Eg-L0012, Norwegian National Archives. For the OECD draft letter, see "Draft Letter from Erik Lykke to Steering Committee of the Long-range Transboundary Air Pollution Project. Cooperative Technical Programme to Measure the Long-range Transport of Air Pollutants: Steering Committee," January 24, 1974, RA-S-2532-2-Dca-L0403, Norwegian National Archives.

32. Brynjulf Ottar, " 'Monitoring møtet.' Overvåking av luftforurensninger i Europa," April 2, 1974, RA-S-2532-2-Dca-L0403, Norwegian National Archives. The request was specifically made to the UN ECE's Senior Advisors on Environmental Problems (SAEP). The SAEP was formed after the UN's 1972 Conference on the Human Environment and consisted of delegates from UN ECE member states. United Nations Economic Commission for Europe, "Measuring Long-Distance Air Pollution: UN ECE Launches Cooperative Interna-

tional Programme," January 18, 1977, BAC 58/1990 No. 2 Pages: 501 Année(s): 1975–1977, Page 0191, European Commission Archives.

33. "Letter from Amasa Bishop to Haakon Hjelde," March 18, 1974, GX 33/1/10, United Nations Archives; "Letter from Amasa Bishop to Dr. K.V. Amanichev, Director of Department State Committee of the USSR Council of Ministers for Science and Technology," May 5, 1974, GX 33/1/10, United Nations Archives.

34. "Letter from Amasa Bishop to Holmoy, Director Smoke Control Council, Oslo, Norway, and Hjelde, Head of Division, Ministry of Environment, Oslo, Norway," July 2, 1974, GX 33/1/10, United Nations Archives.

35. Richard G. Fort, "Sur nedbør problemet: Hva gjør myndighetene?," October 1975, RA-S-2532-2-Dca-L0003, Norwegian National Archives, p. 8. Brodsky was Chief of the Department of State Research Institute of Industry and Sanitary Gas Cleaning (Nauchno-lssledovatel'skiy Institut po Promyshlennoy i Sanitarnoy Ochistke Gazov, or NIIOGAZ) in Moscow. Founded in the 1920s, NIIOGAZ organized several studies into pollution control from power stations. See N. F. Izmerov, *Control of Air Pollution in the USSR* (Geneva: World Health Organization, 1973), pp. 17, 72, 83. Scientists from this Institute later cooperated with US researchers on other pollution problems. See L. I. Kropp, I. N. Shmigol, G. S. Chekanov, et al., "Joint US/USSR Test Program for Reducing Fly Ash Resistivity," *Journal of the Air Pollution Control Association* 29, no. 6 (1979): 665–669.

36. Brynjulf Ottar, "Utkast og økonomiovepslag for øst-vest-konferanse angaende monitoring i Europa," January 25, 1974, RA-S-2532-2-Dca-L0403, Norwegian National Archives; Richard G. Fort, "Internasjonale møter i Oslo i 1974. Et foreløpig kostnadsoverslag," January 29, 1974, RA-S-2532-2-Dca-L0403, Norwegian National Archives; Norwegian Institute for Air Research, "Cooperative Programme for Monitoring and Evaluation of Air Pollutants in Europe: Outline Plan," August 7, 1974, RA-S-2532-2-Dca-L0003, Norwegian National Archives.

37. Brundtland referenced this meeting in her remarks before the Council of Europe on European Environmental Policy in 1975. See Environment Committee, "Statement by Mrs. Gro Harlem Brundtland, Minister for the Environment in the Norwegian Government, at the 26th Session of the Parliamentary Assembly of the Council of Europe on 24th January, 1975," February 14, 1975, ENV(75) 17, OECD Archives, p. 7.

38. Part of the reason for the years that elapsed between the Norwegians' initial proposals in the summer and fall of 1973 and the official start of the work program in the UN Economic Commission for Europe is because the Soviets wanted to wait until the conclusion of the Helsinki talks before beginning formal preparations within the UN Economic Commission for Europe. This may be because they wanted to gain publicity and attention for their support of the UN monitoring program as a détente initiative. See H. C. Christensen, "Referat fra møte i komite for forurensningsspørsmål 1974 tirsdag 22. januar 1974," February 12, 1974, RA-S-1574-E-L0065, Norwegian National Archives.

39. A new environmental policy program had recently been approved by the Comecon. See Council for Mutual Economic Assistance, "Information

Concerning the Activity of the Council for Mutual Economic Assistance (CMEA) and Concerning Cooperation between the Member-Countries of CMEA on Questions Concerning the Prevention of Air Pollution," November 27, 1974, RA-S-2532-2-Dca-Loo03, Norwegian National Archives.

40. Ibid., pp. 4–6.

41. For literature on the history of weather and climate studies, see Paul N. Edwards, *A Vast Machine: Computer Models, Climate Data, and the Politics of Global Warming* (Cambridge, MA: MIT Press, 2010); Kristine C. Harper, *Weather by the Numbers: The Genesis of Modern Meteorology* (Cambridge, MA: MIT Press, 2008); Frederik Nebeker, *Calculating the Weather: Meteorology in the 20th Century* (San Diego: Academic Press, 1995); Robert Marc Friedman, *Appropriating the Weather: Vilhelm Bjerknes and Construction of a Modern Meteorology* (Ithaca, NY: Cornell University Press, 1993); Vladimir Jankovic, Deborah R. Coen, and James Rodger Fleming, eds., *Intimate Universality: Local and Global Themes in the History of Weather and Climate* (Sagamore Beach, MA: Science History Publications/USA, 2006); Mark Monmonier, *Air Apparent: How Meteorologists Learned to Map, Predict, and Dramatize Weather* (Chicago: University of Chicago Press, 1999); James Rodger Fleming, *Meteorology in America, 1800–1870* (Baltimore: Johns Hopkins University Press, 2000); James Rodger Fleming, *Historical Perspectives on Climate Change* (Oxford: Oxford University Press, 2005); James Rodger Fleming, *The Callendar Effect: The Life and Work of Guy Stewart Callendar (1898–1964), the Scientist Who Established the Carbon Dioxide Theory of Climate Change*, illustrated ed. (American Meteorological Society, 2009); James Rodger Fleming, *Fixing the Sky: The Checkered History of Weather and Climate Control* (New York: Columbia University Press, 2010).

42. Hans A. Panofsky, "Air Pollution Meteorology," *American Scientist* 57, no. 2 (July 1, 1969): 269–285.

43. Norwegian Institute for Air Research, "Preliminary Technical Consultation in Connection with the Preparation of a Cooperative Programme for Monitoring and Evaluation of the Transmission of Air Pollutants in Europe: Report of the Meeting of Rapporteurs and Experts in Geneva," October 13–15, 1975, RA-S-2532-2-Dca-Lo292, Norwegian National Archives, pp. 4–7.

44. Ibid., p. 12.

45. Norwegian Institute for Air Research, "Cooperative Programme for Monitoring and Evaluation of Air Pollutants in Europe: Outline Plan," p. 5.

46. Ibid., p. 16.

47. Brynjulf Ottar, "Preliminary Technical Consultation for the Preparation of a European Monitoring and Evaluation System for the Long Range Transmission of Air Pollutants," March 20, 1975, RA-S-2532-2-Dca-Loo03, Norwegian National Archives.

48. The UN Economic Commission for Europe formally approved the monitoring program as part of its work program in February 1975. See Erik Lykke, "Sur nedbør-problemene. Møte i Miljøverndepartementet," May 26, 1975, RA-S-2532-2-Dca-Lo404, Norwegian National Archives, p. 5.

49. Richard G. Fort, "UN ECE. Første møte i task force for opprettelse av et europeisk system for overvåking av luftforurensninger. Geneve, 24.–26. mai 1976," May 20, 1976, RA-S-2532-2-Dca-Lo292, Norwegian National Archives.

50. Britain in particular questioned whether the possible results of the project could justify the scientific and monetary expense given the information that countries would be willing and able to furnish. "Réunion Coordination des 9 E.M. [handwritten notes]," n.d. Likely late 1976 or early 1977, BAC 58/1990 No. 2 Pages: 501 Année(s): 1975–1977, Page 0500, European Commission Archives; United Nations Economic Commission for Europe, "Measuring Long-Distance Air Pollution: UN ECE Launches Cooperative International Programme."

51. Brynjulf Ottar, "Draft Letter for the Preliminary Technical Consultations in Connection with the Preparation of a Cooperative Programme for Monitoring and Evaluation of the Transmission of Air Pollutants in Europe," July 22, 1975, RA-S-2532-2-Dca-Lo292, Norwegian National Archives. Brynjulf Ottar, "European Monitoring System Questionnaire," July 22, 1975, RA-S-2532-2-Dca-Lo292, Norwegian National Archives.

52. "Letter from Brynjulf Ottar to Ekspedisjonssjef Erik Lykke," July 23, 1975, RA-S-2532-2-Dca-Lo292, Norwegian National Archives.

53. United Nations Economic Commission for Europe, "Measuring Long-Distance Air Pollution: UN ECE Launches Cooperative International Programme."

54. Author Translation from Norwegian. "Telegram: Miljøverndepartementet to norske delegasjon Geneve. UN ECE—Monitoring luftforurensninger," August 26, 1975, RA-S-2532-2-Dca-Lo292, Folder UN ECE/EMP 1976–1978, Norwegian National Archives.

55. Lykke did note that his government would be interested in comparing results from any parallel programs between the US and Canada. For the Lykke/Grant correspondence, see "Letter from Lindsey Grant, Environmental and Scientific Affairs, Department of State to Erik Lykke," August 19, 1976, RA-S-2532-2-Dca-Lo292, Norwegian National Archives; "Letter from Erik Lykke to Lindsey Grant, Environmental and Scientific Affairs, Department of State," September 14, 1976, RA-S-2532-2-Dca-Lo292, Norwegian National Archives.

56. Concerns about the impact of Cold War politics may also have played a part in Norway's hesitancy to involve the US. These were not lost on the Norwegian officials, who extensively debated how to deal with the fact that West Germany's Ministry of the Environment had been established in Berlin. The Soviet Union protested this throughout preparations for the 1979 Convention on Long-Range Transboundary Air Pollution. The role that the European Communities was allowed to play in the negotiations was also an object of contention with the Soviets.

57. Brynjulf Ottar, "Forberedelse og igangsettelse av det europeiske monitoring system for luftforurensninger," January 20, 1976, RA-S-2532-2-Dca-Lo292, Norwegian National Archive, p. 2.

58. Ibid. In many ways, these issues mirrored the financial problems that beset the early days of the OECD project, where NILU was forced to shoulder an enormous portion of the startup expenses as well. Ottar hoped that NILU's investment in the UN's monitoring program would pay off by further increasing the visibility of the institute internationally and attracting young, bright scientists to work there. See Brynjulf Ottar, "Notat: planlegging, forberedelse

og igangsetting av det europeiske monitoring program for luftforurensninger (EMP)," March 2, 1976, RA-S-2532-2-Dca-Lo292, Norwegian National Archives, p. 7.

59. "Letter from Brynjulf Ottar to Erik Lykke," July 23, 1975, RA-S-2532-2-Dca-Lo292, Norwegian National Archives.

60. Ottar, "Notat: Planlegging, forberedelse og igangsetting av det europeiske monitoring program for luftforurensninger (EMP)," p. 2.

61. T.A. Lie, "UN ECE/EMEP. Styringsgruppen. Møte i Geneve 31. august–2. september 1977. Norsk forberedende møte i Miljøverndepartementet 18. august 1977," August 26, 1977, RA-S-2532-2-Dca-Lo293, Norwegian National Archives.

62. Brynjulf Ottar, "Programme for Monitoring and Evaluation of the Long Range Transmission of Air Pollutants in Europe," January 22, 1976, RA-S-2532-2-Dca-Lo292, Norwegian National Archives; "Telegram from Brynjulf Ottar to G. Shvedov, State Committee of USSR, Council of Ministers for Science and Technology," August 21, 1975, RA-S-2532-2-Dca-Lo292, Norwegian National Archives.

63. "Letter from Brynjulf Ottar to Miljøverndepartementet. Attn: Byråsjef Hjelde. 'Europeisk Monitoringsystem,'" August 21, 1975, RA-S-2532-2-Dca-Lo292, Norwegian National Archives, p. 1.

64. At this January meeting, Eastern European countries had a more passive role than the Western European counterparts in the discussions. Director Tor Holmøy of the Norwegian Pollution Control Authority (Statens forurensningstilsyn) mused that it was possible the Eastern European countries adopted a more subdued attitude because they were disappointed that the Soviets had not taken on a more prominent role. See Tor Holmøy, "UN ECE—Working Party on Air Pollution Problems, 6. Sesjon 13.-15.1.76," January 19, 1976, RA-S-2532-2-Dca-Lo300, Norwegian National Archives. However, certain governments, such as Poland, were quick to commit meteorological research stations to the program to show their support and interest. If an exclamation point in the margin next to Poland's offer of two stations for the network is any indication, the Norwegians were quite excited by this. See "Letter from Janusz Pawlak, Director, Environment Protection Department, to Erik Lykke," October 21, 1976, RA-S-2532-2-Dca-Lo292, Norwegian National Archives.

65. Author interview with Anton Eliassen, director of the Norwegian Meteorological Institute, May 16, 2011. "Report of the First Meeting of the Task Force for the Development of the Programme for the Monitoring and Evaluation of the Long-range Transmission of Air Pollutants in the UN ECE Region," May 26, 1976, RA-S-2532-2-Dca-Lo292, Norwegian National Archives.

66. Nebeker, *Calculating the Weather*; Harper, *Weather by the Numbers*.

67. Anton spent several of his elementary school years in Princeton, New Jersey, while his father worked on the project. Eliassen joined NILU in 1972 after fulfilling his mandatory military service and receiving a degree in meteorology from the University of Oslo in 1970. Author interview with Øystein Hov, research director of the Norwegian Meteorological Institute, May 19, 2011, Oslo, Norway; Author interview of John Miller, NOAA, Stable Isotope Lab (Institute of Arctic and Alpine Research), October 15, 2010; Author interview with Anton

Eliassen, director of the Norwegian Meteorological Institute, May 16, 2011; Anton Eliassen and Jørgen Saltbones, "Decay and Transformation Rates of SO_2, as Estimated from Emission Data, Trajectories and Measured Air Concentrations," *Atmospheric Environment* 9, no. 4 (April 1975): 425–429.

68. Author interview with Anton Eliassen, Director of the Norwegian Meteorological Institute, May 16, 2011.

69. Nebeker, *Calculating the Weather*. Robert Marc Friedman, *Appropriating the Weather*.

70. Ottar planned to utilize the expertise of research institutes in the other Nordic countries to try out new and improved measurement methods before proposing their inclusion in the UN monitoring program to ensure that their failure did not reflect poorly on the program. Ottar, "Notat: Planlegging, forberedelse og igangsetting av det europeiske monitoring program for luftforurensninger (EMP)," pp. 4, 6.

71. Brynjulf Ottar, "Budsjettforslag EMP 1976 og 1977," February 11, 1976, RA-S-2532-2-Dca-L0292, Norwegian National Archives, p. 2.

72. Brynjulf Ottar, "Planlegging, forberedelse og igangsetting av det europeiske monitoring program for luftforurensninger i UN ECE Task Force," February 23, 1976, RA-S-2532-2-Dca-L0292, Norwegian National Archives. On use of the flux model in the first year of the network's operations, see "The UN ECE Co-operative Programme on Long Range Transmission of Air Pollutants in Europe (EMEP) Technical Meeting 23–25 May 1978, Bilthoven, the Netherlands," May 23, 1978, RA-S-2532-2-Dca-L0293, Norwegian National Archives.

73. For a special meeting between representatives of the US, Soviet Union, and Norway at the UN about this issue, see "Meeting in Oslo 23–24 November 1978 on Air Pollution Emission Data and Monitoring," 1978, RA-S-2532-2-Dca-L0294, Norwegian National Archives.

74. "Report of the First Meeting of the Task Force for the Development of the Programme for the Monitoring and Evaluation of the Long-range Transmission of Air Pollutants in the UN ECE Region," p. 6.

75. Ottar, "Budsjettforslag EMP 1976 og 1977," pp. 20–21.

76. "Report of the First Meeting of the Task Force for the Development of the Programme for the Monitoring and Evaluation of the Long-range Transmission of Air Pollutants in the UN ECE Region."

77. United Nations Economic Commission for Europe, "Measuring Long-Distance Air Pollution: UN ECE Launches Cooperative International Programme."

78. United Nations Economic Commission for Europe, "Co-operative Programme for Monitoring and Evaluation of the Long Range Transmission of Air Pollutants in Europe: Recommendations of the UN ECE Task Force," March 10, 1977, GX 33/1/14, United Nations Archives, pp. 1–7.

79. Erik Lykke, "Pollution Problems across International Boundaries," May 21, 1979, RA-S-2532-2-Dca-L0291, Norwegian National Archives.

80. Author interview with Anton Eliassen, director of the Norwegian Meteorological Institute, May 16, 2011. The modeling aspect of the program began later than the measurement phase. Moscow sent the first modeling results to Oslo beginning in August 1979. See "Co-operative Programme for Monitoring and Evaluation of the Long-range Transport of Air Pollutants in Europe

(EMEP). Summary Report of the Eastern Synthesizing Centre for the First Phase of EMEP," December 1980, RA-S-2532-2-Dca-L0293, Norwegian National Archives.

81. F. B. Smith, "An Assessment of the Cooperative Programme for Monitoring and Evaluation of the Long-range Transmission of Air Pollutants in Europe (EMEP) at the End of the First Stage," January 1981, RA-S-2532-2-Dca-L0293, Norwegian National Archives, pp. 15–16.

82. Ibid., pp. 7–8, 15. Fort and Lykke, "Referat fra SO_2-møtet i Miljøverndepartementet 6. juni 1973."

CHAPTER FIVE

1. Douglas M. Costle, "Remarks by the Honorable Douglas M. Costle, Administrator, U.S. Environmental Protection Agency, Head of the U.S. Delegation, Economic Commission for Europe High-Level Meeting," November 14, 1979, RA/PA-0636–11/D/Dc/L0755, Norwegian National Archives.

2. East European and Soviet Department, UK Foreign Commonwealth Office, "What Do the Russians Mean by, and Seek from, All European Economic Cooperation?," August 16, 1976, FV 61/53, UK National Archives.

3. For example, they coordinated closely with Ellis Cowling, a leading acid rain researcher at North Carolina State, and John Miller, an atmospheric scientist at the Mauna Loa Observatory, on their diplomatic approach to cooperation on acid rain. See "Letter from Ellis B. Cowling, Professor, Plant Pathology and Forest Resources, to Dr. Donald King, Deputy Director for Science, Office of Environmental Affairs, Department of State," October 13, 1976, RA-S-2532-2-Dca-L0292, FN (med særorganisasjoner) ECE—Task force—European Monitoring Programme EMP, Norwegian National Archives.

4. "Letter from Lindsey Grant, Environmental and Scientific Affairs, Department of State to Erik Lykke," August 19, 1976, RA-S-2532-2-Dca-L0292, Norwegian National Archives; "Letter from W. Evan Armstrong, Environment Canada, to Lindsey Grant, State Department," September 15, 1976, RA-S-2532-2-Dca-L0292, FN (med særorganisasjoner) ECE—Task force—European Monitoring Programme EMP, Norwegian National Archives.

5. The sulfur fuel oil directive was originally submitted by the Netherlands. Commission of the European Communities, "Proposal for a Council Directive Relating to the Use of Fuel-oils with the Aim of Decreasing Sulphurous Emissions," December 19, 1975, BAC 156/1990 No. 2417 Année(s) 1975–1976, European Commission Archives, pp. 1–2.

6. The governments who had the most stringent regulations in place at the time were the Netherlands, West Germany, and France. Ibid., p. 5.

7. The City of London did have regulations on the amount of sulfur in fuel oil burned within the city. It was the only area in Britain with such regulations. See ibid., p. 3.

8. Those present at these meetings recognized early on the importance of a cooperative effort between the Swedish and Norwegian governments in this regard. See Erik Lykke, "Sur nedbør-problemene. Møte i Miljøverndepartementet," May 26, 1975, RA-S-2532-2-Dca-L0404, Norwegian National Archives, p. 7.

9. The working group included both government officials and scientific experts. H. C. Christensen, "Norsk-svensk samrådsmøte om sur nedbør problemer. Oslo 6.–7. november 1975," January 7, 1976, RA-S-2532-2-Dca-L0441, Norwegian National Archives.

10. Ibid., p. 5

11. Richard G. Fort, "Møte i den svensk-norske arbeidsgruppen om so$_2$ forsurningsproblemene," February 10, 1976, RA-S-2532-2-Dca-L0441, Norwegian National Archives, p. 5.

12. Commission of the European Communities, "Proposal for a Council Directive Relating to the Use of Fuel-oils with the Aim of Decreasing Sulphurous Emissions," pp. 8–9.

13. In service of this policy, countries were required to have meteorological monitoring networks of air pollution around major power plants. Ibid., pp. 7–18.

14. Hawkins explicitly stated that he feared the European Commission would extend such regulations to coal in his letter. "Letter from Arthur Hawkins to Henri Simonet, Vice-President, Commission of the European Communities," February 26, 1976, BAC 156/1990 No. 2417 Année(s) 1975–1976, European Commission Archives.

15. Ibid., p. 2.

16. The CEGB had previously met with the European Commission to make the case for exempting power plants with high stacks from the low sulfur fuel requirements in April 1975. The wording of some of the articles had been "adjusted" after these meetings, but clearly not enough to completely alleviate the CEGB's concerns about cost. See G. Brondel, "Briefing for Mr. Williams," March 1, 1976, BAC 156/1990 No. 2417 Année(s) 1975–1976, European Commission Archives. For the CEGB's contact with the Department of Energy, see "Letter from D. B. Leason, Generation Studies Engineer, CEGB, to D. Morphet, Department of Energy, UK," January 27, 1976, BAC 156/1990 No. 2417 Année(s) 1975–1976, European Commission Archives. For Hawkins' letter to the European Commission, see "Letter from Arthur Hawkins to Henri Simonet, Vice-President, Commission of the European Communities."

17. "Letter from Arthur Hawkins to Henri Simonet, Vice-President, Commission of the European Communities," p. 3.

18. Brondel, "Briefing for Mr. Williams."

19. A. A. Halliwell, "Mission Report: Visit to C.E.G.B. on 5 and 6 April 1976 to Discuss Matters Arising from the Proposal for a Council Directive Relating to the Use of Fuel Oils with the Aim of Reducing Sulphurous Emissions (doc. COM(75) 681 Final—19.12.1915)," April 8, 1976, BAC 156/1990 No. 2417 Année(s) 1975–1976, European Commission Archives, pp. 1–3.

20. "Attachment 1: SO$_2$ Pollution Control. Letter from D. B. Leason, Generation Studies Engineer, CEGB, to D. Morphet, Department of Energy, UK," January 27, 1976, BAC 156/1990 No. 2417 Année(s) 1975–1976, European Commission Archives.

21. Ibid., p. 2. "Letter from H. Eliasmöller to L. Williams, Director General for Energy," April 13, 1976, BAC 156/1990 No. 2417 Année(s) 1975–1976, European Commission Archives.

22. Communautes Europeenes Le Conseil, "Note. Objet: Proposition de Directive du Conseil Concernant L'utilisation des Fuel-oils en Vue de la Réduction des Émissions de Soufre," May 7, 1976, BAC 156/1990 No. 2417 Année(s) 1975-1976, European Commission Archives, p. 9.

23. The directive was eventually withdrawn from the Council's consideration in 1981 when no agreement could be reached. J. D. Liefferink, *Environment and the Nation-State: The Netherlands, the European Union and Acid Rain* (Manchester: Manchester University Press, 1996), p. 81.

24. European Commission, "Editorial: Pan-European Cooperation and Transfers of Technology," March 26, 1976, FV 61/53, UK National Archives.

25. Halvard P. Johansen, "Norsk-Svensk Møte om SO_2-Problemer. Oslo, 10. Desember 1976," December 22, 1976, RA-S-2532-2-Dca-L0441, Norwegian National Archives.

26. Members of the Central Unit on Environmental Pollution represented Britain on Environmental Committees at the UN, OECD, NATO, and European Communities. It was formed in 1969 and became a part of the Department of the Environment when the latter was created the following year. "Letter from F. S. Feates to Dr. L. E. J. Roberts," October 15, 1974, AB 54/161, UK National Archives; "Letter from N. G. Stewart to Mrs. L. J. Robertson," October 29, 1974, AB 54/161, UK National Archives; "Pollution R & D Co-ordinating Bodies," 1974, AB 54/161, UK National Archives.

27. Harald Dovland and Arne Semb, "Discussion of the Final LRTAP Report, NILU, May 20-21, 1976," July 30, 1976, RA-S-2532-2-Dca-L0405, Norwegian National Archives.

28. Ibid. Author interview with Dick Derwent, retired professor of atmospheric chemistry, Imperial College, London, March 18, 2013. For Scriven's critiques of the OECD report to the British Alkali Inspectorate see Central Electricity Generating Board, "Agreed Notes of the Annual Informal Meeting with the Alkali and Clean Air Inspectorate," December 7, 1976, BT 328/105, UK National Archives.

29. "Letter from Brynjulf Ottar to Leslie Reed," August 2, 1976, RA-S-2532-2-Dca-L0405, Norwegian National Archives; Norwegian Institute for Air Research, "Comments to: 'Notes of a Meeting at NILU on May 20th-21st, 1976, to Discuss the Revision of the Draft OECD Report on Long Range Transport of Air Pollution,'" July 30, 1976, RA-S-2532-2-Dca-L0405, Norwegian National Archives; Leslie Reed, "Notes of a Meeting at NILU on May 20th-21st, 1976 to Discuss the Revision of the Draft OECD Report on Long Range Transport of Air Pollution," July 5, 1976, RA-S-2532-2-Dca-L0405, Norwegian National Archives.

30. Central Electricity Research Laboratories, "Item: British Research on Long Range Pollution," September 23, 1976, RA-S-2532-2-Dca-L0405, Norwegian National Archives.

31. "Letter from Brynjulf Ottar to C. Brosset," February 10, 1977, RA-S-2532-2-Dca-L0292, Norwegian National Archives. There was also considerable pressure on Norway to reach agreement on the report because Britain, France, and the Netherlands had sought to delay the start of the UN's European-wide monitoring program until the OECD report was finished. See Richard G. Fort,

"UN ECE. 'Working Party on Air Pollution Problems' Referat Fra 7. Sesjon I Geneve, 11–14 Januar 1977," January 24, 1977, RA-S-2532-2-Dca-L0300, Norwegian National Archives, p. 3.

32. The Norwegian scientists were amused by their attempts to brave the Norwegian snow in leather shoes. Author interview with Anton Eliassen, director of the Norwegian Meteorological Institute, May 16, 2011, Oslo, Norway; Author interview with Ronald Barnes, retired, former Principal Environmental Adviser, Esso, March 4, 2013. "Telegram from Foreign and Commonwealth Office London to Miljøverndepartementet," May 20, 1976, RA-S-2532-2-Dca-L0292, Norwegian National Archives.

33. Author interview with Anton Eliassen, director of the Norwegian Meteorological Institute, May 16, 2011; author interview with Ronald Barnes, retired, former principal environmental adviser, Esso, March 4, 2013.

34. Author interview with Bernard Fisher, principal scientist, United Kingdom Environment Agency, March 21, 2013.

35. Bernard Fisher, "Comments and Discussion of Chapters on Sector Analysis, Lagrangian Model and Trajectory Model in OECD Report (1977)," January 20, 1977, CEGB Internal Report, copy given to the author by Bernard Fisher; author interview with Bernard Fisher, principal scientist, United Kingdom Environment Agency, March 21, 2013.

36. The undecided portion was subsequently given as one-third of the sulfur deposition entering Norway. Organisation for Economic Co-operation and Development, *The OECD Programme on Long Range Transport of Air Pollutants: Measurements and Findings* (Paris: OECD, 1977), NILU Archives and Library, pp. 9-17–9-19.

37. The rapport they developed was so strong that Fisher made a point of looking after Eliassen when he subsequently travelled to the CERL to give a presentation on atmospheric modeling. Upon his arrival, Eliassen was taken out to lunch by some of his hosts at the lab, who tried to get him to loosen up and imbibe single malt whiskeys to the point of inebriation. Fisher, however, kicked him under the table to warn him not to have anything to drink. Later that afternoon, Eliassen was relieved to be sober during his lecture to Peter Chester and other CERL scientists, who proceeded to tear his presentation to pieces. Author interview with Anton Eliassen, director of the Norwegian Meteorological Institute, May 16, 2011.

38. Author interview with Anton Eliassen, director of the Norwegian Meteorological Institute, May 16, 2011; author interview with Ronald Barnes, retired, former principal environmental adviser, Esso, March 4, 2013.

39. Author interview with Ronald Barnes, retired, former principal environmental adviser, Esso, March 4, 2013, Wantage, England.

40. On the use of science to ease diplomatic tensions after the Second World War, see Clark A. Miller, " 'An Effective Instrument of Peace': Scientific Cooperation as an Instrument of U.S. Foreign Policy, 1938–1950," *Osiris* 21, no. 1 (January 1, 2006): 133–160; Barth, "Catalysts of Change."

41. Erik Lykke appears to have been one of attendees who accused Anton of being a traitor. Author interview with Anton Eliassen, director of the Norwegian Meteorological Institute, May 16, 2011; author interview with Ronald Barnes,

retired, former principal environmental adviser, Esso, March 4, 2013, Wantage, England.

42. Author interview with Anton Eliassen, director of the Norwegian Meteorological Institute, May 16, 2011; author interview with Bernard Fisher, principal scientist, United Kingdom Environment Agency, March 21, 2013.

43. Prior to its completion, the British, French, and West German governments had not officially admitted that long-range transport occurred. The report was approved at the OECD's final meeting of the project's steering committee the following month. For the meeting summary, see Environment Committee, "Cooperative Technical Programme to Measure the Long Range Transport of Air Pollutants," April 22, 1977, OECD Archives, Paris, ENV(77)23.

44. "Letter from Jan Wessel Hegg, Ambassadesekretær, to the Utenriksdepartement and Miljøverndepartement," December 15, 1976, RA-S-2532-2-Daa-Lo506, Norwegian National Archives. An article in the *Guardian* on December 9, 1976 that contained the CEGB's first public statements on acid rain in Norway seems to have spurred the Norwegians to get access to the internal report. See above citation; for the article in question, see Mark Arnold-Forster, "The Norwegian Cry of Stinking Fish," *Guardian*, December 9, 1976.

45. The report began by reviewing the Swedish initiatives on acid rain at the UN in the late 1960s. The CEGB believed Sweden had "backed down" on the issue after realizing that their own oil consumption would make them susceptible to accusations of polluting their own atmosphere. "Notes on the Effect of European Pollutant Emission on Norway," December 7, 1976, RA-S-2532-2-Daa-Lo506, Luftforurensning Forskningsprosjektet "Sur nedbørs virkning på skog og fisk" SO_2-problematikken, Norwegian National Archives, p. 1.

46. The authors were identified at the bottom of the report by their initials: DHL (D. H. Lucas), AJC (A. J. Clarke), and GWB (G. W. Barrett). D. H. Lucas and A. J. Clarke were also quoted by the *Guardian* on the issue. See Arnold-Forster, "The Norwegian Cry of Stinking Fish."

47. Ibid., p 5.

48. "Appendix: Technical Aspects of the Long Range Drift of Sulphur Dioxide," December 7, 1976, RA-S-2532-2-Daa-Lo506, Luftforurensning Forskningsprosjektet "Sur nedbørs virkning på skog og fisk" SO_2-problematikken, Norwegian National Archives. Acid rain is not the only environmental or public health issue to face this type of research manipulation for economic or political interests. For works by other historians that explore this theme on topics ranging from tobacco smoke to lead paint to global warming, see Gerald Markowitz and David Rosner, *Lead Wars: The Politics of Science and the Fate of America's Children* (Berkeley: University of California Press, 2013); Thomas O. McGarity and Wendy Wagner, *Bending Science: How Special Interests Corrupt Public Health Research* (Cambridge, MA: Harvard University Press, 2010); Naomi Oreskes and Erik M. Conway, *Merchants of Doubt: How a Handful of Scientists Obscured the Truth on Issues from Tobacco Smoke to Global Warming* (New York: Bloomsbury Press, 2010); David Michaels, *Doubt Is Their Product: How Industry's Assault on Science Threatens Your Health* (Oxford University Press, 2008).

49. "Appendix: Technical Aspects of the Long Range Drift of Sulphur Dioxide," pp. 2-3.

50. "Britain Gets the Blame for 'Acid Rain,'" *Observer*, July 10, 1977; Economic Commission for Europe, "Rapport de La Cinquième Session (21–25 Février 1977)," March 11, 1977, BAC 58/1990 No. 3 Pages: 382 Année(s): 1977–1978, Page 0174, European Commission Archives, p. 4; Gunnar H. Lindeman, "Note Mellom Miljøvernminister Gro Harlem Brundtland og Secretary of State for Environment Peter Shore, London 26. April 1978," April 28, 1978, RA-S-2532-2-Dca-L0004, Norwegian National Archives.

51. Oddmund Graham, "Telex: fn's oekonomiske kommisjon for europa (ece) og oppfoelgingen av slutt akten fra sikkerhetskonferansen (ksse). oversikt," March 2, 1976, RA-S-1713-D-Dc-L0124, Norwegian National Archives; David Hildyard, "Economic Commission for Europe: 31st Session, Geneva, 30 March—9 April, 1976," April 13, 1976, FV 61/53, UK National Archives; East European and Soviet Department, UK Foreign Commonwealth Office, "What Do the Russians Mean by, and Seek from, All European Economic Cooperation?" For Norway's preparations of the draft at the UN ECE, see Erik Lykke, "OECD—Undersøkelsen Av Langtransport Av Forurensninger," May 6, 1977, RA-S-2532-2-Dca-L0405, Norwegian National Archives.

52. In October 1977 a working group was established under the Nordic Council of Ministers to prepare a possible draft of the Convention text, which was completed and sent to all UN ECE countries in June 1978. The draft was officially submitted to the UN Economic Commission for Europe in July 1978. See Norwegian Institute for Air Research, "UN ECE. Special Group on Long Range Transboundary Air Pollution. Første møte i Geneve 3.-4. juli 1978," July 10, 1978, RA-S-2532-2-Dca-L0294, Norwegian National Archives; Richard G. Fort and Erik Lykke, "Om samtykke til ratifikasjon av en internasjonal konvensjon av 13. november 1979 om langtransportert grenseoverskridende luftforurensning," August 15, 1980, RA-S-2532-2-Dca-L0295, Norwegian National Archives, p. 3.

53. "Statement by the Swedish Delegation," November 3, 1978, RA-S-2532-2-Dca-L0294, Norwegian National Archives.

54. Ibid., p. 5. Also see Richard G. Fort, "Draft: Notes on the Nordic 'Memorandum on Major Elements to Be Considered for Inclusion in an Annex on Emissions of Sulphur Compounds,'" September 9, 1978, RA-S-2532-2-Dca-L0291, Norwegian National Archives.

55. "Nordic Statement on the Draft Convention for the UN ECE Region on Reduction of Emissions Causing Transboundary Air Pollution," October 2, 1978, RA-S-2532-2-Dca-L0294, Norwegian National Archives, pp. 4–5.

56. Armin Rosenkranz, "The Stockholm Conference," *Acid News: The Swedish NGO Seretariat on Acid Rain* 1/83 (1983), p. 7.

57. "Letter from Leslie Reed to Mr. Thompson," October 5, 1977, FCO 76/1553, UK National Archives.

58. Maritime Aviation and Environment Department, "Transboundary Air Pollution: International Aspects," December 15, 1977, FCO 76/1555, UK National Archives.

59. "Letter from LE Reed to Dr. I Cromartie," August 12, 1977, FCO 76/1553, UK National Archives.

60. "Letter from AG Lyall to RJ Chase," October 10, 1977, FCO 76/1553, UK National Archives; "Letter from Leslie Reed to Mr. Thompson."

61. There had been little progress in the negotiations among member governments, with member states each objecting to different aspects of the directive. Generally, the Netherlands, West Germany, and Denmark were in support of the directive, though not in every aspect; for example, West Germany was similarly concerned about the potential for the directive to apply to coal as well as oil given its reserves of the former. Britain was most opposed, with France, Italy, Belgium, and Ireland having some objections as well. See Central Unit on Environmental Pollution, "Interdepartmental Working Group on Community Environment Policies: Report of an Environment Working Group Meeting 16 September 1977. Directive on Sulphur Content of Fuel Oil," September 1977, FCO 76/1550, UK National Archives. These governments' positions on lowering sulfur dioxide emissions were largely consistent with their views on regulations for several other environmental issues debated by European Communities members. See Stanley Johnson and Guy Corcelle, *The Environmental Policy of the European Communities* (Kluwer Law International, 1995).

62. "Letter from RJ Chase to A Lyall," September 28, 1977, FCO 76/1553, UK National Archives, pp. 1–2.

63. Maritime Aviation and Environment Department, "Transboundary Air Pollution: International Aspects," pp. 3–4.

64. Leslie Reed, "Comparison of French, German and UK Positions on Sulphur in the Atmosphere," October 5, 1977, FCO 76/1553, UK National Archives; "Letter from RJ Chase, Maritime Aviation & Environment Dept to Dr. RIT Cromartie, Scientific Counsellor, British Embassy," December 5, 1977, FCO 76/1554, UK National Archives.

65. "Letter from LE Reed to Dr. I Cromartie."

66. "Letter from LE Reed, CUEP to Mr. Thomson," December 14, 1977, FCO 76/1555, UK National Archives.

67. "Telex from Owen to Mr. Sutherland, Mr. Edmonds, Mr. Butler, MAED, WED, ESSO, Dr. L Reed, CUEP, DOE," December 6, 1977, FCO 76/1554, UK National Archives.

68. L. E. Reed, "Emissions of Sulphur Dioxide," December 14, 1977, FCO 76/1555, UK National Archives.

69. "Letter from LE Reed to Dr. I Cromartie."

70. Geoffrey Findlay, "Anglo-French Bilateral Meeting on Long Distance Transfrontier SO_2 Pollution," November 21, 1977, FCO 76/1554, UK National Archives, p. 2.

71. One interesting suggestion to appease the Norwegians that arose in these discussions with the French was to pay for NILU's operating costs rather than agree to sulfur reductions. See "Letter from GWD Findlay to Leslie Reed," September 23, 1977, FCO 76/1550, UK National Archives.

72. "Letter from RIT Cromartie to LE Reed," September 14, 1977, FCO 76/1553, UK National Archives, p. 2.

73. Despite considerable political support for using environmental issues to increase ties across the iron curtain, some officials in the Ministry of the Interior and the Foreign Affairs Ministry had raised concerns about Eastern European countries being able to judge West Germany's environmental policies. Tor C. Hildan, "Notat for Statsråd Harlem Brundtland: Samtale med Willy Brandt i

Bonn 9. november 1978," November 13, 1978, RA-S-2532-2-Dca-L0295, Norwegian National Archives; Tor C. Hildan, "Referat fra samtale mellom Miljøvernminister Gro Harlem Brundtland og Innenriksminister Rudolf Baum i Bundesministerium des Innern, Bonn, 9. november 1978," November 13, 1978.

74. For a detailed discussion of West Germany's Ostpolitik, see Werner D. Lippert, *The Economic Diplomacy of Ostpolitik: Origins of NATO's Energy Dilemma* (New York: Berghahn Books, 2010).

75. "Letter from RIT Cromartie to LE Reed," p. 4.

76. Ibid.

77. "Letter from Robert Chase, Maritime Aviation and Environment Department to Geoffrey Findlay, British Embassy, Paris," November 30, 1977, FCO 76/1554, UK National Archives; Maritime Aviation and Environment Department, "Transboundary Air Pollution: International Aspects," p. 4.

78. "Letter from AG Lyall to RJ Chase"; Reed, "Emissions of Sulphur Dioxide." This "secret" alliance ended up being not so secret, as the three governments continually aligned themselves during European Communities discussions of the Convention. See Council of the European Communities, "Environment: Council Expected to Favour Air-Pollution Convention Signature," September 18, 1979. BAC 495/2005 No. 450 Pages: 751 Année(s): 1980, Page 0149, European Commission Archives.

79. "Letter from LE Reed, CUEP to Mr. Thomson."

80. "Telex All European Environment Conference: Tactics for UN ECE Plenary," March 20, 1978, FV 61/54, UK National Archives, p. 1.

81. J. Smeets, "Note à L'attention de Monsieur M. Carpentier. Objet: Projet de Convention UN ECE-Genève Concernant la Réduction des Émissions Pouvant Avoir pour Conséquences une Pollution Transfrontière," May 17, 1978, BAC 58/1990 No. 4 Pages: 250 Année(s): 1977–1979, Page 0139, European Commission Archives, p. 1.

82. Communautes Europeenes Le Conseil, "Note. Objet: Concertation au Sujet des Activités Internationales dans le Domaine de l'Environnement (UN ECE-Genève)," September 18, 1978, BAC 48/1984 No. 109 Pages: 401 Année(s): 1978–1980, Page 0014–0019, European Commission Archives, pp. 3–8.

83. Smeets, "Note à L'attention de Monsieur M. Carpentier. Objet: Projet de Convention UN ECE-Genève Concernant la Réduction des Émissions Pouvant Avoir pour Conséquences une Pollution Transfrontière," pp. 3–4; Commission des Communautes Europeenes, Service de l'environnement et de la protection des consommateurs, "Document de Travail des Services de la Commission sur la Définition de la Position de la Communauté lors de la 33ème Session de la Commission Economique pour l'Europe au Sujet de la Tenue Éventuelle d'une Réunion à Haut Niveau sur la Protection de l'Environnement," March 7, 1978, BAC 58/1990 No. 4 Pages: 250 Année(s): 1977–1979, Page 0183–0186, European Commission Archives, pp. 1–4.

84. Author translation of quote. "Telex: UN ECE—Miljoevern—Forberedelser Til Hoeynivaamoete," July 6, 1978, RA-S-2532-2-Dca-L0294, Norwegian National Archives, p. 6. The Soviet Union and East Germany were early supporters of the Nordic proposal. See Erik Lykke, "Forhandlingene om en luftforurensningsavtale for UN ECE-regionen. Momenter for besøk i Wien og Bonn

uke 45," November 3, 1978, RA-S-2532-2-Dca-L0005, Norwegian National Archives.

85. The UN Economic Commission for Europe resolution does state that it was "hoped" the meeting could be held in 1979. See Economic Commission for Europe, "Draft Resolution on the Work and Future Activities of the Commission and the Proposal in Regard to the Holding of All-European Congresses or Interstate Conferences on Co-operation in the Field of Protection of the Environment, Development of Transport, and Energy," April 22, 1978, BAC 58/1990 No. 4 Pages: 250 Année(s): 1977–1979, Page 0161–0165, European Commission Archives, pp. 2–6.

86. Economic Commission for Europe, "Senior Advisers to UN ECE Governments on Environmental Problems. Special Group on Long-range Transboundary Air Pollution. Geneva, 3–4 July 1978," July 6, 1978, RA-S-2532-2-Dca-L0294, Norwegian National Archives.

87. "Telex from Delegasjonen i Geneve to Miljoeverndepartementet," July 5, 1978, RA-S-2532-2-Dca-L0294, Norwegian National Archives.

88. Ibid. Also see Economic Commission for Europe, "Senior Advisers to UN ECE Governments on Environmental Problems. Special Group on Long-range Transboundary Air Pollution. Geneva, 3–4 July 1978."

89. It is unclear who suggested Barnes should be selected as the consultant, but it was likely Leslie Reed, who chaired the committee and, as his former boss, knew Barnes had recently departed the Central Unit on Environmental Pollution. See "Telex from Delegasjonen i Geneve to Miljoeverndepartementet."

90. Janez Stanovnik, "Annex I. Invitation to UN ECE Member Governments to Provide Information Related to Long-range Transboundary Air Pollution," July 17, 1978, RA-S-2532-2-Dca-L0294, Norwegian National Archives.

91. "Letter from RJ Chase to LE Reed," December 7, 1977, FCO 76/1555, UK National Archives.

92. The level of accuracy in emissions data and the need to simplify trajectory transport by examining only the mixing layer were particularly criticized, as was the "large degree of uncertainty" (±50%) in the estimated contributions from individual source areas. Central Unit on Environmental Pollution, "Doc A. Long Range Transport of Sulphur Compounds: A Paper by the United Kingdom," August 24, 1978, RA-S-2532-2-Dca-L0294, Norwegian National Archives, pp. 7, 17.

93. This became known later as the principle of nonlinearity. Ibid., p. 13.

94. The report did mention that land use practices and agriculture might have contributed to increasing acidity, a statement which likely has its roots in consultations between Ivan Rosenqvist and the CEGB during 1976. Ibid., pp. 1, 15–16.

95. Ibid. The report went on to detail the economic costs of reducing sulfur dioxide emissions, especially from flue gas desulfurization. Norway consulted Swedish experts on these cost estimates since they had been researching the issue for several years, and they advised the Ministry of the Environment that they believed the British estimates were greatly overstating the economic impact. However, Norway chose to focus on the British critiques about the scientific research on atmospheric transport and environmental impacts in their rebuttal to the

British. For the Swedish response regarding the economic costs, see Bo Assarssen, "Kommenterer till UK's Rapport till UN ECE: Long Range Transport of Sulphur Compounds," September 21, 1978, RA-S-2532-2-Dca-L0294, Norwegian National Archives.

96. Ibid., pp. 2–3.

97. Ibid., p. 4.

98. Richard G. Fort, "UN ECE. Spesialgruppen for Luftforurensninger. Referat Fra Byråmøte i Geneve 11.-13. September 1978," September 14, 1978, RA-S-2532-2-Dca-L0294, Norwegian National Archives, pp. 3–4.

99. Delegation of Norway, "UN ECE. Special Group on Long-Range Transboundary Air Pollution. Norwegian Comments to 'Doc. A', Sent Out by the UK to All Special Group Participants 25 August 1978," October 1978, RA-S-25 32-2-Dca-L0294, Norwegian National Archives.

100. Ibid., pp. 2–4.

101. Ibid., p. 1.

102. Ibid.

103. Fort, "UN ECE. Spesialgruppen for Luftforurensninger. Referat Fra Byråmøte i Geneve 11.-13. September 1978," pp. 3–4.

104. It is not clear who decided Barnes should be accompanied by Brown, but it was likely someone at the CEGB, Central Unit on Environmental Pollution, or Department of Energy. Brown had not yet published any scientific research on acid rain. It's also unknown who told the Norwegians he would be accompanying Barnes. There are records that show the SNSF ecologists were informed about a week ahead of time of his attendance. See Budsjettkontoret, "UN ECE—Spesialgruppen for luftforurensninger besøk av Dr. Barnes," October 9, 1978, RA-S-2532-2-Dca-L0294, Norwegian National Archives; Arne Tollan, "Møte i forbindelse med UN ECE," October 9, 1978, RA-S-2532-2-Dca-L0294, Norwegian National Archives.

105. Arne Tollan, "Meeting with Dr. Ron Barnes and Dr. David Brown, Oslo, 17. Oct. 1978. Summary of Presentations," October 24, 1978, RA-S-2532-2-Dca-L0294, Norwegian National Archives; "Letter from Arne Tollan to H. Dovland, A. Semb, A. Henriksen, I. P. Muniz, G. Abrahamsen, A. Stuanes, H. M. Seip, L. N. Overrein," October 9, 1978, RA-S-2532-2-Dca-L0291, Norwegian National Archives; Richard G. Fort, "UN ECE—Referat fra 2. møte i spesialgruppen for luftforurensninger," October 9, 1978, RA-S-2532-2-Dca-L0294, Norwegian National Archives.

106. European Communities, "Proposition Présentée au Nom de La Communauté Éonomique Européenne. Lignes Directices Qui Pourraient Résulter d'une Réunion à Haut Niveau," November 2, 1978, RA-S-2532-2-Dca-L0294, Norwegian National Archives.

107. The European Communities also touted its proposals to reduce sulfur from fuel oils and referenced these efforts as proof of their commitment to reducing sulfur dioxide pollution. Commission des Communautes Europeenes, "Proposition pour une Position Communautaire à l'Égard du Projet de Convention des Pays Nordiques sur la Réduction de la Pollution Atmosphérique Transfrontière," September 15, 1978, BAC 58/1990 No. 4 Pages: 250 Année(s): 1977–1979, Page 0018, European Commission Archives, p. 6; Commission des

Communautes Europeenes, Service de l'environnement et de la protection des consommateurs, "33ème Session de la Commission Economique pour l'Europe: Intervention au Nom de la Communauté Economique Européenne sur le Programme de Travail des Conseillers Principaux des Gouvernements de la CEE sur les Problèmes d'Environnement," April 15, 1978, BAC 58/1990 No. 4 Pages: 250 Année(s): 1977–1979, Page 0168–0172, European Commission Archives, pp. 1–5; "Statement by the Federal Republic of Germany on Behalf of the European Economic Community," November 2, 1978, RA-S-2532-2-Dca-L0294, Norwegian National Archives, Economic Commission for Europe, "Third Meeting of the Special Group on Long-range Transboundary Air Pollution. Geneva, 2–3 November. Draft Report," November 2, 1978, RA-S-2532-2-Dca-L0294, Norwegian National Archives.

108. Emphasis in original. Oddmund Graham, "UN ECE Special Group on Long Range Transboundary Air Pollution, 3rd Session, Geneva, 2–3 November: Statement by the Norwegian Delegation," November 3, 1978, RA-S-2532-2-Dca-L0294, Norwegian National Archives, p. 1.

109. Ibid.

110. Ibid., p. 4.

111. This point is even more evident in a published, shortened version of the original report, which Barnes completed the following year. See Ronald Barnes, "The Long Range Transport of Air Pollution: A Review of the European Experience," *Journal of the Air Pollution Control Association* 29, no. 12 (December 1979): 1223–1227. Historian Nils Roll-Hansen has noted that Rosenqvist was the lone advocate for his argument that land use changes were the major cause of acidification of surface water ecosystems, rather than acid rain from transboundary pollution. See Nils Roll-Hansen, "Ideological Obstacles to Scientific Advice in Politics? The Case of 'Forest Death' from 'Acid Rain,'" Makt- og demokratiutredningens rapportserie, Rapport 48, November 2002, p. 7. As one example of the misleading claims Barnes made about other Norwegian ecological research, he stated that H. M. Seip and A. Tollan, two Norwegian scientists who worked on SNSF projects, agreed with Rosenqvist that land use changes in forestry and agriculture could have increased what was known as "weak acid production" in soils, accounting for the changes in pH to rivers and lakes during water runoff. Seip and Tollen unequivocally disagreed with Rosenqvist on this point. See H. M. Seip and A. Tollan, "Acid Precipitation and Other Possible Sources for Acidification of Rivers and Lakes," *Science of the Total Environment* 10, no. 3 (November 1978): 253–270; I. Th. Rosenqvist, "Acid Precipitation and Other Possible Sources for Acidification of Rivers and Lakes," *Science of the Total Environment* 10, no. 3 (November 1978): 271–272.

112. R. A. Barnes, "The Long Range Transboundary Transport of Air Pollution: Summary, Discussion and Conclusions of the Synthesis Report Prepared for the Senior Advisers to UN ECE Governments on Environmental Problems," November 26, 1978, RA-S-2532-2-Dca-L0294, Norwegian National Archives; R. A. Barnes, "The Long Range Transboundary Transport of Air Pollution: Summary, Discussion and Conclusions of the Synthesis Report Prepared for the Senior Advisers to UN ECE Governments on Environmental Problems," n.d. BAC 58-1990 No. 4 Pages 250 Années 1977–1979, Page 0030–0039, European

Commission Archives. The European Communities archives include only pages 1–2 and 93–99 of the report. A full copy was also obtained by the author from Ronald Barnes. Author interview with Ronald Barnes, retired, former principal environmental adviser, Esso, March 4, 2013, Wantage, England.

113. P. Stief-Tauch, "Participation à la 4ème Réunion du Groupe Special de l'UN ECE/Genève 'Pollution Atmosphérique Transfrontière à Longue Distance' (Genève, 28 Novembre au 1er Décembre 1978)," December 7, 1978, BAC 58/1990 No. 4 Pages: 250 Année(s): 1977–1979, Page 0022, European Commission Archives, p. 2; "Some Preliminary Remarks by the Norwegian and Swedish Delegation on the Report Submitted by Dr. Barnes on the Long Range (Transboundary) Transport of Air Pollution," November 30, 1978, RA-S-2532-2-Dca-L0294, Norwegian National Archives. Handwritten on the top of this document is the notation "Ikke Distribuert" (not distributed) so this document was likely never circulated.

114. Barnes denies that anyone at the CEGB influenced the report, but it is perplexing why David Brown, an ecologist at the CERL, attended Barnes' meeting in the Scandinavian countries if not to assist him with the report. Barnes does reference personal communications with CERL scientists in the report, such as J. Bettelheim and D. H. Lucas. Author interview with Ronald Barnes, retired, former principal environmental adviser, Esso, March 4, 2013, Wantage, England.

115. "UN ECE. Special Group on Transboundary Air Pollution. Meeting in Geneva 28.11–2.12.1978. Statement by Delegation of Norway," November 29, 1978, RA-S-2532-2-Dca-L0294, Norwegian National Archives; Richard G. Fort, "Telex: for eksp.sjef Lykke. 1. nordisk moete 27.11," November 28, 1978, RA-S-2532-2-Dca-L0294, Norwegian National Archives.

116. Richard G. Fort, "UN ECE. Spesialgruppen for luftforurensninger møte i Geneve 15.—19. januar 1979. Referat fra forberedende nordisk møte i København 12. januar 1979," January 15, 1979, RA-S-2532-2-Dca-L0291, Norwegian National Archives.

117. Richard G. Fort, "Besøk av M. Michel Leclerc til Oslo 12. februar 1979," February 15, 1979, RA-S-2532-2-Dca-L0291, Norwegian National Archives. Fort and Lykke, "Om samtykke til ratifikasjon av en internasjonal konvensjon av 13. november 1979 om langtransportert grenseoverskridende luftforurensning."

118. Telegram from US State Department, attached to Memo "Betr.: UN ECE Vorbereitung einer hochrangigen Umwelttagung (HRT). Hier: Weitraumige grenzüberschreitende Luftverschmutzung (WGL)," January 23, 1979, German National Archives, Koblenz, B 106 65850, Band 9.

119. Ibid., pp. 4–5.

120. Ibid.

121. Richard G. Fort, "Notat: UN ECE. Forberedelser til høynivåmøte om miljøvern. Referat fra nordisk samrådsmøte i Stockholm, 16. februar 1979," February 23, 1979, RA-S-2532-2-Dca-L0291, Norwegian National Archives.

122. West Germany had some initial reservations about the wording that were subsequently resolved. Ibid.

123. In lieu of including any recommendations for reducing sulfur in the Convention itself, the parties consented to have the UN Economic Commission

for Europe's Senior Advisors on Environmental Problems use a revised version of the Annex as guidelines for implementing articles 3 and 4 of the Convention. The senior advisers requested information on actual and planned measures for sulfur emissions control to be submitted by each signatory to the Convention, as well as information on total national sulfur emissions for the most recent year available and projected estimates for the years from 1985 through 1990. "Letter from Erik Lykke to Janez Stanovnik," June 13, 1980, RA-S-2532-2-Dca-L0295, Norwegian National Archives; "Annex: Guidelines for the Completion of a Document Setting Out the Strategies and Policies for Abatement of Air Pollution Caused by Sulphur Compounds," 1980, RA-S-2532-2-Dca-L0295, Norwegian National Archives. The UK continued to insist that "proof" of harm would be required before it would agree to reduce emissions. See "2. Coordination SAEP," February 19, 1979. BAC 58/1990 No. 8 Pages: 331 Année(s): 1979, Page 0181, European Commission Archives, p.1; Fort and Lykke, "Om samtykke til ratifikasjon av en internasjonal konvensjon av 13. november 1979 om langtransportert grenseoverskridende luftforurensning," p. 4.

124. Fort and Lykke, "Om samtykke til ratifikasjon av en internasjonal konvensjon av 13. november 1979 om langtransportert grenseoverskridende luftforurensning," p. 3. A subsequent meeting to review the signatories' strategies and policies for air pollution abatement was scheduled for October 1980. UN ECE Secretariat, "Matters Arising from the Thirty-fifth Session of the Economic Commission for Europe and the Eighth Session of the Senior Advisers to UN ECE Governments on Environmental Problems," June 25, 1980, RA-S-2532-2-Dca-L0295, Norwegian National Archives, p. 4.

125. Marcus Fox, "UN ECE High Level Meeting," November 13, 1979, RA-PA-0636-11-D-Dc-L0755, Norwegian National Archives, p. 3.

126. Ibid.

127. Ibid., p. 4.

128. G. R. Baum, "Federal Republic of Germany," November 13, 1979, RA-PA-0636–11-D-Dc-L0755, Norwegian National Archives.

129. Rolf Hansen, "UN ECE High Level Meeting on Environment Geneva, 13–16 November 1979 Statement by Rolf Hansen, Norwegian Minister of Environment," November 13, 1979, RA-PA-0636–11-D-Dc-L0755, Norwegian National Archives, pp. 2–3.

130. E. Fouéré, "File Note—Subject: Convention on Long-range Transboundary Air Pollution—First Meeting of the Interim Executive Body (Geneva 27–31/10/1980)," November 14, 1980, BAC 495/2005 No. 450 Pages: 751 Année(s): 1980, Page 0003–0004, European Commission Archives, pp. 1–2.

131. Ibid., page 0006 of the file, page 5 of the document.

CHAPTER SIX

1. "Letter from Don A. Rolt, British Embassy to Leslie Reed, Chief Alkali Inspector," January 15, 1982, AT 82/203, UK National Archives.

2. H. M. Seip, "Comments to Letter and Paper by P. F. Chester," December 16, 1981, AT 82/203, UK National Archives, pp. 2–3.

3. Central Electricity Generating Board, "Agreed Notes of the Annual In-

formal Meeting with HM Alkali and Clean Air Inspectorate," January 8, 1980, BT 328/105, UK National Archives.

4. Ibid.

5. Hans C. Christensen, "Notat (justert 27.1.77) til møte 15/76 i styringsutvalget for SNSF-prosjektet. Vannkvalitet, tanker om målsetting for fase II," January 27, 1977, RA-S-2532-2-Dca-Loo15, Norwegian National Archives.

6. Ibid., p. 2. The second phase of the project also expanded the geographic scope of the research to Northern Norway when it was discovered that additional lakes in that region were also increasing in acidity. George Perkins Marsh, *Man and Nature* (New York: Charles Scribner, 1864).

7. Lars Walløe, "SNSF prosjektet: Arbeidsprogram og budsjett for fase II, 1977–1979," April 15, 1977, RA-S-2532-2-Dca-Loo15, Norwegian National Archives, pp. 2, 46, 50–80.

8. Christensen, "Notat (justert 27.1.77) til møte 15/76 i styringsutvalget for SNSF-prosjektet. Vannkvalitet, tanker om målsetting for fase II."

9. Walløe, "SNSF prosjektet: Arbeidsprogram og budsjett for fase II, 1977–1979," p. 3; H. C. Christensen, "Referat fra møte 1/77 i styringsutvalget for SNSF-prosjektet 12. januar 1977," February 11, 1977, RA-S-2532-2-Dca-Loo15, Norwegian National Archives, p. 2.

10. Attachment to Hans M. Seip, "Angående brev fra K. M. Bennett, EPA, om samarbeid innen sur nedbør forskningen," August 17, 1982, RA-S-2532-2-Dca-L6048, Norwegian National Archives, p. 5.

11. Nils Roll-Hansen and Geir Hestmark, *Miljøforskning mellom vitenskap og politikk: En studie av forskningspolitikken omkring andre fase av storprosjektet "sur nedbørs virkning på skog og fisk" (SNSF), 1976–1980* (Oslo: NAVF, 1990), pp. 64–65.

12. Author interview with Hans Martin Seip, professor emeritus, University of Oslo, January 15, 2014. On Seip's relationship with SNSF program director Lars Walløe, see ibid., p. 64.

13. D. Drabløs and Arne Tollan, eds., *Ecological Impact of Acid Precipitation: Proceedings of an International Conference, Sandefjord, Norway, March 11–14, 1980* (Oslo: SNSF, 1980), pp. 358–375; Lars N. Overrein, Hans Martin Seip, and Arne Tollan, *Acid Precipitation: Effects on Forest and Fish: Final Report of the SNSF-Project 1972–1980*, Fagrapport (Oslo: SNSF-prosjektet, 1980), pp. 105–110.

14. "Letter from Hans C. Christensen, NTNF to Mr. Peter D. Burgess, Department of the Environment," January 26, 1982, AT 82/203, UK National Archives.

15. Arne Henriksen and Hans Martin Seip, "Strong and Weak Acids in Surface Waters of Southern Norway and Southwestern Scotland" (SNSF-prosjektet, January 1980), National Library of Norway.

16. A. Henriksen and H. M. Seip, "Strong and Weak Acids in Surface Waters of Southern Norway and Southwestern Scotland," *Water Research* 14, no. 7 (1980): 809–813.

17. "International Conference on the Ecological Impact of Acid Precipitation March 11–14 1980: Press Release," March 14, 1980, RA-S-2939-D-Db-Dbk-Dbke-Lo605, Norwegian National Archives; Brynjulf Ottar, "Air Pollution

Research in Norway," April 8, 1980, RA-S-2532-2-Dca-L0005, Norwegian National Archives, p. 5; Drabløs and Tollan, *Ecological Impact of Acid Precipitation: Proceedings of an International Conference, Sandefjord, Norway, March 11–14, 1980.*

18. Drabløs and Tollan, *Ecological Impact of Acid Precipitation: Proceedings of an International Conference, Sandefjord, Norway, March 11–14, 1980,* p. 10.

19. Ellis B. Cowling, "Acid Precipitation in Historical Perspective," *Environmental Science & Technology* 16, no. 2 (February 1, 1982): 110A–123A.

20. John E. Carroll, "Acid Rain: An Issue in Canadian-American Relations, Prepared for the Acid Rain Task Force," August 1981, RA-S-2532-2-Dca-L6051, Norwegian National Archives, p. 60.

21. Naomi Oreskes and Erik M. Conway, *Merchants of Doubt: How a Handful of Scientists Obscured the Truth on Issues from Tobacco Smoke to Global Warming* (New York: Bloomsbury Press, 2010), pp. 73–74.

22. "Washington Report: Federal Acid Rain Research," *EPRI Journal* 8, no. 9 (November 1983), p. 56.

23. National Research Council, *Atmosphere-Biosphere Interactions: Toward a Better Understanding of the Ecological Consequences of Fossil Fuel Combustion* (Washington, DC: National Academies Press, 1981); "SNSF Prosjektet Statusrapport Med Regnskap 2. Kvartal 1979," 1979, RA-S-2532-2-Dca-L0015, Norwegian National Archives, p. 2.

24. National Research Council, *Atmosphere-Biosphere Interactions.*

25. Thomas Korosec, "Coal-Fired Plants Are Acid Rain's Source, Report Says," *Boston Globe,* September 11, 1981, p. 8; "The Evidence Justifies Action," *Hartford Courant,* September 25, 1981, p. A22; "Dropping Acid," *Washington Post,* October 16, 1981, p. A26; Leo H. Carney, "Acid Rain Held Peril to Crops," *New York Times,* November 8, 1981, p. NJ20.

26. Robert Reinhold, "Acid Rain Issue Creates Stress between Administration and Science Academy," *New York Times,* June 8, 1982, p. C1.

27. Jonathan Harsh, "EPA Chief Charts Austere Course," *Christian Science Monitor,* November 20, 1981, p. 4; Cass Peterson, "Acid Rain Research Budget Cut by EPA: Funding Slash Ordered by Administration," *Boston Globe,* December 5, 1982, p. 62.

28. "U.N. Environmental Conference Ends Feeling Pinch of Reduced U.S. Role," *Washington Post,* May 19, 1982, p. A19.

29. D. Fowler, J. N. Cape, I. D. Leith, et al., "Rainfall Acidity in Northern Britain," *Nature* 297, no. 3 (June 3, 1982): 383–386, p. 383.

30. House of Commons, "Evidence Submitted by the Institute of Terrestrial Ecology" (Session 1983–1984, June 27, 1984), HC-CL-JO-10–1794, British Parliamentary Archives, pp. 240–241.

31. Central Unit on Environmental Pollution, "Sulphur Dioxide Review: Meeting Held at 2:30 pm on Wednesday 5 November 1975 in Room N19/13A Marsham Street," November 28, 1975, AT 34–10, UK National Archives.

32. "Progress in Research on Air Pollution in the United Kingdom: Report by the Interdepartmental Committee on Air Pollution Research to the Clean Air Council," September 1975, AT 34–86, UK National Archives.

33. Author interview with David Fowler, professor, University of Edinburgh, January 15, 2014.

34. R. Harriman, B. R. S. Morrison, L. A. Caines, et al., "Long-Term Changes in Fish Populations of Acid Streams and Lochs in Galloway South West Scotland," *Water, Air, and Soil Pollution* 32, no. 1–2 (January 1, 1987): 89–112.

35. Fowler et al., "Rainfall Acidity in Northern Britain"; author interview with David Fowler, January 15, 2014.

36. Marek Mayer, "Britain Slashes Research on Air Pollution," *New Scientist* (April 29, 1982), p. 271; Fred Pearce, "Research Cuts Corrode Britain's Acid Rain Strategy," *New Scientist* (July 15, 1982), p. 141.

37. House of Commons, "Minutes of Evidence Taken Before the Environment Committee: Natural Environment Research Council" (Session 1983–1984, June 27, 1984), HC-CL-JO-10–1794, British Parliamentary Archives, p. 259; "Letter from PJW Saunders, Science Division, Natural Environment Research Council, to MW Holdgate, Department of the Environment," May 28, 1982, AT 82/203, UK National Archives.

38. Michael Kilian, "Protection of the U. S. Environment Endangered," *Chicago Tribune*, December 30, 1980, p. D2.

39. Andy Pasztor, "Interior Choice James Watt Seems Poised to Confront Opponents to Confirmation," *Wall Street Journal*, January 7, 1981, p. 6; Edward Flattau, "Political Appointee May Soil the EPA," *Chicago Tribune*, March 11, 1981, p. A4.

40. Office of the Press Secretary, "White House Press Release," December 10, 1982, RG 412—Studies and Other Related Records, Acid Rain, IT FAP RCC MTS, US National Archives, Box 14.

41. Ibid.

42. "Nomination of Dr. James R. Mahoney to be Assistant Secretary for Oceans and Atmosphere and Deputy Administrator for the National Oceanic and Atmospheric Administration," Committee on Commerce, Science, and Transportation, United States Senate (Washington, DC: US GPO, 2003), p. 6.

43. P. F. Chester, G. D. Howells, and R.A. Scriven, "The Present Position on Acid Rain and Its Relation to Power Station Emissions," November 9, 1983, HC-CP-9443, British Parliamentary Archives, p. 3.

44. "Another Look at Acid Rain," *Chicago Tribune*, June 25, 1983, p. 8.

45. "Statement of John Hernandez, Ph.D, Deputy Administrator, U.S. Environmental Protection Agency, before the Subcommittee on Health and the Environment of the Committee on Energy and Commerce, U.S. House of Representatives," September 23, 1981, Acid Rain Papers, Testimony Thru US Canada Work Group #2 Phase 1, Box 1, US National Archives; Laura Westra and Bill E. Lawson, *Faces of Environmental Racism: Confronting Issues of Global Justice* (Lanham, MD: Rowman & Littlefield, 2001), p. 14.

46. Electric Power Research Institute, "1982 Annual Report," *EPRI Journal* 8, no. 3 (1983), p. 53; "Risk in the Pursuit of Benefit," *EPRI Journal* 3, no. 9 (1978), p. 14.

47. "William Nierenberg: Blockbuster," *EPRI Journal* 2, no. 10 (December 1977), p. 19.

48. Oreskes and Conway, *Merchants of Doubt*, p. 88. Oreskes and Conway note Nierenberg's communications with Starr but do not discuss his deeper history with the organization. For their discussion of his role in the Reagan Administration's manipulation of acid rain studies, see pp. 95–101.

49. Ronald Reagan, "Nomination of Kathleen M. Bennett to Be an Assistant Administrator of the Environmental Protection Agency," July 2, 1981. Published online by Gerhard Peters and John T. Woolley, *The American Presidency Project*, http://www.presidency.ucsb.edu/ws/?pid=44033.

50. "Letter from Dick Funkhouser to Kathleen Bennett, Subject: US-Canada Phase III Work Groups," December 7, 1981, AT 82/203, UK National Archives; "Acid Rain's Causes and Effects: The Answers Are In," *New York Times*, August 25, 1983, p. A22.

51. Philip Shabecoff, "U.S. Holds Up Report on Global Environment," *New York Times*, March 28, 1982, p. 4.

52. "Letter from Alvin Trivelpiece, Department of Energy, to Richard Funkhouser, EPA," February 3, 1982, RA-S-2532-2-Dca-L0648, Norwegian National Archives; "Acid Precipitation: Report on the Director's Workshop on Selected Global Environment and Resource Issues, Held at OECD, Paris 27th & 28th October, 1981," n.d., RA-S-2532-2-Dca-L0648, Norwegian National Archives. This report is marked up, likely by Trivelpiece and others in the Department of Energy. Also see the attached "Comments on OECD Paper on Acid Rain," which details their criticisms. On Funkhouser's broader opposition to US support for international environmental cooperation, see Philip Shabecoff, "U.S. Goes to Ecology Parley under Cloud of Doubt," *New York Times*, May 5, 1982, p. A2.

53. Norsk Institutt for Vannforskning, "Sekretariatsoppgaver Vedr. Sur Nedbor: Beretning Desember 1981-Oktober 1982," October 22, 1982, RA-S-2532-2-Dba-L0061, Norwegian National Archives.

54. "Letter from Jack Siegal, Director, Office of Environment, Department of Energy to Lowell Smith, U.S. Chairman, Work Group 3A, Environmental Protection Agency," January 30, 1981, RG 412—Studies and Other Related Records, Acid Rain, Folder US Canada Work Group 3A, U.S. National Archives and Record Administration II, Box 1.

55. Ibid., p. 2.

56. "Letter from Raymond Wilhour, Chief, Air Pollution Effects Branch to Gary Foley, Director, Acid Deposition Work Group," April 29, 1982, RG 412—Studies and Other Related Records, Acid Rain, Folder DOE Comments, U.S. National Archives and Record Administration II, Box 2.

57. He met these government officials, who worked for the Royal Norwegian Council for Scientific and Industrial Research, at a conference sponsored by Exxon Mobil. "Letter from Dr. P. F. Chester, CEGB to Dr. J. M. Doderlein, Royal Norwegian Council for Scientific and Industrial Research," October 28, 1981, RA-S-2532-2-Dca-L6051, Norwegian National Archives.

58. Ibid.

59. Hans C. Christensen, "Utspill fra Dr. P.F. Chester, CERL," November 13, 1981, RA-S-2532-2-Dca-L6051, Norwegian National Archives.

60. "Letter from Hans C. Christensen, NTNF to Mr. Peter D. Burgess, Department of the Environment."

61. Seip, "Comments to Letter and Paper by P.F. Chester," pp. 2–3.

62. Ibid.

63. "Letter from JPG Rowcliffe to RB Wilson, CEGB Note on Acid Rain Etc: Norwegian Comments," February 9, 1982, AT 82/203, UK National Archives; "Letter from MW Holdgate to Mr. Rowcliffe, 'Acid Rain,' " February 16, 1982, AT 82/203, UK National Archives.

64. "Letter from MW Holdgate to Dr. Everest, 'Acid Rain and Acidified Lakes,' " May 7, 1982, AT 82/203, UK National Archives.

65. Senior Advisers to UN ECE Governments on Environmental Problems, "Working Party on Air Pollution Problems: Tenth Session, 28–30 April 1980. Draft Report," April 29, 1980, RA-S-2532-2-Dca-L0300, Norwegian National Archives, p. 2; "Norsk innlegg 29/4. vedl. 3," April 29, 1980, RA-S-2532-2-Dca-L0300, Norwegian National Archives, p. 1.

66. Review Group on Acid Rain, Warren Spring Laboratory, "Confidential Draft: Rainfall Acidity in the United Kingdom. Draft Report No. 3," 1982, AT 82/203, UK National Archives, p. 19.

67. "Letter from MW Holdgate to Mr. Wedd," March 1, 1982, AT 82/203, UK National Archives.

68. David Fowler has said he was not the source of the leak and believes it may have been someone either in the Department of the Environment or a university-based consultant who reviewed the report draft before it was published. Author interview with David Fowler, professor, University of Edinburgh, January 15, 2014. For the report, see Review Group on Acid Rain, Warren Spring Laboratory, "Confidential Draft: Rainfall Acidity in the United Kingdom. Draft Report No. 3." For the *Observer* article, see Geoffrey Lean and Marek Mayer, "Britain Moving towards Acid Rain Disaster," *Observer*, June 13, 1982, p. 2. On the DOE's response to the leak, see "Letter from MW Holdgate to Mr. Wedd, 'Acid Rain Disaster,' " June 17, 1982, AT 82/203, UK National Archives; "Letter from JPG Rowcliffe to Mr. King, 'Acid Rain: Radio Scotland Program,' " April 21, 1982, AT 82/203, UK National Archives.

69. For some examples of Parliament's dramatic increase in attention to acid rain following the literature survey, which was led by members of the Warren Spring Laboratory, as well as an additional increase in interest after the 1982 Stockholm Conference at the end of June, see "Acid Rain. HC Deb 17 June 1982 Vol 25 cc340–1W" (Hansard, June 17, 1982); "Acid Rain. HC Deb 21 June 1982 Vol 26 cc33–4W" (Hansard, June 21, 1982); "Acid Rain. HC Deb 08 July 1982 Vol 27 c207W" (Hansard, July 8, 1982); "Lake District National Park. HC Deb 21 July 1982 Vol 28 c199W" (Hansard, July 21, 1982); "Acid Rain. HC Deb 21 July 1982 Vol 28 cc384–6" (Hansard, July 21, 1982); "Meteorological Office (Rainfall Measurement). HC Deb 29 July 1982 Vol 28 cc696–7W" (Hansard, July 29, 1982); "Acid Rain: Warren Spring Report. HL Deb 02 August 1982 Vol 434 c666WA" (Hansard, August 2, 1982); "Acid Rain: Areas Most Affected. HL Deb 02 August 1982 Vol 434 c666WA" (Hansard, August 2, 1982); "Rainfall (Acid Content). HC Deb 15 November 1982 Vol 32 c76W" (Hansard, November 15, 1982).

70. "Letter from MW Holdgate, Director General of Research, Department of Environment to PJW Saunders, Science Division, Natural Environment Research Council," June 9, 1982, AT 82/203, UK National Archives.

71. "Letter from MW Holdgate, Chief Scientist and Deputy Secretary, Department of Environment, to FT Last, Institute of Terrestrial Ecology," April 21, 1982, AT 82/203, UK National Archives.

72. On Last's description of the key areas needing attention in acid rain research, particularly acid rain's effects on vegetation and soils in Britain, see "Letter from FT Last, Institute of Terrestrial Ecology, to MW Holdgate, Chief Scientist and Deputy Secretary, Department of Environment," May 26, 1982, AT 82/203, UK National Archives.

73. After detailing possible areas of research, such as acid rain's pathway through soils to surface and ground waters in rural areas in Britain, Saunders concluded his letter with a terse "Please don't bother to reply." See "Letter from PJW Saunders, Science Division, Natural Environment Research Council, to MW Holdgate, Department of the Environment."

74. "Letter from MW Holdgate to Mr Wedd, 'Acid Rain,'" July 5, 1982, AT 82/203, UK National Archives.

75. John Jones and Jack Smith (pseudonyms), "Critics of E.P.A. Are Right," *New York Times*, September 1, 1982, p. A23; John Jones and Jack Smith (pseudonyms), "Critics of the EPA Are Right," *Chicago Tribune*, September 8, 1982, p. 19.

76. Bertil Hägerhäll, "Kommittén 10-år efter Stockholm—arbetsuppgifter m.m.," January 14, 1981, SE/RA/323591/~/1, Kommittén med uppgift att samordna aktiviteter i samband med uppföljningen av Stockholmskonferensen 1972 om den mänskliga miljön, Swedish National Archives, Marieberg.

77. Bertil Hägerhäll, "Anteckningar från kommittens sammanträde 1981–12–22," December 22, 1981, SE/RA/323591/~/1, Kommittén med uppgift att samordna aktiviteter i samband med uppföljningen av Stockholmskonferensen 1972 om den mänskliga miljön, Swedish National Archives, Marieberg.

78. "Stockholmskonferensen 28–30 juni 1982 om försurning av miljön," February 1982, SE/RA/323591/~/5, Kommittén med uppgift att samordna aktiviteter i samband med uppföljningen av Stockholmskonferensen 1972 om den mänskliga miljön, Swedish National Archives, Marieberg, p. 3.

79. "Memorandum: The 1982 Stockholm Conference on Acidification of the Environment, Stockholm, June 21–30 1982," February 24, 1982, SE/RA/323591/~/5, Kommittén med uppgift att samordna aktiviteter i samband med uppföljningen av Stockholmskonferensen 1972 om den mänskliga miljön, Swedish National Archives, Marieberg.

80. "Letter from MW Holdgate to Mr Rowcliffe, 'Acidification,'" January 25, 1982, AT 82/203, UK National Archives, p. 1.

81. CDEP, "Acidification Conference in Stockholm, June 1982. Points Arising Out of a Meeting Held in Room N19/20 on 29 January 1982," January 1982, AT 82/203, UK National Archives.

82. "Letter from MW Holdgate to Mr Rowcliffe, 'Acidification,'" p. 1.

83. "Letter from LF Rutterford to Miss McConnell, 'UK Representation at the Expert Meetings Preceding the Stockholm High Level Conference on Acidification,'" January 12, 1982, AT 82/203, UK National Archives.

84. "Letter from LF Rutterford to Mr. Rowcliffe, 'Acidification,'" January 21, 1982, AT 82/203, UK National Archives.

85. "Letter from MW Holdgate to Mr. Rowcliffe, 'International Conference on Acidification,' " January 7, 1982, AT 82/203, UK National Archives, pp. 1–2.

86. "Letter from MW Holdgate, Chief Scientist and Deputy Secretary, Department of Environment, to PL Gregson, Cabinet Office," June 14, 1982, AT 82/203, UK National Archives.

87. Ibid, p. 2.

88. Ibid. Unfortunately, Holdgate had been unable to attend the meeting where Cabinet officers, the Department of Energy, and the Department of Interior had discussed these issues.

89. "Letter from Dr. Nicholson to Mr. Gregson," June 18, 1982, AT 82/203, UK National Archives. On Nicholson, see Peter Large, "Taxpayer to Fund More of R&D Bill," *Guardian*, December 12, 1984, p. 18.

90. "Letter from Mr. Shaw, Undersecretary of State, Department of the Environment, to Secretary of State and Mr. King," undated, handwritten, AT 82/203, UK National Archives.

91. "Letter from JA Catterall, Head, Energy Technology Division, Department of Energy, to MW Holdgate, Chief Scientist, Department of Environment," June 22, 1982, AT 82/203, UK National Archives.

92. "Letter from PL Gregson, Deputy Secretary, Cabinet Office, to MW Holdgate, Chief Scientist and Deputy Secretary, Department of the Environment," June 23, 1982, AT 82/203, UK National Archives.

93. "Letter from JA Catterall, Head, Energy Technology Division, Department of Energy, to MW Holdgate, Chief Scientist, Department of Environment." Holdgate was quick to try to smooth things over with his colleagues, writing to the Cabinet Office that he was happy to have the matter settled and that further conversations between the government's chief scientists could take place based on the outcome of the Stockholm meeting. See "Letter from MW Holdgate, Chief Scientist and Deputy Secretary, to PL Gregson, Deputy Secretary, Cabinet Office," July 2, 1982, AT 82/203, UK National Archives.

94. Dr. James L. (Larry) Regens headed the US acid rain task force group at the EPA from 1980 to 1983. Regens also participated in the US-Canadian negotiations on acid rain as part of the technical working groups, as well as serving as a US delegate to the OECD and UN Economic Commission for Europe on acid rain discussions. See James L. Regens and Robert W. Rycroft, *The Acid Rain Controversy* (Pittsburgh: University of Pittsburgh Press, 1988), pp. xvii–xviii. On Regens' discussions with Holdgate, including a reference to Regens saying he had "heard enough off the wall questions from [Dick] Funkhouser," see "Letter from Don A. Rolt, Attache (Science) to Ms. Fiona McConnell, Head of EPIE Division, Department of the Environment," February 19, 1982, AT 82/203, UK National Archives.

95. "Letter from Don Rolt, Attache (Science) to Leslie Reed, Chief Alkali Inspector," January 15, 1982, AT 82/203, UK National Archives.

96. "Letter from MW Holdgate to Miss McConnell, 'Acid Precipitation,' " February 17, 1982, AT 82/203, UK National Archives.

97. Michael Oppenheimer made the accusation. See Chris Mosey, "US Pelted over Acid Rain," *Christian Science Monitor*, July 2, 1982, p. 2.

98. National Swedish Environment Protection Board, "Ecological Effects of Acid Deposition: Report and Back Ground Papers 1982 Stockholm Conference

on the Acidification of the Environment. Expert Meeting I," January 1983, SE/
RA/323591/~/5, Kommittén med uppgift att samordna aktiviteter i samband
med uppföljningen av Stockholmskonferensen 1972 om den mänskliga miljön,
Swedish National Archives, Marieberg, p. 13; Fred Pearce, "The Menace of Acid
Rain," *New Scientist* 95, no. 1318 (August 12, 1982), p. 422.

    99. Pearce, "The Menace of Acid Rain," p. 424.

    100. "Draft Letter for Mr Shaw to Send to David Mellor," 1982, AT 82/203,
UK National Archives. Though America and Britain's "unrepentant" attitudes did
attract attention from the international scientific community, including coverage in
Nature and *New Scientist*, the Stockholm conference received minimal mainstream
media coverage in both countries. In the US, coverage was limited to several articles
in the *New York Times* without much mention of acid rain. See Shabecoff, "U.S.
Goes to Ecology Parley under Cloud of Doubt." In Britain, the only two articles that
appeared at the time were very brief. See Lean and Mayer, "Britain Moving towards
Acid Rain Disaster," p. 5; Chris Mosey, "Pledge on Acid Rain," *Observer*, June 20,
1982, p. 13; Jasper Becker, "Acid Rain: UK Unrepentant," *Nature* 298 (July 8,
1982): 112; Fred Pearce, "Science and Politics Don't Mix at Acid Rain Debate,"
New Scientist 95, no. 1312 (July 1, 1982), p. 3; Fred Pearce, "It's an Acid Wind That
Blows Nobody Any Good," *New Scientist* 95, no. 1313 (July 8, 1982), p. 80.

    101. "Acidification Conference, Stockholm 28–30 June 1982: Delegation
Report," July 1982, AT 82/203, UK National Archives.

    102. There are several drafts of this letter to David Mellor in the archival file.
The early draft is more heavily descriptive concerning the domestic arguments
for the need to review acid rain research, while the final version points to the
international tensions at the Stockholm conference and in the European Communities. This seems likely to have stemmed from concerns about interagency
conflict over the acid rain issue, which had developed during the preparations for
Stockholm. Had the Department of the Environment signaled that they were considering regulating British emissions because of domestic acid rain, it may have provoked a greater backlash from the Department of Energy as well as the National
Coal Board and CEGB. See "Draft Letter for Mr Shaw to Send to David Mellor."

    103. "Draft Letter for Mr. Shaw to Send to David Mellor," 1982, AT 82/203,
UK National Archives, pp. 1–2.

    104. "Letter from Nigel Lawson, Secretary of State, Department of Energy,
to Norman Siddall, Chairman, National Coal Board," January 28, 1983, COAL
96/85, UK National Archives.

    105. Ibid. Regarding the need to notify the CEGB of the importance of using
abatement technologies going forward, see "Draft Letter for Mr Shaw to Send to
David Mellor," p. 2.

    106. A. D. Dainton, "Report of a Meeting between NCB and CEGB," Febru-
ary 23, 1983, COAL 96/85, UK National Archives, p. 1.

    107. National Research Council, *Acid Deposition. Atmospheric Processes in
Eastern North America: A Review of Current Scientific Understanding* (Wash-
ington, DC: National Academies Press, 1983). For the public response, see Philip
Shabecoff, "Acid Rain Panel Urges Curb on Pollutants in East," *New York
Times*, June 30, 1983.

    108. National Research Council, *Acid Deposition,* pp. 7–9.

109. "Acid Rain, 1983," US Senate, Committee on the Environment and Public Works (Washington, DC: US GPO, 1983), pp. 351–353.

110. "Acid Rain," US Senate, Committee on the Environment and Public Works (Washington, DC: US GPO, 1982), p. 53.

111. Glen E. Gordon, "A Decade of Acid Rain Research," in *The Chemistry of Acid Rain*, vol. 349, ACS Symposium Series 349 (American Chemical Society, 1987), 2–9, p. 5; Dale W. Johnson, J. M. Kelly, W. T. Swank, et al., *A Comparative Evaluation of the Effects of Acid Precipitation, Natural Acid Production, and Harvesting on Cation Removal from Forests, Publication No. 2508* (Oak Ridge, TN: Oak Ridge National Laboratory, Environmental Sciences Division, 1985).

112. René Malès, "R & D Status Report: Energy Analysis and Environment Division," *EPRI Journal* 3, no. 9 (1978), p. 55.

113. René Malès, "R & D Status Report: Energy Analysis and Environment Division," *EPRI Journal* 4, no. 3 (1979), p. 50.

114. They also worked on the environmental effects of nuclear power. For the acid rain grant, see "Removal of Pollution from Power Plant Plumes by Precipitation," in "EPRI Negotiates 62 Contracts," *EPRI Journal* 2, no. 2 (1977), p. 25; René Malès, "R & D Status Report: Energy Analysis and Environment Division," *EPRI Journal* 3, no. 9 (1978), p. 55.

115. Mark McIntyre, "Students Get a Short Course in Acid Rain," *Newsday*, January 14, 1983, p. 25.

116. Environmental Research and Technology, Inc., "Design of the Sulfate Regional Experiment (SURE)" (EPRI EC-125, February 1976), p. iii.

117. Peter E. Coffey, "Author's Reply," *Journal of the Air Pollution Control Association* 35, no. 2 (1985): 129–130; F.W. Lipfert, E. Kaplan, and M. Daum, "Statistical Analysis of Relationships between Precipitation Chemistry and Air Quality in New Hampshire and Precursor Emissions," *Proceedings of the Air Pollution Control Association Annual Meeting*, 1984.

118. F. Lipfert, "Exposure to Acidic Sulfates in the Atmosphere: Review and Assessment: Final Report" (EPRI, 1988), Report number EPRI-EA-6150; F Lipfert, "The Role of Dry Deposition in Acidification of Waters," (Brookhaven, 1984), Report number BNL-35667; Frederick W. Lipfert, H. Mitchell Perry Jr, J. Philip Miller, et al., "The Washington University-EPRI Veterans' Cohort Mortality Study: Preliminary Results," *Inhalation Toxicology* 12 Suppl 4 (2000): 41–73; National Research Council, Commission on Engineering and Technical Systems, and Committee to Review DOE's Office of Fossil Energy's Research Plan for Fine Particulates, *Review of the U.S. Department of Energy Office of Fossil Energy's Research Plan for Fine Particulates* (Washington, DC: National Academies Press, 1999), p. 42; Gary Polakovic, "Air Pollution Harmful to Babies, Fetuses, Studies Say," *Los Angeles Times*, December 16, 2001, http://articles.latimes.com/2001/dec/16/news/mn-15433; Laura Johannes, "Pollution Study Sparks Debate over Secret Data," *Wall Street Journal*, April 7, 1997, p. B1.

119. Lawrence Kleinman, "A Regional Scale Modeling Study of the Sulfur Oxides with a Comparison to Ambient and Wet Deposition Monitoring Data," *Atmospheric Environment* 17, no. 6 (1983): 1107–1121.

120. Stephen E. Schwartz, "Both Sides Now," *Annals of the New York Academy of Sciences* 502, no. 1 (July 1, 1987): 83–144; Stephen E. Schwartz and

Leonard Newman, "Processes Limiting Oxidation of Sulfur Dioxide in Stack Plumes," *Environmental Science & Technology* 12, no. 1 (January 1, 1978): 67–73; Stephen Eugene Schwartz, *Trace Atmospheric Constituents: Properties, Transformations, and Fates* (Hoboken, NJ: John Wiley & Sons, 1983).

121. "Letter from Bernard Markowitz to Senator Max Baucus," November 10, 1983, Hearings before the Committee on Environment and Public Works, US Senate (Washington, DC: US GPO, 1983), p. 530.

122. Ronald Kotulak, "Skies Raining Death on Lakes: The Poisons above Us; Acid Rain Turning the World's Lakes into 'Deserts,'" *Chicago Tribune*, March 23, 1982, p. 1.

123. On the European linearity research prior to the publication of the National Academy of Sciences 1983 report, see J. R. Mitchell and M. L. Williams, "The Significance of Non-linearity between Emissions and Depositions of Acidic Pollutants," 1985, FV 12/67, UK National Archives. The work of Dick Derwent in Britain during the early 1980s was also crucial in resolving this question. For Peter Chester's attacks on Derwent's conclusion that there was a broadly linear relationship between emissions and deposition, see "CEGB Scientist Attacks Plans to Curb Acid Rain" (ENDS Report 99, April 1983), COAL 96/91, UK National Archives; author interview with Dick Derwent, retired professor of Atmospheric Chemistry, Imperial College, London, March 18, 2013, Newbury, England.

124. "Acid Rain, 1983," US Senate, Committee on the Environment and Public Works, p. 392.

125. Stephen E. Schwartz, "Acid Deposition: Unraveling a Regional Phenomenon," *Science* 243 (February 10, 1989): 753–763.

126. James Fay and Dan Golomb, "Acid Rain Models: Response," *Science* 244, no. 4901 (April 14, 1989): 127–128.

127. "Congress to Weigh Curb on 'Acid Rain,'" *New York Times*, January 17, 1982, p. CN5.

128. "CEGB Confidential: Report on a Visit to the USA May 9th to 23rd 1985 by WD Halstead," August 1, 1985, COAL 29/655, UK National Archives.

CHAPTER SEVEN

1. Anthony Tucker, "Britain's Two Million Tonne Chemical Warfare Onslaught," *Guardian*, September 22, 1983, p. 19.

2. Hywel Davies, who initiated discussions on the general community policies on acid rain, was currently deputy director general of the European Communities' Directorate for Science, Research and Development. See D. H. Davies, "Note for the Attention of Mr. Ph. Bourdeau. Subject: Acid Rain," October 7, 1983, BAC 107/1993 893, European Commission Archives.

3. "Million-Dollar Problem—Billion-Dollar Solution?" *Nature* 268, no. 5616 (July 14, 1977): 89.

4. Erik Lykke, "Europe versus Itself," *Nature* 269, no. 5627 (September 1977): 372.

5. Fred Singer seems to have been the first person to use this phrase in the US while working on the Nierenberg panel on acid rain. See S. Fred Singer, "Appendix 5," in William A. Nierenberg, "Report of the Acid Rain Review Panel,"

Office of Science Technology and Policy (Washington, DC: US G.P.O., 1984), pp. A5–10; Carl Bagge, "The Case Has Yet to Be Proven," *New York Times*, October 28, 1984, p. F2; S. Fred Singer, "Acid Rain: A Billion-Dollar Solution to a Million-Dollar Problem?," *Policy Review*, no. 27 (Winter 1984): 56. For a discussion of Singer's concerns about the costs and benefits of acid rain control, see Naomi Oreskes and Erik M. Conway, *Merchants of Doubt: How a Handful of Scientists Obscured the Truth on Issues from Tobacco Smoke to Global Warming* (New York: Bloomsbury Press, 2010), pp. 84–94.

6. Concerns about damage to German forests from industrial activities can be traced at least to the nineteenth century. See George Perkins Marsh, *Man and Nature* (New York: Charles Scribner, 1864).

7. It was modeled after the International Geophysical Year of 1957. On the International Biological Program, see Sharon E. Kingsland, *The Evolution of American Ecology, 1890–2000* (Baltimore: Johns Hopkins University Press, 2005), p. 221; Joel Bartholemew Hagen, *An Entangled Bank: The Origins of Ecosystem Ecology* (New Brunswick, NJ: Rutgers University Press, 1992), pp. 164–186; Elena Aronova, Karen S. Baker, and Naomi Oreskes, "Big Science and Big Data in Biology: From the International Geophysical Year through the International Biological Program to the Long Term Ecological Research (LTER) Network, 1957–Present," *Historical Studies in the Natural Sciences* 40, no. 2 (May 1, 2010): 183–224.

8. A pure beech stand on a high plateau in the area was selected for the studies. Michael Bredemeier and Martin Jansen, "Interdisciplinary Forest Ecosystem Experiments at Solling, Germany–from Plot Scale to Landscape Level Integration," *Forest Snow and Landscape Research* 78, no. 1 (2004): 33–52.

9. Ibid., pp. 36–37.

10. Informasjonsgruppen mot sur nedbør, "Stopp sur nedbør," October 1986, Norwegian National Archives, Oslo, Norway, RA-S-2939-D-Db-Dbk-Dbkj-Lo738.

11. "Säure-Regen: 'Da Liegt Was in Der Luft,'" *Der Spiegel* 49 (1981): 174–190, Norwegian National Archives, RA-S-2532-2-Daa-Lo509. Many of his colleagues in West Germany shared similar views, although there were others who did not believe the situation was as dire as Ulrich claimed, including Scandinavian scientists who had studied the issue since the seventies. For Ulrich's comments, see Patricia Clough, "Germany's Progress Is Killing Its Trees," *Irish Times*, January 12, 1982, p. 1. Historian Nils Roll-Hansen has shown how the media misrepresented the current scientific consensus on acid rain's forest impacts, instead playing up the warnings of Ulrich and other West German scientists. See Nils Roll-Hansen, "Science, Politics, and the Mass Media: On Biased Communication of Environmental Issues," *Science, Technology & Human Values* 19, no. 3 (July 1, 1994): 324–341.

12. B. Ulrich, R. Mayer, and P. K. Khanna, "Chemical Changes due to Acid Precipitation in a Loess-Derived Soil in Central Europe," *Soil Science* 130, no. 4 (October 1980): 193–199, p. 199.

13. Ulrich's initial hypothesis was published in 1980 in the journal *Soil Science*. See ibid. For other explanations of his thinking, see B. Ulrich, "Eine ökosystemare Hypothese über die Ursachen des Tannensterbens (Abies alba Mill.),"

Forstwissenschaftliches Centralblatt 100, no. 1 (January 1, 1981): 228–236; B. Ulrich, "A Concept of Forest Ecosystem Stability and of Acid Deposition as Driving Force for Destabilization," in B. Ulrich and Jürgen Pankrath, *Effects of Accumulation of Air Pollutants in Forest Ecosystems* (New York: Springer Science & Business Media, 1983), pp. 1–29.

14. Heike Faller, "Prognoseopfer Wald: Lange Wurde Er Totgesagt, Dann Schien Es Ihm Wieder Besser Zu Gehen, Nun Wird Ihm Zum Ersten Mal Seit Langem Wieder Eine Schlechte Zukunft Prophezeit: 'Schon in Den Nächsten Jahren Werden in Deutschland Großflächig Wälder Absterben,'" *Die Zeit*, December 2003. Most scientists involved in acid rain studies in Norway stated that there was little evidence to suggest Norwegian forests were at risk of death from acid rain and other air pollutants. The other suspected causes of West Germany's forest decline included ozone, drought, and disease in addition to sulfur dioxide and nitrogen oxides. See Gunnar Abrahamsen, "Skogdød i Tyskland og risikoen for skader i Norge," November 18, 1983, RA-S-2532-2-Dca-L0297, Norwegian National Archives. On consultations between the European Communities and Norwegian environmental officials regarding the West German initiative, see Erik Lykke, "Referat fra 2. arlige kontaktmøte om miljøvern med EF-kommisjonen, Brussel 22. september 1982," September 28, 1982, RA-S-2532-2-Daa-L0047, Norwegian National Archives.

15. West Germany's Minister of the Interior, Gerhart Baum, spearheaded the regulations. "Säure-Regen: 'Da Liegt Was in Der Luft.'" Also see Morten Wetland, "Artikkel i 'Der Spiegel' om sur nedbør-del III," December 1, 1981, RA-S-2532-2-Daa-L0509, Norwegian National Archives; Abrahamsen, "Skogdød i Tyskland og risikoen for skader i Norge." The general surge in attention to environmental issues in Germany as well as the rise of the Green Party's influence in politics during the late 1970s and early 1980s contributed to this focus on acid rain as well. See Miranda A. Schreurs, *Environmental Politics in Japan, Germany, and the United States* (Cambridge: Cambridge University Press, 2003), pp. 96–98; Ion Bogdan Vasi, *Winds of Change: The Environmental Movement and the Global Development of the Wind Energy Industry* (Oxford: Oxford University Press, 2011), pp. 59–60.

16. "Chairman's Report for Adoption by Air Sector Group: Working Paper in relation to item No. 4 of Agenda, for 7th AMRG Meeting," March 15, 1971, OECD Archives, NR/ENV "Divers" E.43547; "Council: Meeting of the Environment Committee at Ministerial Level (Note by the Environment Committee)," February 27, 1974, OECD Archives, CE/M(74)6(Prov.), p. 6. The Nordic countries viewed these efforts as the most promising avenue for future negotiations on acid rain in the 1980s. See "Statsradens besøk i DDR—12.–14. juli 1982—arbeidet i FN/UN ECE om virkninger av svovelutslipp," July 5, 1982, RA-S-2532-2-Dca-L0005, Norwegian National Archives; Richard G. Fort, "Status i det internasjonale samarbeide for a begrense utslipp av So₂," September 15, 1981, RA-S-2532-2-Dca-L0296, Norwegian National Archives.

17. "Environment Directorate: Monthly Report October to December," December 17, 1971, OECD Archives, ENV/D/574 E.49834, p. 3.

18. Organisation for Economic Co-operation and Development, *Environment Policies for the 1980s* (Paris: Organisation for Economic Co-operation and Development, 1980), p. 52.

19. OECD, *Coal and Environmental Protection: Costs and Costing Methods* (Paris: OECD, 1983). Also see "Conference on Saving the Forests: International Pollution and Environmental Responsibility, Sponsored by the Aspen Institute Berlin," September 16, 1983, RA-S-2939-D-Db-Dbk-Dbke-L0605, Norwegian National Archives. For more on the OECD's involvement in environmental economics in this period, see chapter 9 in Matthias Schmelzer, *The Hegemony of Growth: The OECD and the Making of the Economic Growth Paradigm* (Cambridge: Cambridge University Press, 2016).

20. For newspaper articles discussing the OECD's analysis of the economic costs versus benefits of regulating pollution in the context of acid rain, see Franco Goy, "Le Piogge Acide: Una Gravissima Forma Di Inquinamento Che Minaccia La Natura E La Salute Umana," *VIA*, October 1983, BAC 107/1993 897, European Commission Archives. Also see Leopold Lukschanderl, "Saurer Regen: Malaria Der Biosphäre," *Organ Der Osterreichischen Gesellschaft Für Natur-Und Umweltschutz*, November 1983, BAC 107/1993 897, European Commission Archives, p. 17; Derek Brown, "Acid Rain Damage Costs EEC £33bn a Year," *Guardian*, January 20, 1984, p. 6.

21. Richard G. Fort, "Rapport fra UN ECE—Første møte i styringsorganet for konvensjonen om langtransportert grenseoverskridende luftforurensning Geneve 6.-10. juni 1983," June 27, 1983, RA-S-2532-2-Dca-L0297, Norwegian National Archives. The need to incorporate cost-benefit analyses was further emphasized by the European Communities and West Germany during the 1983 Karlsruhe conference on acid rain, which was held to achieve consensus not only on the environmental effects of the problem but also its "socio-economic" implications. See D. W. Gill, "Acid Deposition—A Challenge for Europe. Impressions of a Symposium Organised by the Commission of the European Communities. Karlsruhe, Germany 19–21 September 1983," October 4, 1983, COAL 96/91, UK National Archives.

22. Organisation for Economic Co-operation and Development, *Environment Policies for the 1980s*, p. 13.

23. On the importance of weighing future potential environmental risks of nuclear fallout, waste, and war against national security needs of weapons testing and nuclear power see Michael Egan, *Barry Commoner and the Science of Survival: The Remaking of American Environmentalism* (Cambridge, MA: MIT Press, 2007). Also see Rachel Rothschild, "Environmental Awareness in the Atomic Age: Radioecologists and Nuclear Technology," *Historical Studies in the Natural Sciences* 43, no. 4 (September 1, 2013): 492–530. On the need for caution regarding environmental policies toward DDT as voiced by the Presidential Scientific Advisory Committee, see Zuoyue Wang, *In Sputnik's Shadow: The President's Science Advisory Committee and Cold War America* (New Brunswick, NJ: Rutgers University Press, 2008), pp. 200–218.

24. On the initial use of "vorsorge" in German environmental policy, see François Ewald, Christian Gollier, and Nicholas De Sadeleer, *Le principe de précaution* (Paris: Presses Universitaires de France, 2001). Also see Andrew Jordan and Timothy O'Riordan, "The Precautionary Principle: A Legal and Policy History," in *The Precautionary Principle: Protecting Public Health, the Environment and the Future of Our Children*, ed. Marco Martuzzi and Joel A. Tickner

(Geneva: World Health Organization, 2004). As scholar Kerry Whiteside notes, though Ewald and Gollier have identified the use of the word vorsorge [precaution] in German environmental law in the 1970s, it was not until the mid-1980s that the word came to be associated with a lack of scientific knowledge even within West Germany, though she does not posit an explanation for this evolution. See Kerry Whiteside, *Precautionary Politics: Principle and Practice in Confronting Environmental Risk* (Cambridge, MA: MIT Press, 2006).

25. United Nations Economic Commission for Europe, "Strategies et Politiques Visant á Reduire la Pollution Atmosphèrique Provquée par les Composés Sulfureux. Renseignements Communiqués par le Gouvernment de la Republique Federale d'Allemagne," July 21, 1980, BAC 495/2005 No. 450 Pages: 751 Année(s): 1980 Page 0314, European Commission Archives, p. 1.

26. As examples, see S. Lehringer, "EG-Symposium in Karlsruhe. Saure Niederschläge—Eine Herausforderung Für Europa," 1983, BAC 107/1993 897, European Commission Archives; "Gemeinsam Gegen Den Sauren Regen," *Dienstag* 217 (September 20, 1983), BAC 107/1993 897, European Commission Archives.

27. Davies, "Note for the Attention of Mr. Ph. Bourdeau. Subject: Acid Rain."

28. "Letter from Stanley Clinton Davis, European Commission, to Sir Hugh Rossi, Member of Parliament, House of Commons," July 31, 1985, HC-CP-12066, British Parliamentary Archives, p. 3.

29. Davies, "Note for the Attention of Mr. Ph. Bourdeau. Subject: Acid Rain."

30. European Communities, "Acid Rain—A Challenge for Europe," 1983, BAC 107/1993 897, European Commission Archives, p. 7.

31. Ibid., p. 9.

32. European Communities, "Resolution on the Combating of Acid Rain, Adopted by the European Parliament at Its Session of 20 January 1984," January 25, 1984, BAC 107/1993 893, European Commission Archives.

33. Ibid., p. 11.

34. F. C. Widdowson, Legal Department, British Coal Corporation, "Air Pollution: CBI Environmental and Technical Legislation Committee," January 23, 1984, COAL 96/91, UK National Archives, p. 1.

35. Dr. A. J. Dainton and M. J. Edwards, "National Coal Board—General Purpose Committee: Sub-committee on Research & Development. Memorandum: Acid Rain," November 17, 1982, COAL 96/85, UK National Archives; "Letter from Ray Beasley to P. M. Moullin Esq, Deputy Secretary National Coal Board," November 24, 1982, COAL 96/85, UK National Archives; "Letter from Deputy Secretary P. M. Moullin to Mr. L. J. Mills," February 1983, COAL 96/85, UK National Archives; "Letter from Dr. A. Challis, CBE Chief Scientist, Department of Energy, to Mr. O. G. Tregelles, National Coal Board," January 4, 1983, COAL 96/85, UK National Archives.

36. A. D. Dainton, National Coal Board, "Coal Research Establishment. The Environmental Problem," January 20, 1983, COAL 96/85, UK National Archives.

37. Dr. A. D. Dainton and M. J. Edwards, "DRAFT. National Coal Board—General Purpose Committee: Sub-committee on Research and Development.

Memorandum: Acid Rain," November 11, 1982, COAL 96/85, UK National Archives.

38. "Letter from P. M. Moullin to Dr. A. D. Dainton, Mr. F. L. Edwards, Mr. W. G. Jensen, Mr. R. Omerrod, Mr. P. G. Tregelles, Mr. R. W. C. Wheatley, Mr. F. C. Widdowson," December 9, 1982, COAL 96/85, UK National Archives, p. 2.

39. "Letter from R. J. Ormerod to Dr. A. D. Dainton," January 5, 1983, COAL 96/85, UK National Archives; "Letter from A. D. Dainton to Mr. R. J. Ormerod," January 10, 1983, COAL 96/85, UK National Archives; "Letter from A. J. Hustings to Mr. R. J. Ormerod," January 13, 1983, COAL 96/85, UK National Archives; Central Planning Unit, National Coal Board, "Draft: The NCB's Response to the Threat of Acid Rain Legislation," January 20, 1983, COAL 96/85, UK National Archives.

40. Central Planning Unit, National Coal Board, "Draft: The NCB's Response to the Threat of Acid Rain Legislation"; A. D. Dainton, "Report of a Meeting between NCB and CEGB," February 23, 1983, COAL 96/85, UK National Archives.

41. "Letter from Sir Walter Marshall to Sir Andrew Huxley, President, The Royal Society," June 3, 1983, COAL 96/91, UK National Archives.

42. The inclusion of the Scandinavian academies was not mentioned in Chester's initial proposal. On the isolation of Britain and the US regarding the scientific evidence for the need to reduce acidic pollution emissions, see Fred Pearce, "It's an Acid Wind That Blows Nobody Any Good," *New Scientist* 95, no. 1313 (July 8, 1982): 80; Jasper Becker, "Acid Rain: UK Unrepentant," *Nature* 298 (July 8, 1982): 112.

43. "Note of the Conclusions of a Meeting Held in Room 370 Hobart House at 2.30 pm 1st March 1983 to Discuss Acid Rain," March 7, 1983, COAL 96/91, UK National Archives, p. 2.

44. This research included such cost saving methods as rerouting high sulfur coal to plants with flue gas desulfurization technology, which was estimated at the time to reduce emissions equivalently to retrofitting three to ten additional plants with the technology at half to a third of the cost. W. D. Halstead, "CEGB Confidential. Report on a Visit to the USA. May 9th to 23rd 1985," August 1, 1985, COAL 29/655, UK National Archives, p. 1; K. Gregory, "Operational Research Executive. Acid Rain Section. Concentration of High Sulphur Coal to Power Stations Fitted with Flue Gas Desulphurisation," January 1985, COAL 29/655, UK National Archives.

45. Ibid.

46. Walter Marshall, "Fund for Independent Research on the Effect of Acid Deposition on Surface Waters and Fisheries," June 2, 1983, COAL 96/91, UK National Archives.

47. Ibid., p. 3.

48. D. Fishlock and L. E. J. Roberts, "Walter Charles Marshall, C. B. E., Lord Marshall of Goring. 5 March 1932–20 February 1996," *Biographical Memoirs of Fellows of the Royal Society* 44 (November 1, 1998): 299–312; Ian Mogford, "CERL: From Shed to Watershed" (National Power, 1993), 7802/1/7–13, Surrey History Centre Archives, pp. 15, 30.

49. "Letter from P. F. Chester to John Mills, National Coal Board," June 23, 1983, COAL 96/91, UK National Archives.

50. Kjell Herlofsen, Norske Videnskaps-Akademi, "Virkningen av sur nedbør på sammensetning av overflatevann og fiskerier i det sydlige Skandinavia," August 5, 1983, RA-S-2532-2-Dca-L6052, Norwegian National Archives.

51. In discussions with the Ministry of the Environment, the Norwegian academy scientists eventually expressed serious trepidation about the potential "strings attached" to the money. See Hans C. Christensen, "Referat fra møte hos Statssekretær Levy Tirsdag L. november 1983," November 3, 1983, RA-S-5806-D-Lo196, Norwegian National Archives, p. 2.

52. Their skepticism was shared by British environmental groups, who openly accused the CEGB of using the study to "purchase" time before they had to make reductions and questioned the objectivity of the research. See Des Wilson, chairman, Friends of the Earth, "Blinded with Science when Peering into the Acid Rain," *Guardian*, September 7, 1983, COAL 74/1716, UK National Archives. These sentiments received sharp rebukes from the British Royal Society. See "Letter from Sir Morris Sugden, Vice President, Royal Society, to the Editor, The Guardian," September 7, 1983, COAL 74/1716, UK National Archives.

53. Erik Lykke, "Notat: Henvendelse til det Videnskaps-Akademiet om norsk deltakelse i et britisk finansiert sur nedbør forskningsprogram," August 9, 1983, RA-PA-0641-E-Eg-Lo013, Norwegian National Archives.

54. The one exception was Ivan Rosenqvist, a longtime critic of the SNSF research. See ibid.

55. Ibid.

56. Arne Semb-Johansson, "Utviklingen av surt vann-prosjektet," December 1, 1983, RA-S-5806-Lo196, Norwegian National Archives.

57. Christensen, "Referat fra møte hos Statssekretær Levy Tirsdag L. november 1983," pp. 1–2.

58. L. U. Mole, "The Royal Society. Surface Water Acidification Programme Management Group," November 30, 1983, RA-S-5806-D-Lo196, Norwegian National Archives.

59. Andrew Moncur, "£5m Research on Ways to Curb Pollution by Acid Rain," *Guardian*, September 6, 1983, p. 2.

60. Georg Parmann, "Britisk prosjekt om sur nedbør startes," *Aftenpostens*, September 5, 1983, RA-S-5806-D-Lo196, Norwegian National Archives.

61. Godfrey Brown, "Britain Is Blamed as Norway Counts Cost of Acid Rain," *Telegraph*, September 16, 1983, FT 48/35, UK National Archives.

62. Parmann, "Britisk prosjekt om sur nedbør startes."

63. Erik Lykke, "Notat: Britisk henvendelse til Videnskaps-Akademiet om deltakelse i et nytt sur nedbør forskningsprogram. (Det vises til notat av 9.8.1983.)," August 26, 1983, RA-PA-0641-E-Eg-Lo013, Norwegian National Archives.

64. Nils Christophersen, "Reiserapport fra ekspertmøte i Geneve 17–19/10.1983," November 3, 1983, RA-S-2532–2-Dca-Lo297, Norwegian National Archives.

65. Semb-Johansson, "Utviklingen av surt vann-prosjektet."

66. Translation by the author. Christophersen, "Reiserapport fra ekspertmøte i Geneve 17–19/10.1983."

67. There were also concerns about the fact that the director of the program was to have broad power and authority, leading Christensen to warn against the potential for him to act as a British "commissioner" for the Scandinavian environment. See Hans C. Christensen, "Samarbeide om surnedbørforskning mellom Norge, Sverige og Storbritannia," December 23, 1983, RA-S-5806-L0196, Norwegian National Archives, pp. 1–2.

68. Ibid., p. 2.

69. Mole, "The Royal Society. Surface Water Acidification Programme Management Group"; G. E. Hemmen, "Surface Water Acidification Programme. Minutes At a Meeting of the Surface Water Acidification Programme Management Group Held at the Royal Society, London on 17 and 18 December 1984," January 14, 1985, RA-S-2939-D-Db-Dbk-Dbke-L0605, Norwegian National Archives.

70. Carl Olaf Tamm, "Anteckningar från sammanträde med management group i SWAP I Kristiansand 19 juni och i haugesund 21 juni 1984," June 25, 1984, RA-S-2532-2-Dca-L0652, Norwegian National Archives, pp. 1- 3.

71. "Letter from Carl Olaf Tamm to J. H. Oxley, Battelle," June 25, 1984, RA-S-2532-2-Dca-L0652, Norwegian National Archives.

72. Southwood took over as chairman after the death of the initial chair, Maurice Sugden, in 1984.

73. "Letter from Carl Olaf Tamm to J. H. Oxley, Battelle," p. 1. Tamm appears to have developed a rapport with Southwood that likely made him feel comfortable expressing his honest appraisal of SWAP's problems, noting at the conclusion of his letter that he had very much enjoyed conversations with Southwood on the role of science in society as well as scientific issues more generally. See p. 2.

74. Catherine Caufield, "Acid Rain Study under Fire," *New Scientist*, September 27, 1984, p. 3.

75. John Mason and Hans Martin Seip, "The Current State of Knowledge on Acidification of Surface Waters and Guidelines for Further Research," *Ambio* 14, no. 1 (January 1, 1985): 45–51.

76. Fred Pearce, "Norwegians Protest over Gag on Research," *New Scientist* 109, no. 1500 (March 20, 1986): 24.

77. Hemmen, "Surface Water Acidification Programme. Minutes At a Meeting of the Surface Water Acidification Programme Management Group Held at the Royal Society, London on 17 and 18 December 1984," p. 7.

78. "Statement of the United Kingdom Delegation, Munich Air Pollution Conference, 24–27 June 1984. Annex B, House of Commons Environment Committee, Minutes of Evidence. Department of the Environment," July 9, 1984, HC-CL-JO-10–1794, British Parliamentary Archives, pp. 299–300.

79. "House of Commons Environment Committee. Session 1983–1984. Acid Rain: Minutes of Evidence. National Coal Board," July 2, 1984, HC-CL-JO-10–1794, British Parliamentary Archives, p. 296.

80. House of Commons Environment Committee, "Acid Rain," December 16, 1983, HC-CP-9073, British Parliamentary Archives.

81. Their initial notes on gathering background information on acid rain specifically mention the need to investigate possible legislation at the level of the European Communities. See ibid., pp. 1–2.

82. The House of Commons Environment Committee consisted of seven Conservatives, three Labour party members, and one Liberal. Its composition is striking given its outspoken critiques of Thatcher. See Armin Rosencranz, "The Acid Rain Controversy in Europe and North America: A Political Analysis," *Ambio* 15, no. 1 (January 1, 1986): 49.

83. The Canadian Minister of the Environment spearheaded the meeting after growing frustrated with the Reagan administration's refusal to discuss a bilateral agreement. The nations that attended were Canada, West Germany, France, the Netherlands, Norway, Sweden, Denmark, Finland, Austria, and Switzerland. "Canada, 9 Other Countries Move to Combat Acid Rain," *Sun*, March 22, 1984, p. A2; "9 Countries to Confer in Ottawa on Acid Rain," *New York Times*, February 7, 1984, p. A5.

84. "Accord on Acid Rain," *Washington Post*, March 22, 1984, p. A36; "Acid Rain Moves to World Forums," *Christian Science Monitor*, April 2, 1984, p. 23.

85. For more on the Thirty Percent Club, see Alex Scott, "Britain Pressed on Acid Rain," *Guardian*, November 29, 1985, p. 8.

86. Erik Lykke, "Sur nedbør: Ottawa-møtet. Spørsmålet om bindende forpliktelse til a redusere So_2-utslippene med minst 30 prosent innen 1993," January 17, 1984, RA-S-2532-2-Daa-Lo511, Norwegian National Archives.

87. The House of Commons Environment Committee solicited the assistance of several academic scientists with expertise in pollution problems, most notably Nigel Bell of Imperial College London and Alan Williams of Leed University. Bell, Williams, and other university scientists provided private and nearly always critical assessments of the CEGB's scientific claims during the House of Commons hearings.

88. "Letter from the Central Electricity Generating Board to the Environment Committee. House of Commons Environment Committee. Session 1983–1984," March 7, 1984, HC-CL-JO-10–1794, British Parliamentary Archives, p. 3.

89. Parliamentary Debates, House of Commons, *Acid Rain* (London, UK: Hansard, January 11, 1985), cc. 1023–1024. Walter Waldegrave, who represented the Department of the Environment at the hearing, also asserted that there was no scientific consensus on acid rain during the debate. It was amid this dispute that Dr. David Clark, a Labour Party politician for over two decades, accused his government of "waging chemical warfare not only on our neighbors but on our country." Dr. Clark prefaced these remarks with an avowal that he was choosing his words very carefully. Representatives from the Department of the Environment repeatedly asked Clark to take back his statement, but he refused.

90. "Minutes of Evidence. Central Electricity Generating Board. Acid Rain. House of Commons Environment Committee. Session 1983–1984," May 21, 1984, HC-CL-JO-10–1794, British Parliamentary Archives, pp. 20–21.

91. Ibid., p. 36.

92. Secretary of State for the Environment, by Command of Her Majesty, "Department of the Environment: Acid Rain. The Government's Reply to the Fourth Report from the Environment Committee," December 1984, HL-PO-JO-10–11–2264, British Parliamentary Archives.

93. Ibid., pp. 5-8.

94. It was the first Environmental Committee Report debated on the floor of the House of Commons since the 1970s. See Parliamentary Debates, House of Commons, *Acid Rain*, cc. 1012.

95. House of Commons Environment Committee, "Acid Rain. Volume I. Report Together with the Proceedings of the Committee Relating to the Report," July 30, 1984, HC-CL-JO-10-1794, British Parliamentary Archives.

96. Ibid., p. xiv.

97. Alexander MacLeod, "British Government and Parliament at Odds over Acid Rain," *Christian Science Monitor*, September 12, 1984, sec. International, p. 15.

98. "Letter from Sir Hugh Rossi, Chairman of the House of Commons Environment Committee, to Sir Walter Marshall, Chairman of the Central Electricity Generating Board," November 14, 1984, HC-CP-9068, British Parliamentary Archives, p. 1.

99. Ibid., p. 3.

100. For the Rossi/Marshall debate, see Central Electricity Generating Board, "Environment Committee Report on Acid Rain Statement by CEGB," September 18, 1984, HC-CP-9073, British Parliamentary Archives; House of Commons Environment Committee, "Press Notice Sir Hugh Jousts Sir Walter," November 14, 1984, HC-CP-9073, British Parliamentary Archives; "Letter from Sir Walter Marshall to Sir Hugh Rossi," December 12, 1984, HC-CP-9073, British Parliamentary Archives.

101. "Letter from Carl-Dieter Spranger, Parliamentary Secretary, Ministry of the Interior, Deutschen Bundestages to Sir Hugh Rossi, Chairman of the Environment Committee, House of Commons," September 5, 1984, HC-CP-9073, British Parliamentary Archives. The Environment Committee was discouraged to note that only its research recommendations appeared to have any influence on government allocation of funding and resources to study the problem. R. H. Hobden, Clerk to Committee, "Acid Rain: Government's Response," December 3, 1984, HC-CP-9073, British Parliamentary Archives.

102. Secretary of State for the Environment, by Command of Her Majesty, "Department of the Environment: Acid Rain. The Government's Reply to the Fourth Report from the Environment Committee," p. 21. Although the government report repeatedly stated that it had made considerable reductions in pollution since 1970, almost none of this reduction had come from the CEGB, despite it being the largest source of sulfur dioxide in the country. House of Commons Environment Committee, "Acid Rain. Volume I. Report Together with the Proceedings of the Committee Relating to the Report," p. xi.

103. United Nations, *Protocol to the 1979 Convention on Long-Range Transboundary Air Pollution on the Reduction of Sulphur Emissions or Their Transboundary Fluxes by at least 30 Per Cent*, opened for signature July 8, 1985, Treaty Series: Treaties and International Agreements Registered of Filed and Recorded with the Secretariat of the United Nations, vol. 1480, p. 215, https://treaties.un.org/doc/Publication/UNTS/Volume%201480/v1480.pdf.

104. John Ardill and Donald Fields, "New Target for Cutting Pollution," *Guardian*, July 9, 1985, p. 2. The Helsinki Protocol specified that countries

could substitute reductions in "transboundary fluxes" instead of total domestic emissions to comply with the agreement. This was negotiated to bring the Soviet Union and Eastern Europe onboard. See Article 5 of United Nations, *Protocol to the 1979 Convention on Long-Range Transboundary Air Pollution on the Reduction of Sulphur Emissions or Their Transboundary Fluxes by at least 30 Per Cent*. On Britain's decision to abstain from signing the Helsinki Protocol, see Tony Samstag, "Britain Likely to Break Ranks on Acid Rain; Helsinki Meeting on National Sulphur Emissions," *Times* (London), July 8, 1985, p. 2; Scott, "Britain Pressed on Acid Rain."

105. A. J. Hustings, "Central Electricity Generating Board/National Coal Board. Joint Study of Sulphur in Coal Record of the Eleventh Meeting Held on Friday 7th June, 1985 at Hobart House," June 24, 1985, COAL 29/655, UK National Archives.

106. Walter G. Jensen, "European Community: Developments in Coal and Energy Affairs," May 1985, COAL 29/655, UK National Archives, pp. 1–3.

107. "Letter from Martin Holdgate, Chief Scientist, Department of the Environment to JD Rimington, Director General, Health and Safety Executive," April 22, 1986, BT 328/152, UK National Archives.

108. The UK reserved its agreement to the decision. "Draft EC Large Combustion Plant Directive Declaration Agreed at the Environment Council, 6 March 1986. Annex I," March 1986, BT 328/152, UK National Archives.

109. Department of the Environment, "Draft EC Large Combustion Plant Directive: Background Note on the Directive," April 22, 1986, BT 328/152, UK National Archives.

110. Ibid., p. 2.

111. Specific country targets were still to be negotiated. Department of the Environment, "Cabinet Steering Committee on European Questions. Draft EC Large Combustion Plant Directive. Note by the Department of the Environment," May 7, 1986, BT 328/152, UK National Archives.

112. Ibid., pp. 2–3.

113. Cabinet Office, "Cabinet Steering Committee on European Questions. Minutes of a Meeting Held in Conference Room C, Cabinet Office on Friday 9 May 1986 at 2.30 PM," May 12, 1986, BT 328/152, UK National Archives.

114. Ibid., pp. 5–6.

115. "Letter from R. J. Perriman, Chief Industrial Air Pollution Inspector to J. R. Rimington, Director General," May 14, 1986, BT 328/152, UK National Archives.

116. "Central Electricity Generating Board/British Coal. Joint Study of Sulphur in Coal. A Record of the Thirteenth Meeting Held on Wednesday, 11 June, 1986 at Bretby," July 14, 1986, COAL 29/655, UK National Archives, p. 3.

117. P. F. Chester, "Acid Lakes in Scandinavia—The Evolution of Understanding" (Central Electricity Research Laboratories, July 1986), personal papers, Richard Skeffington, obtained during interview.

118. Ibid., p. 14.

119. Margaret Thatcher, who studied chemistry at the University of Oxford, had a reputation for wanting to engage directly with scientists and government officials about the intricacies of scientific debates on several issues, including acid

rain. Author interview with Richard Skeffington, professor, University of Reading, former biologist at the CERL, March 15, 2013; author interview with Anthony Kallend, retired, head of environmental chemistry at CERL, March 25, 2013.

120. "Cabinet. Ministerial Steering Committee on Economic Strategy. Sub Committee On Economic Affairs. Acid Rain: The Scientific Evidence: Note by Secretaries of State for the Environment and for Energy," September 4, 1986, BT 328/152, UK National Archives.

121. "Sulphur Emissions, Acid Deposition and Freshwater Ecosystems: Current Scientific Understanding. Memorandum by the Chief Scientist, Department of Energy, Chief Environment Scientist, Department of the Environment, and Director, Environment, Technology, Planning and Research Division, Central Electricity Generating Board," September 4, 1986, BT 328/152, UK National Archives.

122. Ibid.

123. CEGB Press Information, "CEGB to Reduce Power Station Emissions. Recommendation to Government Based on New Research Results," September 11, 1986, COAL 29/655, UK National Archives.

124. John Ardill and Jonathan Steele, "Move to Curb Acid Rain Fails to Ease PM's Path," Guardian, September 12, 1986, p. 1; Reuters, "Maggie's Norway Trip Off to Stormy Start," New Straits Times, September 13, 1986, p. 13.

125. CEGB Press Information, "CEGB to Reduce Power Station Emissions. Recommendation to Government Based on New Research Results," p. 3. After Britain agreed to begin retrofitting three power plants with flue gas desulfurization technologies in September 1986, the CEGB and National Coal Board ceased its regularly scheduled meetings on acid rain and sulfur in coal. See P. Cammack, "Coal Use Environment Committee. Acid Rain. NCB/CEGB Joint Study on Sulphur in Coal," November 4, 1986, COAL 29/655, UK National Archives, pp. 5–6.

126. Margaret Thatcher, "Press Conference Visiting Norway," September 12, 1986, Thatcher Archive: COI transcript, Margaret Thatcher Foundation. Scholars have largely taken Thatcher at her word on the role of SWAP in this decision. For example, Maarten Hajer states that it was scientific results from SWAP that persuaded Britain to introduce flue gas desulfurization in 1986. See Maarten A. Hajer, "Discourse Coalitions and the Institutionalization of Practice: The Case of Acid Rain in Britain," in The Argumentative Turn in Policy Analysis and Planning, ed. Frank Fischer and John Forester (Durham: Duke University Press, 1993). For news reports on the reversal in British policy, see "Government Sets Up £600m Project to Combat Acid Rain," Glascow Herald, September 12, 1986, p. 8; David Hencke and John Ardill, "Britain May Join 'Club' for Acid Rain Reduction," Guardian, September 4, 1986, p. 2.

127. In comparison, other member states committeed to seventy percent reductions. Matthieu Glachant, Implementing European Environmental Policy: The Impacts of Directives in the Member States (Cheltenham, UK: Edward Elgar Publishing, 2001), p. 60.

128. The explicit use of the precautionary principle in grappling with pollution problems was part of a growing need to address environmental threats on longer and longer timescales with even more costly implications for industry. Global warming and the ozone hole, which both became matters of political interest toward the end of the heyday of debates on acid rain, are the clearest

examples of this. On similarities between acid rain, ozone, and global warming, see William C. Clark, *Learning to Manage Global Environmental Risks: A Comparative History of Social Responses to Climate Change, Ozone Depletion, and Acid Rain* (Cambridge, MA: MIT Press, 2001).

129. The United Nations Conference on Environment and Development, "Rio Declaration on Environment and Development," Rio de Janeiro, June 3–14, 1992.

CHAPTER EIGHT

1. "The Next Generation of Poison: An Issue for the No-Issue Campaign," *New York Times*, October 9, 1988, p. E22.

2. "Treasury Vetoes Action on Acid Rain," *New Scientist*, September 15, 1983, COAL 74/1716, UK National Archives; "Acid Rain: More Thunder than Enlightening," *Economist*, September 17, 1983, COAL 74/1716, UK National Archives.

3. Andy Pasztor, "Reagan to Request Added Funds to Probe Acid Rain but Won't Seek Pollution Curbs," AMAX Daily Press Clippings, January 25, 1984, COAL 96/91, UK National Archives.

4. According to Likens, the Edison Electric Institute had earlier solicited scientists for proposals to show that Likens' research on acid rain was incorrect. Author interview with Gene Likens, August 18, 2017; "Is Industry Poisoning the Forests?," *Wall Street Journal*, June 2, 1984, COAL 96/91, UK National Archives.

5. "Acid Rain," *Wall Street Journal*, June 30, 1980, p. 14.

6. Naomi Oreskes and Erik Conway have also made note of the disparity of coverage of acid rain. See Naomi Oreskes and Erik M. Conway, *Merchants of Doubt: How a Handful of Scientists Obscured the Truth on Issues from Tobacco Smoke to Global Warming* (New York: Bloomsbury Press, 2010), pp. 101–102.

7. The British journal *Atmospheric Environment* published many of these articles by power industry scientists. Scandinavian, British, and American scientists viewed the journal as mishandling the publication process on acid rain articles and sometimes having an industry bias. For examples, see D. A. Hansen and G. M. Hidy, "Review of Questions Regarding Rain Acidity Data," *Atmospheric Environment* 16, no. 9 (January 1, 1982): 2107–2126; Charles V. Cogbill, Gene E. Likens, and Thomas A. Butler, "Uncertainties in Historical Aspects of Acid Precipitation: Getting It Straight," *Atmospheric Environment* 18, no. 10 (January 1, 1984): 2261–2268; D. A. Hansen and G. M. Hidy, "Comments from Hansen and Hidy," *Atmospheric Environment* 18, no. 10 (January 1, 1984): 2268–2270.

8. For a debate between Gene Likens and Michael Oppenheimer against Laurence Kulp, Ralph Perhac, and George Hidy about the need to reduce emissions at one such conference, see James C. White, *Acid Rain: The Relationship between Sources and Receptors* (New York: Springer Science & Business Media, 1988).

9. *Acid Rain, 1984: Hearings before the Committee on Environment and Public Works, Senate, Ninety-Eighth Congress, Second Session* (Washington, DC: US G.P.O., 1984), p. 393.

10. "Review the Effects of Acid Deposition and Other Air Pollutants on Forest Productivity; Forest Ecosystems and Atmospheric Pollution Research Act of 1985; and the Endangered Forest Research Act of 1985," Subcommittee on Forests, Family Farms, and Energy, Committee on Agriculture, United States House

of Representatives, Ninety-Ninth Congress, Second Session (Washington, DC: US G.P.O., 1986), pp. 73–84, 314.

11. Leslie R. Alm, *Crossing Borders, Crossing Boundaries: The Role of Scientists in the U.S. Acid Rain Debate* (Westport, CT: Praeger, 2000), pp. 70–73; Oreskes and Conway, *Merchants of Doubt*, p. 103; Oran R. Young, *The Effectiveness of International Environmental Regimes: Causal Connections and Behavioral Mechanisms* (Cambridge, MA: MIT Press, 1999), p. 245.

12. Associated Press, "U.S. Acid-Rain Aide Resigns," *New York Times*, September 26, 1987, p. 36; Philip Shabecoff, "Government Acid Rain Report Comes under Sharp Attack," *New York Times*, September 22, 1987, p. C1.

13. Leslie Roberts, "Fresh Look at Acid Rain," *Science* 240 (May 6, 1988): 715.

14. Ron Doel, "Oral History: J. Laurence Kulp—Session II," April 12, 1996, American Institute of Physics Niels Bohr Library and Archives. Gene Likens debated Kulp over the interim report on television and stated that Kulp lied several times about the issue during its filming. Author interview with Gene Likens, August 18, 2016.

15. "New Technical Reports," *EPRI Journal* 15, no. 9 (1990): 51; J. Laurence Kulp, "Acid Rain: Causes, Effects and Control," *CATO Review of Business and Government* 13, no. 1 (1990).

16. Donald Rheem, "Environmental Action: A Movement Comes of Age," *Christian Science Monitor*, January 14, 1987, p. 16.

17. John Fitzgerald, "Environmentalists Reinforced in Congress," *Hartford Courant*, February 2, 1987, p. A1.

18. John Fitzgerald, "Water Bill Enacted by Senate: Presidential Veto Overridden, 86–14," *Hartford Courant*, February 5, 1987, p. A1.

19. Philip Shabecoff, "U.S. Agrees to Limit Pollutant Linked to Acid Rain," *New York Times*, November 2, 1988, sec. US, p. A24.

20. "Reagan's U-Turn on Acid Rain Seen as Election Strategy," *Nature* 334, no. 6182 (August 11, 1988): 460. Concern about ozone depletion also might have played a part in President Reagan's softening on environmental issues, but there were efforts by antiregulatory forces to derail negotiations on that issue as well. See Richard Elliot Benedick, *Ozone Diplomacy: New Directions in Safeguarding the Planet*, enl. ed. (Cambridge, MA: Harvard University Press, 1991), pp. 46, 58–67.

21. Michael Weisskopf, "Reagan Agrees to Freeze U.S. Emissions of Pollutant: Way Is Cleared for Acid-Rain Treaty; Reagan Order Clears Way for Treaty on Acid Rain," *Washington Post*, August 6, 1988, p. A1.

22. Philip Shabecoff, "U.S. and Soviet to Study Ways to Save the Ozone," *New York Times*, December 19, 1986, p. A14.

23. "The Next Generation of Poison," p. E22.

24. The change in media coverage of the two issues together in 1988 is dramatic compared to several years prior, especially in the summer prior to the fall election. See "The Greenhouse Effect? Real Enough," *New York Times*, June 23, 1988, p. A22; Ronald Kotulak, "Pollution Expert Warns of Rise in Deaths, Disease," *Chicago Tribune*, June 28, 1988, p. 3; Philip Shabecoff, "Parley Urges Quick Action to Protect Atmosphere," *New York Times*, July 1, 1988, p. A3; Warren Leary, "Reagan, in Switch, Agrees to a Plan on Acid Rain," *New York Times*, August 7, 1988, p. 1.

25. Clifford May, "Pollution Ills Stir Support for Environment Groups," *New York Times*, August 21, 1988, p. 30.

26. "Environment and Election," *Los Angeles Times*, September 6, 1988, p. E6.

27. Philip Shabecoff, "Democrats Assail G.O.P. on Pollution: Presidential Hopefuls Gather to Discuss Environment—Republicans Absent," *New York Times*, November 2, 1987, p. B8.

28. Timothy McNulty, "Bush Scrubs GOP's Image on Pollution: Campaign '88," *Chicago Tribune*, September 1, 1988, p. 1.

29. Barbara Rosewicz and Michel McQueen, "Bush, Resolving Clash in Campaign Promises, Tilts to Environment: How He Settled on Proposals That Irk Business Shows Break with Reagan Style," *Wall Street Journal*, June 13, 1989, p. A1.

30. Richard Conniff, "The Political History of Cap and Trade," *Smithsonian Magazine*, August 2009, http://www.smithsonianmag.com/science-nature/the-political-history-of-cap-and-trade-34711212/.

31. Gabriel Chan, Robert Stavins, Robert Stowe, and Richard Sweeney, "The SO_2 Allowance-Trading System and the Clean Air Act Amendments of 1990: Reflections on 20 Years of Policy Innovation," *National Tax Journal* 65, no. 2 (June 2012): 419–452.

32. Volker A. Mohnen, "The Challenge of Acid Rain," *Scientific American* 259, no. 2 (August 1988): 30–38.

33. Henry A. Waxman, "Overview and Critique: An Overview of the Clean Air Act Amendments of 1990," *Environmental Law* 21 (July 1, 1991): 1721–1816, p. 1793.

34. Richard Stevenson and Thomas Hayes, "Bush Clean-Air Plan Assessed," *New York Times*, June 14, 1989, p. A1.

35. Edward J. Markey, "Give Global Warming the Acid Rain Test," *New York Times*, November 21, 1989, p. A24.

36. Kathy Sawyer, "Groups Say Clean Air Bill Needs Work," *Washington Post*, June 16, 1989, p. A9.

37. For example, Canada agreed to further cuts in mobile source emissions that mirrored those the US imposed through the 1990 Clean Air Act amendments. See Jeffrey Roelofs, "United States-Canada Air Quality Agreement: A Framework for Addressing Transboundary Air Pollution Problems," *Cornell International Law Journal* 26, no. 2 (April 1, 1993): 421–454.

38. Waxman, "Overview and Critique: An Overview of the Clean Air Act Amendments of 1990," p. 1794.

39. Matthew L. Wald, "Utility Is Selling Right to Pollute," *New York Times*, May 12, 1992, p. A1.

40. Richard A. Kerr, "Acid Rain Control: Success on the Cheap," *Science* 282, no. 5391 (November 6, 1998): 1024–1024.

41. Andy Coghian, "EEC to Corner Britain over Sulphur Emissions," *Chemistry and Industry*, November 18, 1985, BT 308/152, UK National Archives. On differences between lowering emissions through a bubble concept versus other forms of distributing reduction requirements, see Cecilia Albin, "Rethinking Justice and Fairness: The Case of Acid Rain Emission Reductions," *Review of International Studies* 21, no. 2 (April 1, 1995): 119–143.

42. For an example of such a model developed for the United Nations Economic Commission for Europe, see Jean-Paul Hettelingh and Leen Hordijk, "Environmental Conflicts: The Case of Acid Rain in Europe," *Annals of Regional Science* 20, no. 3 (November 1986): 38–52.

43. The concept of critical loads, while apparently discussed between American and Canadian diplomats, never seems to have taken hold in scientific and policy circles in North America to the extent that occurred in Europe. For some background discussion on the US and Canadian use of this concept, see Samuel P. Hays, *Explorations in Environmental History: Essays* (Pittsburgh: University of Pittsburgh Press, 1998), pp. 281–282. Kenneth Wilkening claims the concept of critical loads emerged in Canada in 1985, but this is questionable given the long-standing interest of Scandinavian scientists in developing such a tool for policymakers. See Kenneth E. Wilkening, *Acid Rain Science and Politics in Japan: A History of Knowledge and Action toward Sustainability* (Cambridge, MA: MIT Press, 2004), pp. 190–191.

44. William C. Clark, *Learning to Manage Global Environmental Risks: A Comparative History of Social Responses to Climate Change, Ozone Depletion, and Acid Rain* (Cambridge, MA: MIT Press, 2001), p. 327.

45. *Critical Loads for Sulphur and Nitrogen: Report from a Workshop Held in Skokloster, Sweden, 19–24 March, 1988* (Skokloster, Sweden: Nordic Council of Ministers, 1988), p. 8.

46. The IIASA was initially hesitant to incorporate work on global systems modeling because of controversy over such methods following the publication of *The Limits of Growth* by the Club of Rome. Nonetheless, it sponsored a series of conferences on global systems modeling shortly after its founding, and this work became a focus of its resident scientists. These scientific attachés often stayed at the IIASA for a term of approximately two years, but this was variable. Regarding the controversy over incorporating global modeling into the IIASA's work, see Paul N. Edwards, *A Vast Machine: Computer Models, Climate Data, and the Politics of Global Warming* (Cambridge, MA: MIT Press, 2010), p. 370.

47. Roger Levien, "RAND, IIASA, and the Conduct of Systems Analysis," in *Systems, Experts, and Computers: The Systems Approach in Management and Engineering, World War II and After*, ed. Thomas Hughes and Agatha C. Hughes (Cambridge, MA: MIT Press, 2011).

48. "Møterapport fra Ivar Isaksen, Institutt for Geofysikk, i forbindelse med deltagelse I IEA/OECD Workshop on Carbon Dioxide Research and Assessment, 11th-12th February 1981," 1981, RA-S-2532-2-Dca-L6048, Norwegian National Archives; "Letter from Lars Björkbom, Ministry for Foreign Affairs, Sweden to Mr. C. Lopez-Polo, Director, Environment and Human Settlement Division, ECE," August 31, 1983; "Rapport: Møte i ECE om kostnadsnyttevurderinger av tiltak mot utslipp av So_2 13.-15. desember 1982," February 28, 1983, RA-S-2532-2-Dca-L0297, Norwegian National Archives; "Statement by WHO/EURO, First Session of the Executive Body for the Convention on Long-Range Transboundary Air Pollution," June 8, 1983, RA-S-2532-2-Dca-L0297, Norwegian National Archives; "Draft Report: Work Plan for the Third Phase of EMEP, 1984–1986 Chemical Programme," November 11, 1983, RA-S-2532-2-Dca-L0293, Norwegian National Archives, p. 7.

49. Petter Talleraas and Kristin Magnussen, "Invitasjon til møte i Miljøvernde-partementet torsdag 4. desember. Presentasjon av modeljæn for langtransport-erte luftforurensninger i Norge," November 21, 1986, RA-S-2939-D-Db-Dbk-Dbke-L0605, Norwegian National Archives.

50. J. Alcamo, R. Shaw, and L. Hordijk, eds., *The RAINS Model of Acid-ification: Science and Strategies in Europe*, 1990 edition (Dordrecht: Springer, 1990), p. vi; Leen Hordijk, "Use of the RAINS Model in Acid Rain Negotiations in Europe," *Environmental Science & Technology* 25, no. 4 (1991): 596–603, p. 598.

51. Harvey Brooks and Alan McDonald, "The International Institute for Applied Systems Analysis, the TAP Project, and the RAINS Model," in *Systems, Experts, and Computers: The Systems Approach in Management and Engineer-ing, World War II and After*, ed. Agatha C. Hughes and Thomas Hughes (Cam-bridge, MA: MIT Press, 2011), pp. 420–423.

52. The atmospheric portions of the IIASA model were largely based on the work of Ivar Isaksen at the Norwegian Meteorological Institute in Oslo. Geir Taugbøl, "Reiserapport. First Meeting of Acidification Research Coordinators (Marc. 1) 17–19 November 1986. Bilthoven, Nederland," December 1, 1986, RA-S-3281-A-L0001, Norwegian National Archives.

53. Brynjulf Ottar, "Rapport fra møte i Geneve angående global spredning av luftforurensninger og mulighetene for å begrense deter igjennom lovbestem-melser," August 1986, NILU RR 17/86, NILU Library and Archives. For examples of such work by Ottar, see B. Ottar, J. M. Pacyna, and T. C. Berg, "Aircraft Mea-surements of Air Pollution in the Norwegian Arctic," December 1984, NILU Re-port 59/84, NILU Library and Archives; Jozef M. Pacyna, Brynjulf Ottar, and Val Vitols, "The Occurrence of Air Pollutants in the Arctic Measured by Aircraft," De-cember 1985, NILU Report 66/85, NILU Library and Archives. Ottar retired from his position as director of NILU in 1988 and died the following year at age 71.

54. Willemijn Tuinstra, Leen Hordijk, and Markus Amann, "Using Com-puter Models in International Negotiations: The Case of Acidification in Eu-rope," *Environment: Science and Policy for Sustainable Development* 41, no. 9 (November 1, 1999): 32–42.

55. Scholars have been highly critical of the idea of consensus group assess-ment as a tool in policy negotiations, especially regarding climate change and the IPCC. See Keynyn Brysse, Naomi Oreskes, Jessica O'Reilly, et al., "Climate Change Prediction: Erring on the Side of Least Drama?," *Global Environmental Change* 23, no. 1 (February 2013): 327–337.

56. In fact, some of the IIASA modelers involved in the negotiation process themselves attributed the model's success in guiding negotiations to their close relationships with the delegates rather than the superiority of their modeling ap-proach in terms of accuracy. See Markus Amann, "Some Lessons from the Use of the RAINS Model in International Negotiations," in *Diplomacy Games*, ed. Pro-fessor Dr Rudolf Avenhaus and Professor I. William Zartman (Berlin: Springer, 2007), pp. 197–210.

57. Tuinstra et al., "Using Computer Models in International Negotiations."

58. Ibid., p. 38. For the per country percentage reductions, see United Na-tions, *Protocol to the 1979 Convention on Long-Range Transboundary Air Pol-lution on Further Reduction of Sulphur Emissions*, opened for signature June 14,

1994, Treaties and International Agreements Registered of Filed and Recorded with the Secretariat of the United Nations, vol. 2030, p. 122, https://treaties.un.org/doc/Publication/UNTS/Volume%202030/v2030.pdf.

59. For press coverage, see as examples Jessica Mathews, "Clean Sweeps: Two Success Stories for the Environment," *Washington Post*, December 18, 1995, p. A23; Dallas Burtraw, "Call It 'Pollution Rights,' But It Works: At Last, an Innovative Environmental Policy We Can Breathe Easier Over," *Washington Post*, March 31, 1996, p. C3; Paul Brown, "The Dilemma That Confronts the World: Water Shortages, Global Warming and Nitrogen Pollution Threaten Planet's Future Unless Politicians Act Now, Says UN Environment Report," *Guardian*, September 16, 1999, p. 3; "The Invisible Green Hand," *Economist*, July 4, 2002, http://www.economist.com/node/1200205. The success of the cap-and-trade system of acid rain has also been heavily promoted by the Environmental Defense Fund, and still is today. See "Acid Rain Pollution Solved Using Economics," *Environmental Defense Fund*, accessed July 24, 2017, https://www.edf.org/approach/markets/acid-rain.

60. Gun Lövblad, Leonor Tarrasón, Kjetil Tørseth, and Sergey Dutchak, eds., *EMEP Assessment. Part I: European Perspective* (Oslo: Norwegian Meteorological Institute, 2004), pp. 17, 162.

61. Chan et al., "The SO_2 Allowance-Trading System and the Clean Air Act Amendments of 1990."

62. Kerr, "Acid Rain Control," p. 1024.

63. Lauraine G. Chestnut and David M. Mills, "A Fresh Look at the Benefits and Costs of the US Acid Rain Program," *Journal of Environmental Management* 77, no. 3 (November 1, 2005): 252–266.

64. The cessation of the SWAP research project was especially significant in this regard. See "Money Runs Out for Acid Rain Research," *New Scientist*, March 24, 1990, https://www.newscientist.com/article/mg12517090-900-and-money-runs-out-for-acid-rain-research/. On US cuts over the course of the 1990s, see *U.S. Congressional Record. Proceedings and Debates of the 106th Congress, Second Session* (Washington, DC: US GPO, 2000), p. 11751.

65. Lövblad et al., *EMEP Assessment*, p. 160.

66. Oreskes and Conway, *Merchants of Doubt*, pp. 197–215.

67. Shannon Hall, "Exxon Knew about Climate Change Almost 40 Years Ago," *Scientific American*, October 26, 2015, https://www.scientificamerican.com/article/exxon-knew-about-climate-change-almost-40-years-ago/.

68. Thomas A. Clair and Atle Hindar, "Liming for the Mitigation of Acid Rain Effects in Freshwaters: A Review of Recent Results," *Environmental Review* 13 (2005): 91–128.

69. Per Warfvinge, Maria Holmberg, Maximilian Posch, and Richard F. Wright, "The Use of Dynamic Models to Set Target Loads," *Ambio* 21, no. 5 (August 1, 1992): 369–376.

70. Hans Løkke, Jesper Bak, Ursula Falkengren-Grerup, et al., "Critical Loads of Acidic Deposition for Forest Soils: Is the Current Approach Adequate?," *Ambio* 25, no. 8 (December 1, 1996): 510–516.

71. Tryge Hesthagen, Iver H. Sevaldrud, and Hans M. Berger, "Assessment of Damage to Fish Populations in Norwegian Lakes Due to Acidification," *Ambio* 28, no. 2 (March 1, 1999): 112–117.

72. Rob Edwards, "Revealed: The 200 Scottish Lochs Polluted by Acid Rain," *Herald Scotland*, October 20, 2012.

73. US Government Accountability Office, "Water Quality: EPA Faces Challenges in Addressing Damage Caused by Airborne Pollutants" (Washington, DC, January 2013), http://www.gao.gov/products/GAO-13-39.

74. Douglas A. Burns, Mark E. Fenn, Jill S. Baron, et al., "National Acid Precipitation Assessment Program Report to Congress 2011: An Integrated Assessment" (Executive Office of the President, National Science and Technology Council, December 2011).

75. Martin Lorenz, "Air Pollution Impacts on Forests in a Changing Climate," in *Forests and Society—Responding to Global Drivers of Change*, ed. Gerardo Mery, Pia Katila, Glenn Galloway, Rene I. Alfaro, Markku Kanninen, Maxim Lobovikov, and Jari Varjo (Vantaa, Finland: International Union of Forest Research Organizations, 2010).

76. Per Gundersen, "Air Pollution and Acid Rain," in *Sceptical Questions and Sustainable Answers*, ed. Christian Ege and Jeanne Lind Christiansen (Danish Ecological Council, 2002); Fredric C. Menz and Hans M. Seip, "Acid Rain in Europe and the United States: An Update," *Environmental Science & Policy 7*, no. 4 (August 1, 2004): 253–265.

77. For the media argument, see Nils Roll-Hansen, "Science, Politics, and the Mass Media: On Biased Communication of Environmental Issues," *Science, Technology & Human Values* 19, no. 3 (July 1, 1994): 324–341. For an example of distorting the forest death issue to claim acid rain does not harm forests, see Bjørn Lomborg, *The Skeptical Environmentalist: Measuring the Real State of the World* (Cambridge: Cambridge University Press, 2001). For a response to Lomborg, see Naomi Oreskes, "Science and Public Policy: What's Proof Got to Do with It?," *Environmental Science & Policy* 7, no. 5 (October 2004): 369–383.

78. Daniel Markewitz, Daniel D. Richter, H. Lee Allen, and Byron J. Urrego, "Three Decades of Observed Soil Acidification in the Calhoun Experimental Forest: Has Acid Rain Made a Difference?," *Soil Science Society of America Journal* 62, no. 5 (September 1998): 1428–1439.

79. Ibid.

80. Christopher G. Reuther, "Winds of Change: Reducing Transboundary Air Pollutants," *Environmental Health Perspectives* 108, no. 4 (April 1, 2000): A170–175.

EPILOGUE

1. Erik Lykke, "Pollution Problems across International Boundaries" (Twelfth Rochester International Conference on Environmental Toxicity, University of Rochester, NY, May 21, 1979), RA-S-2532-2-Dca-L0291, Norwegian National Archives.

2. Bert Bolin, "Världskonferens—Nödvändig! -Effektiv?," *Framsteg*, 1969, sec. no. 5, SE/RA/322619/~/6. Nationalkommittén för 1972 års FN-konferens om den mänskliga miljön. Kommittén för forskning och andra sakfrågor, protokoll jämte bilagor nr 1–49, Swedish National Archives, Marieberg. Bert Bolin, "Case Study 'Nederbördens Försurning,'" November 2, 1970, SE/RA/322619/~/8. Na-

tionalkommittén för 1972 års FN-konferens om den mänskliga miljön. Kommittén för forskning och andra sakfrågor, protokoll jämte bilagor nr 91–141, Swedish National Archives, Marieberg.

3. David G. Hirst, "Controlling the Agenda," in *International Organizations and Environmental Protection: Conservation and Globalization in the Twentieth Century*, ed. Wolfram Kaiser and Jan-Henrik Meyer (New York: Berghahn Books, 2016). This was also the lesson many diplomats had drawn from the ozone depletion issue. See Richard Elliot Benedick, *Ozone Diplomacy: New Directions in Safeguarding the Planet*, enl. ed. (Cambridge, MA: Harvard University Press, 1991).

4. For the IPCC reports, see Intergovernmental Panel on Climate Change, "Reports," https://www.ipcc.ch/publications_and_data/publications_and_data _reports.shtml.

5. Bolin was asked to accept the award on behalf of the IPCC but was sadly too ill with stomach cancer to receive it in person. Paul Brown, "Obituary: Bert Bolin," *Guardian*, January 9, 2008, sec. Environment, http://www.theguardian .com/environment/2008/jan/09/climatechange.mainsection.

6. Michael Le Page, "Was Kyoto Climate Deal a Success? Figures Reveal Mixed Results," *New Scientist*, June 14, 2016, https://www.newscientist.com /article/2093579-was-kyoto-climate-deal-a-success-figures-reveal-mixed -results/; Glen P. Peters, Jan C. Minx, Chrisopher L. Weber, and Ottmar Edenhofer, "Growth in Emission Transfers via International Trade from 1990 to 2008," *Proceedings of the National Academy of Sciences* 108, no. 21 (May 24, 2011): 8903–8908.

7. Jonathan Watts and Kate Connolly, "World Leaders React after Trump Rejects Paris Climate Deal," *Guardian*, June 2, 2017, sec. Environment, http:// www.theguardian.com/environment/2017/jun/01/trump-withdraw-paris-climate -deal-world-leaders-react.

8. Kathy Mulvey and Seth Shulman, "The Climate Deception Dossiers: Internal Fossil Fuel Industry Memos Reveal Decades of Corporate Disinformation," Union of Concerned Scientists, July 2015, http://www.ucsusa.org/sites/default /files/attach/2015/07/The-Climate-Deception-Dossiers.pdf.

9. On the limited influence of scientific expertise in the diplomatic process for negotiating the Montreal Protocol, see Leen Hordijk, "Use of the RAINS Model in Acid Rain Negotiations in Europe," *Environmental Science & Technology* 25, no. 4 (1991): 596–603. For a discussion on the important political, social, and cultural factors that facilitated the Montreal Protocol, see Reiner Grundmann, "Ozone and Climate: Scientific Consensus and Leadership," *Science, Technology, & Human Values* 31, no. 1 (January 1, 2006): 73–101. The degree of scientific involvement is disputed by diplomat Richard Benedick, head of the US delegation, who has argued they played a more prominent role in the negotiations. Further scholarship on this question may shed more light on the issue. For Benedick's account, see Benedick, *Ozone Diplomacy*.

10. For arguments about the need to better understand the scientific process, and the hope that this might aid in breaking an impasse over climate change, see Naomi Oreskes and Erik M. Conway, *Merchants of Doubt: How a Handful of Scientists Obscured the Truth on Issues from Tobacco Smoke to Global Warming* (New York: Bloomsbury Press, 2010), pp. 266–274.

11. Grundmann, "Ozone and Climate."

12. Roger Harrabin, "G20: Merkel's Mission Is to Co-opt Saudis and Russia to Embarrass US," *BBC News*, July 7, 2017, sec. Europe, http://www.bbc.com /news/world-europe-40529447.

13. Shehab Khan, "Brexit Could 'Derail' the EU's Attempts to Tackle Climate Change," *Independent*, February 8, 2017, http://www.independent.co.uk/envi ronment/brexit-latest-news-derail-eu-attempts-climate-change-greenhouse-gas -emissions-meps-ets-a7569896.html; Andrew Simms, "The Curious Disappearance of Climate Change, from Brexit to Berlin | Andrew Simms," *Guardian*, March 30, 2017, sec. Environment, http://www.theguardian.com/environment/2017/mar/30 /the-curious-disappearance-of-climate-change-from-brexit-to-berlin.

14. Simms, "The Curious Disappearance of Climate Change, from Brexit to Berlin | Andrew Simms."

15. Markus Becker and Christoph Seidler, "The Threat of Pollution Tariffs: Economists Warn of a Climate Trade War," *Spiegel Online*, December 24, 2009, sec. International, http://www.spiegel.de/international/business/the-threat-of -pollution-tariffs-economists-warn-of-a-climate-trade-war-a-668635.html.

16. Oliver Milman, "Why Has Climate Change Been Ignored in the US Election Debates?," *Guardian*, October 19, 2016, sec. US news, http://www .theguardian.com/us-news/2016/oct/19/where-is-climate-change-in-the-trump -v-clinton-presidential-debates; Marianne Lavelle, "Climate Change Treated as Afterthought in Second Presidential Debate," *Inside Climate News*, October 10, 2016, https://insideclimatenews.org/news/10102016/presidential-debate-town -hall-donald-trump-hillary-clinton-climate-change-global-warming.

17. Coral Davenport, "Counseled by Industry, Not Staff, E.P.A. Chief Is Off to a Blazing Start," *New York Times*, July 1, 2017, sec. Politics, https://www .nytimes.com/2017/07/01/us/politics/trump-epa-chief-pruitt-regulations-climate -change.html.

Sources

Archival sources

In the endnotes, all manuscript boxes and folders are labeled with their record numbers only and the name of the archive. Full citations, including the titles of the boxes, their date range, and the locations of the archives are listed here. If boxes or folders within the same subseries contain documents having a different range of dates, a combined range is given in the reference below.

British Parliamentary Archives, London, United Kingdom
Note: The files below were viewed after submitting a Freedom of Information Act Request to the House of Commons. They should now be open to the public.

House of Commons Environment Committee.

HC-CP-9068 Environment Committee.
4th Report. Inquiry on Acid Rain.
1983–1984.
HC-CP-9073 Environment Committee.
4th Report. Inquiry on Acid Rain.
1983–1984.
HC-CP-12066. Environment Committee.
Inquiry on Acid Rain. 1987–1988.

House of Commons Journal Office.

HC-CL-JO-10–1794. House of Commons Unprinted Papers. 1983–1984.

House of Lords.

HL/PO/JO/10/11/2264. HL Main Papers. 1984–1985. 30 Nov 1984 to 10 Dec 1984. Nos. 30G–2G5.

Bundesarchiv Berlin (German National Archives), Berlin, Germany

Ministerium für Umweltschutz und Wasserwirtschaft.

Konvention über die Zusammenarbeit beim Schutz der Atmosphäre vor Verschmutzung.

DK 5/4573 Bd. 1. 1974–1976.
DK 5/3741 Bd. 2. 1975.

Meteorologische Aspekte der Verunreinigung der Atmosphäre.—Beratungen im RGW.

DK 5/57 Bd. 1. 1974–1983.

Umweltkongress der ECE über die Reduzierung der Schwefel-Emissionen oder deren grenzüberschreitender Ströme um mindestens 30%.

DK 5/1525 1985

Zusammenarbeit mit der BRD, Dänemark, Frankreich, Finnland, Norwegen, Schweden, Großbritannien, Iran, USA und Japan über Maßnahmen und Technologien zur Luftreinhaltung

DK 5/3394 1985–1989.

Arbeitsgruppe Luftverunreinigung der ECE.–4. Tagung vom 7.–11. 1.1974 in Genf.

DK 5/3721 Bd. 2. 1974.

Bundesarchiv Koblenz (German National Archives), Koblenz, Germany

Bundesministerium des Innern.- Umweltpolitik. Gesamteur-
opäischer Kongress über Umweltfragen.-Weiträumige
grenzüberschreitende Luftverschmutzung.

B 106 86364 Bandnummer: 7–9. 1978.
B 106 86365 Bandnummer: 10–11. 1978–1979.

Bundesministerium des Innern.- Umweltpolitik. Umweltschutz
bei weltweiten und anderen internationalen Organisationen
und Einrichtungen. Wirtschaftskommission für Europa
(Economic Commission for Europe—ECE). Arbeitsgruppe
"Luftverschmutzung".- Übereinkommen über weiträumige
Luftverschmutzung; ECE-Programm "Überwachung und
Beurteilung des weiträumigen Transports von Luftverun-
reinigungen in Europa."

B 106 69305 Bandnummer: 1–2. 1973–1976.
B 106 69306 Bandnummer: 3. 1976–1977.
B 106 69307 Bandnummer: 4–6. 1977–1982.

Bundesministerium des Innern.- Umweltpolitik. Umweltschutz
bei weltweiten und anderen internationalen Organisationen
und Einrichtungen. Wirtschaftskommission für Europa
(Economic Commission for Europe—ECE). Hochrangige
Tagung über Umweltfragen im Rahmen der ECE. Weiträumige, grenzüberschreitende Luftverschmutzung.

B 106 65843 Bandnummer: 5–7. 1978.
B 106 65846 Bandnummer: 3–4. 1978.
B 106 65849 Bandnummer: 1–2. 1977–1978.
B 106 65850 Bandnummer: 8–9. 1978–1979.
B 106 65851 Bandnummer: 10. 1979.
B 106 65852 Bandnummer: 11–12. 1979.
B 106 65853 Bandnummer: 14–15. 1979.

Bundesministerium für Umwelt, Naturschutz und Reaktorsi-
cherheit. Zusammenarbeit mit dem Ausland in Fragen des
Umweltschutzes. Zusammenarbeit mit einzelnen Staaten.
Zusammenarbeit mit Norwegen.

B 295 2621. Bandnummer: 1. 1972–1981.

B 295 2622. Bandnummer: 1–2. 1976–1990.

European Commission Archives, Brussels, Belgium

BAC 25/1983 (No. 1782–1787) Mise en oeuvre des actions de la Coopération européenne dans le domaine de la recherche scientifique et technique (COST). Période: 1971–1978.

BAC 70/1984 (No. 369–370) Pollution de l'air par le soufre; Pollution de l'air au sein de l'Organisation de Coopération et de Développement Economiques (OCDE). Période: 1963–1974.

BAC 49/1987 (No. 1439, 1695–1696) Mise en oeuvre des actions Coopération européenne dans le domaine de la recherche scientifique et technique (COST). Période: 1982.

BAC 82/1988 (No. 48) Coopération européenne dans le domaine de la recherche scientifique et technique (COST). Période: 1982.

BAC 156/1990 (No. 2382–2390) Activités de l'Organisation de Coopération et de Développement Economiques (OCDE) dans le domaine de l'environnement et en matière de recherche dans le domaine pollution de l'air et de l'eau. Période: 1967–1979.

BAC 156/1990 (No. 2415–2419) Pollution de l'air par le soufre. Période: 1972–1979.

BAC 2531/1991 (No. 560) Pollution de l'air, compte rendu et documents préparatoires à la réunion de l'Organisation de Coopération et de Développement Economiques (OCDE) à Paris (France) le 28 mai 1970. Période: 1967–1970.

BAC 2627/1991 (No. 457) Mesures de la pollution atmosphérique par air sulfureux—documents de travail. Période: 1974–1979. Période: 1978–1982.

BAC 107/1993 (No. 470–477) Mise en oeuvre des actions COST (coopération européenne dans le domaine de la recherche scientifique et technique). Période: 1978–1982.

BAC 133/1993 (No. 1624) Coopération européenne dans le domaine de la recherche scientifique et technique (COST) action 61a et 61a bis: polluants atmosphériques. Période: 1980.

BAC 133/1993 (No. 1878) Acte de notification de l'approbation par la Communauté de la convention sur la pollution atmosphérique transfrontière à longue distance. Période: 1982.

BAC 79/1996 (No. 373) Mise en oeuvre des actions de la Coopération européenne dans le domaine de la recherche scientifique et
technique (COST)—Accord de cooperation européen sur l'action
no 61a: "Recherches sur le comportement physico-chimique de
l'anhydride sulfureux dans l'atmosphère". Période: 1971–1978.

BAC 275/1998 (No. 52–53) Coopération européenne dans le
domaine de la recherche scientifique et technique (COST)
61a. Période: 1970–1981.

BAC 549/1998 (No. 627) Coopération européenne dans le
domaine de la recherche scientifique et technique (COST)
action 61a. Période: 1975.

BAC 495/2005 (No. 450) Convention de Genève (Suisse)—
Pollution atmosphérique transfrontalière à longue distance.
Période: 1980.

BAC 495/2005 (No. 454) Convention de Genève (Suisse)—
Pollution atmosphérique transfrontalière à longue distance.
Période: 1979.

Note: The following files on acid rain were made available to the author with
special permission from the Directorate General for Research and Innovation
(DG RTD). They had not yet been processed by the European Commission Archives and were viewed on the premises of the DG RTD in Brussels, Belgium.

BAC 82/1988 (No. 49)
BAC 82/1988 (No. 65)
BAC 82/1989 (No. 574–575)
BAC 82/1989 (No. 84)
BAC 107/1993 (No. 893–897)

Nasjonalbiblioteket (Norwegian National Library), Oslo, Norway
Newspaper Collections

A-Magasinet
Aftenpostens
Stavanger Aftenblad
Trønder-Avisa

**Sur nedbørs virkning på skog og fisk prosjektet (Acid rain's effects on forests
and fish project)**

Interne rapporter (IR). 1973–1982.

The National Archives, Kew Gardens, United Kingdom

AB 58/53 United Kingdom Atomic Energy Authority: Atomic Energy Research Establishment, Harwell: Senior Management, Correspondence and Papers. Overall policy on collaboration with Department of the Environment (including information on DOE organization). January 1980–December 1985.

AB 54/161 United Kingdom Atomic Energy Authority: Atomic Energy Research Establishment, Harwell: Health and Safety, Correspondence and Papers. Relations with external organizations on atmospheric pollution work. January 1, 1971–December 31, 1976.

AB 88/113 United Kingdom Atomic Energy Authority: Atomic Energy Research Establishment, Harwell: Diversification Projects, Correspondence and Papers. Atmospheric pollution program at Harwell: ministry requirements. January 1, 1972–December 31, 1972.

AB 89/26 United Kingdom Atomic Energy Authority: Atomic Energy Research Establishment, Harwell: Reactor Development, Correspondence and Papers. Central Electricity Generating Board: collaboration. January 1972–December 1975.

AIR 20/12566 Air Ministry, and Ministry of Defence: Papers accumulated by the Air Historical Branch. Long-range pollution. 1972–1974.

AT 34/10 Department of the Environment: Noise, Clean Air and Coast Protection Division: Registered Files (NPCA, NP and CACP Series). Air Pollution Monitoring Management Group: discussion papers and related correspondence; copied publications. January 1974–December 1975.

AT 34/12 Department of the Environment: Noise, Clean Air and Coast Protection Division: Registered Files (NPCA, NP and CACP Series). Air Pollution Monitoring Management Group: discussion papers; minutes of meetings; official and departmental correspondence. January 1, 1974–December 31, 1975.

AT 34/86 Department of the Environment: Noise, Clean Air and Coast Protection Division: Registered Files (NPCA, NP and CACP Series). Air Pollution Monitoring Management Group: minutes of meetings; agendas; discussion papers; official and interdepartmental correspondence. January 1974–December 1975.

AT 34/89 Department of the Environment: Noise, Clean Air
and Coast Protection Division: Registered Files (NPCA, NP
and CACP Series). Clean Air Council: Standing Technical
Committee; minutes of meetings; discussion papers; correspondence. January 1, 1975–December 31, 1976.

AT 34/90 Department of the Environment: Noise, Clean Air
and Coast Protection Division: Registered Files (NPCA, NP
and CACP Series). Clean Air Council: Standing Information Committee; minutes of meetings; discussion papers;
agendas.

AT 82/203 Department of the Environment: Directorate General of Research; Registered Files (DGR and DGRP Series).
Review Group on Acid Rain; reports on rainfall acidity in
the UK; Stockholm Acidification Conference, 1982. January
1981–1982.

BT 328/105 Board of Trade and successors: Alkali Inspector-
ate and successors: Unregistered Records. Chief Inspector's
Correspondence with Trade Associations and Industry
Councils, Including Minutes of Meetings, Industry Re-
search Reports and Statistics. Central Electricity Generating
Board: includes annual meeting papers. 1959–1979.

BT 328/106 Board of Trade and successors: Alkali Inspector-
ate and successors: Unregistered Records. Chief Inspector's
Correspondence with Trade Associations and Industry
Councils, Including Minutes of Meetings, Industry Re-
search Reports and Statistics. Central Electricity Generating
Board: includes annual meeting papers. 1980–1986.

BT 328/152 Board of Trade and successors: Alkali Inspectorate and successors: Unregistered Records. Reports and
Correspondence from Industry. Sulphur dioxide emission
in the UK: problems of sulphuric acid and EEC policy.
1968–1988.

BT 328/162 Board of Trade and successors: Alkali Inspectorate and successors: Unregistered Records. Reports and
Correspondence from Industry. Control of industrial air
pollution: proposals for legislation to control industrial
discharge to the air; correspondence with local authorities.
1982–1988.

COAL 29/655 National Coal Board and British Coal Corporation: Production Department and successors. Coal
Preparation. NCB/CEGB joint study group: Sulphur in coal:

Agendas, minutes, papers, and reports. November 1, 1984–
July 31, 1987.

COAL 74/1028 National Coal Board: Minutes Series, Headquarters. Coal Use Environment Committee—Atmospheric
Pollution—Acid Rain Formation: Headquarters. 1984.

COAL 74/1716 National Coal Board: Minutes Series,
Headquarters. Fuel Technology—Atmospheric Pollution—
Agricultural Use of Pesticides Etc. Acid Rain. Formation:
Headquarters. 1972–1983.

COAL 96/85 National Coal Board and British Coal Corporation:
Central Planning Unit: Registered Files. Acid Rain: Papers
relating to Legislation and Sizewell B Inquiry. October 10,
1982–February 28, 1983.

COAL 96/91 National Coal Board and British Coal Corporation: Central Planning Unit: Registered Files. Acid Rain:
Papers. March 1, 1983–November 30, 1986.

COAL 97/541 National Coal Board and British Coal Cor-
poration: Coal Research Establishment: Registered Files.
Fluidised Combustion Files. Nitrogen and Sulphur Oxides
Pollutants. August 1970–January 1972.

DSIR 23/38676 Department of Scientific and Industrial Research: Aeronautical Research Council: Reports and Papers.
Atmospheric Environment Committee: Harwell air pollution project. 1971.

FCO 76/1550 Foreign and Commonwealth Office: Marine
and Transport Department: Registered Files (MR Series).
Environment. EEC directive on sulphur content of fuel oil.
January 1, 1977–December 31, 1977.

FCO 76/1553 Foreign and Commonwealth Office: Marine and
Transport Department: Registered Files (MR Series). Environment. Transboundary pollution: ECE high-level meetings on the environment. January 1, 1977–December 31,
1977.

FCO 76/1554 Foreign and Commonwealth Office: Marine
and Transport Department: Registered Files (MR Series).
Transboundary pollution: ECE high-level meetings on the
environment. January 1, 1977–Dcember 12, 1977.

FCO 76/1555 Foreign and Commonwealth Office: Marine
and Transport Department: Registered Files (MR Series).
Transboundary pollution: ECE high-level meetings on the
environment. January 1, 1977–December 31, 1977.

FT 8/103 Nature Conservancy and Nature Conservancy Council: Information and Advice Matters, Registered Files. Visits overseas: includes Educational Use of Living Organisms (Netherlands); Woodlands and Forests of France; visit to Iran; Barnacle Goose project (Svalbard); Western Palearctic Migratory Bird Management (Paris); study tour of East Coast of United States; Wildlife and Oil Pollution in North Sea; Ecological Impact of Acid Precipitation (Norway); Study of Falkland Islands; World Conferences on Birds of Prey (Greece). January 1, 1972–December 31, 1982.

FT 48/35 Nature Conservancy Council and its successors: Energy Advisory Office Registered Files (Series D and UFD). Pollution other than oil—acid rain: press coverage; reports, including "Acid deposition and its implications for nature conservation in Britain." January 1, 1982–December 31, 1987.

FV 12/644 Ministry of Technology and successors: Warren Spring Laboratory: Laboratory Reports (LR Series). The measurement and dispersion of sulphur dioxide in the Forth Valley of Scotland with specific reference to emissions from high chimneys. AWC Keddie, JS Bower, RA Maughan, and GH Roberts. January 1, 1981–December 31, 1981.

FV 61/52/2 Department of Trade and Industry and Department of Trade: Europe, Industry and Technology Division: Registered Files (ECT Series). Science and technology: international cooperation of Britain in the forum of the United Nations Economic Commission for Europe. January 1, 1975–December 31, 1976.

FV 61/54 Department of Trade and Industry and Department of Trade: Europe, Industry and Technology Division: Registered Files (ECT Series). Science and technology: international cooperation of Britain in the forum of the United Nations Economic Commission for Europe. January 1, 1977–December 31, 1978.

FV 61/63 Department of Trade and Industry and Department of Trade: Europe, Industry and Technology Division: Registered Files (ECT Series). Import controls: planning and draft regulations for import licence controls. January 1, 1975–December 31, 1976.

HLG 120/1571 Ministry of Housing and Local Government, Local Government Division, and Department of the Environment:

Local Government General Policy and Procedure, Registered Files (LG Series). Sulphur dioxide: presence of sulphur in atmosphere and residual fuel oil; official, interdepartmental and Parliamentary correspondence, background papers, notes of meetings and press cuttings. January 1, 1971–December 31, 1973.

HLG 120/1579 Ministry of Housing and Local Government, Local Government Division, and Department of the Environment: Local Government General Policy and Procedure, Registered Files (LG Series). Interdepartmental Committee on Air Pollution Research: consideration of future of The National Survey of Smoke and Sulphur Dioxide; draft report of the working party on air pollution monitoring and surveillance, with comments. January 1, 1972–December 31, 1972.

The National Archives and Records Administration II, College Park, Maryland, USA

RG 412—Studies and Other Related Records, Acid Rain. National Archives Identifier: 22123773.
Office of Research and Development; Studies and Other Related Records Regarding Acid Rain, 1979–1983 (Entry A1 59). Boxes 1–15.

Norsk institutt for luftforskning, Archives and Library, Kjeller, Norway

Papers of Brynjulf Ottar. Scientific Reports (OR), Technical Reports (TR), Lectures (F), Travel Reports (RR). 1973–1989.

Organisation for Economic Co-operation and Development Archives, Paris, France.
Note: The citations in the main text contain the reference number for the exact document viewed, per the classification system of the OECD archives. Below are the general reference codings of all the series cited in the main text as well as the call numbers for the microfilms and microfiches viewed.

ENV/AIR Environment Directorate. Air Management Group.

ENV/AIR 76 1–21 1976
ENV/AIR 77 1–20 1977

ENV/AIR 78.8 1978
ENV/AIR 78.10–78.13 1978
ENV/AIR 78.15 1978
ENV/AIR 78.19 1978
ENV/AIR/M 78.21 1979

ENV/M Environment Committee. Reunions. Ministerial.

ENV/M 1975
ENV/M 1976 1–3
ENV/M 1977 1–2

ENV/TFP Environment Committee. Transfrontier Pollution
Group.

ENV/TFP/1976/1/19 1976
ENV/TFP/M/1976/4/7 1976
ENV/TFP/DIVERS/1976
ENV/TFP/1977/DIVERS 1977
ENV/TFP/77/1/24 1977
ENV/TFP/78.1/38.961 1978
ENV/TFP/78.1/39.701 1978
ENV/TFP/78.2/38.630 1978
ENV/TFP/78.3/38.482 1978
ENV/TFP/78.3/42.777 1978
ENV/TFP/78.5/39.444 1978
ENV/TFP/78.6/39.707 1978
ENV/TFP/78.9/41.093 1978
ENV/TFP/78.10/ANNEXE/42.825 1978
ENV/TFP/78.11/42.805 1978
ENV/TFP/78.11/ANNEXE/43.805 1978
ENV/TFP/78.12 1978
ENV/TFP/78.14/42.833 1978
ENV/TFP/78.15/45.207 1978
ENV/TFP/78.17/46.516 1978
ENV/TFP/78.17/48.450 1978
ENV/TFP/78.18/44.716 1978
ENV/TFP/41.123 1978
ENV/TFP/A/78.11/E2333 1978
ENV/TFP/M/78.11/40.673 1978
ENV/TFP/78.1/40.724 1978

ENV/TFP/7817/48.805 1979
ENV/TFP/78.3/45.071 1979

NR/ENV. Environment Directorate.

NR/ENV 1–62 and Divers 1973
NR/ENV 1–80 1974

Riksarkivet (Norwegian National Archives), Oslo, Norway
Note: In some cases the boxes (Stykke) contained several folders of many different subjects. In these cases, I have noted the specific folder (Mappe) which was viewed and subsequently cited in the main text. Some of the files below are not currently listed in the public catalogue but can be located with the assistance of the archivists in Oslo. These are the Ministry of Environment (S-2532) records: Daa series (Records on Pollution/ Sakarkiv Forurensningsavdelingen) and Dca series (Records of the Department's Organizational Unit/ Sakarkiv Organisasjonsavdelingen).

RA/PA-0636/II/D/Dc/Lo755 Norges Industriforbund. Mappe:
Sur nedbør. 1975–1984.
RA/PA-0636/II/D/Dc/Lo754 Norges Industriforbund. Mappe:
Sur nedbør. 1975–1983.
RA/PA-0641/E/Eg/Loo12 Norges Naturvernforbund. Stykke:
Sur Nedbør. 1971–1981.
RA/PA-0641/E/Eg/Loo13 Norges Naturvernforbund. Stykke:
Sur Nedbør. 1975–1981.
RA/S-1574/E/Loo25 Norges teknisk-naturvitenskapelige
forskningsråd. Mappe: Komite for forurensningsspørsmål.
1970–1984.
RA/S-1574/E/Loo65 Norges teknisk-naturvitenskapelige
forskningsråd. Mappe: Komite for forurensningsspørsmål.
1970–1975.
RA/S-1574/E/Lo405 Norges teknisk-naturvitenskapelige
forskningsråd. Mappe: Modern Society ENV. The Environment Committee. 1966–1971.
RA/S-2477/D/Dc/Loo74 Norges fiskeriforskningsråd. Mappe:
Koordinering og stimulering av sur nedbørforskning. 1980–
1982.
RA/S-1713/D/Dc/Lo124 Samferdselsdepartementets sentralarkiv. Mappe: FN- Den økonomiske kommisjon for Europa, ECE. 1976.

RA/S-2532/2/Daa/Loo48 Miljøverndepartementet. Stykke: FN.
ECE. UNEP. Diverse. 1973–1980.

RA/S-2532/2/Daa/Loo54 Miljøverndepartementet. Stykke: FN
(med særorganisasjoner). Internasjonalt forsvarssamarbeid
NATO. 1984.

RA/S-2532/2/Daa/Lo506. Miljøverndepartementet. Luftforurensning Forskningsprosjektet "Sur nedbørs virkning på
skog og fisk" SO2/problematikken. 1972–1977.

RA/S-2532/2/Daa/Lo512. Miljøverndepartementet. Stykke:
Luftforurensning. Generelt. NILU- prosjekter. 1978–1984.

RA/S-2532/2/Daa/Lo509 Miljøverndepartementet. Stykke:
Luftforurensning. Generelt. 1980–1984.

RA/S-2532/2/Dca/Loo03. Miljøverndepartementet. Stykke:
Statsråd Gro H. Brundtlands Besøk i DDR, Egypt Nederland, USA, Sverige, Sjekkoslov. 1972–1979.

RA/S-2532/2/Dca/Loo04. Miljøverndepartementet.
Stykke: Statsrådens og statssekretærens besøk i Sovjet
1978. Miljøvernminister Hansens besøk til Svalbard.
Statssekretærbesøk fra Finland. Besøk av DDRs miljøvern-
minister september 1978. Statsråd Brundtlands besøk i
Storbritannia. 1976–1978.

RA/S-2532/2/Dca/Loo05. Miljøverndepartementet. Stykke:
Statsbesøk fra utlandet. Statsrådenes besøk til Finland,
DDR, USA og Østerrike. 1977–1983.

RA/S-2532/2/Dca/Loo14. Miljøverndepartementet. Stykke:
Oppgave over proposisjoner som aktes fremmet Forskningsprosjektet Sur nedbørs virkning på skog og fisk. 1974–
1979.

RA/S-2532/2/Dca/Loo15. Miljøverndepartementet. Stykke:
Oppgave over proposisjoner som aktes fremmet Forskningsprosjektet Sur nedbørs virkning på skog og fisk. 1976–
1979.

RA/S-2532/2/Dca/Lo253. Miljøverndepartementet. Stykke:
EEC, EF (Fellesmarkedet). Miljøvern. 1984.

RA/S-2532/2/Dca/Lo291. Miljøverndepartementet. Stykke:
FN (med særorganisasjoner). ECE—Høynivåmøte I Geneve
(1979). 1978–1981.

RA/S-2532/2/Dca/Lo292. Miljøverndepartementet. Stykke:
FN (med særorganisasjoner) ECE / Task force / European
Monitoring Programme EMP. 1976–1978.

RA/S-2532/2/Dca/Lo293. Miljøverndepartementet. Stykke:

FN (med særorganisasjoner). ECE / Task force / European
Monitoring Programme EMP. 1978–1984.

RA/S-2532/2/Dca/Lo294. Miljøverndepartementet. Stykke: FN
(med særorganisasjoner). ECE—Spesialgruppe for luftforurensning. 1976–1978.

RA/S-2532/2/Dca/Lo295. Miljøverndepartementet. Stykke:
FN (med særorganisasjoner). ECE/ IEB. Interimskomiteen
for gjennomføring av konvensjonen om langtransporterte
grenseoverskridende luftforurensning. 1980–1982.

RA/S-2532/2/Dca/Lo297. Miljøverndepartementet. Stykke:
FN (med særorganisasjoner). ECE/ IEB. Interimskomiteen
for gjennomføring av konvensjonen om langtransporterte
grenseoverskridende luftforurensning. 1983–1984.

RA/S-2532/2/Dca/Lo300. Miljøverndepartementet. Stykke: FN
(med særorganisasjoner). ECE: Working party on air pollution (WPAP). Ad hoc Committee for energy. 1976–1984.

RA/S-2532/2/Dca/Lo404. Miljøverndepartementet. Stykke:
Internasjonalt samarbeid om handel og økonomi. OECD.
LRTAP (Langtransport av luftforurensninger). 1975–1976.

RA/S-2532/2/Dca/Lo405. Miljøverndepartementet. Stykke:
Internasjonalt samarbeid om handel og økonomi. OECD.
LRTAP (Langtransport av luftforurensninger). 1975–1977.

RA/S-2532/2/Dca/Lo409. Miljøverndepartementet. Stykke:
Internasjonalt samarbeid om handel og økonomi. OECD.
Miljøvernminister 1974. SO2 Virkning av sur nedbør—
Konferanse. 1974–1975.

RA/S-2532/2/Dca/Lo437. Miljøverndepartementet. Stykke:
Internasjonalt samarbeid om handel og økonomi. OECD.
Miljøvern/turisme. Luftforurensningsgruppen ENV/AIR. Grup-
pen for forurensning over landegrensene TFP. 1977–1984.

RA/S-2532/2/Dca/Lo441. Miljøverndepartementet. Stykke:
Internasjonalt samarbeid om handel og økonomi. OECD.
Svensk-norsk arbeidsgruppe om sur nedbør. OECD/Env
bureau. OECD/Council. OECD/Group of experts on the
state of the Environment. 1974–1984.

RA/S-2532/2/Dca/L6048. Miljøverndepartementet. Stykke:
Luftforurensning. Sur nedbør. 1982.

RA/S-2532/2/Dca/L6051. Miljøverndepartementet. Stykke:
Luftforurensning. Sur nedbør. 1983.

RA/S-2939/D/Db/Dbk/Dbke/Lo604. NAVF, Rådet for natur
vitenskapelig forskning. Mappe: Sur nedbør. 1980–1985.

RA/S-2939/D/Db/Dbk/Dbke/L0605. NAVF, Rådet for natur
vitenskapelig forskning. Mappe: Kontaktgruppe sur nedbør-
forskning. 1985.
RA/S-2939/D/Db/Dbk/Dbkj/L0734. NAVF, Rådet for natur-
vitenskapelig forskning. Mappe: Sur nedbør. 1972–1984.
RA/S-2939/D/Db/Dbk/Dbkj/L0738. NAVF, Rådet for natur-
vitenskapelig forskning. Mappe: Sur nedbør. 1972–1984.
Mappe: Sur nedbør-forskning. 1975–1984.
RA/S-5806/D/L0196. Kultur- og vitenskapsdepartementet.
Mappe: Miljøverndepartementet. 1980–1986.

Riksarkivet Arninge (Swedish National Archvies), Arninge, Sweden

SE/RA/1211/22/E 1 A/620. Jordbruksdepartementet. 9 okt.
I någon/några av volymerna till beslutsdatum 9 okt ingår
akt till Prop. nr 31 om godkännande av konvention om
långväga gränsöverskridande luftföroreningar. 1980.
SE/RA/730289/L/L 2/L 2 d/75. Svenska Naturskyddsförenin-
gen. Naturens villkor—stoppa försurningen. 1980.
SE/RA/730289/L/L 2/L 2 d/70. Svenska Naturskyddsföreningen.
1980. Försurningen en miljökatastrof. Vattenvärnet/SNF.
SE/RA/730289/L/L 2/L 2. Svenska Naturskyddsföreningen. 1984.
Handbok för kretsarbete "Stoppa försurningen!" 1984.
SE/RA/730289/L/L2/L2 d/82. Svenska Naturskyddsföreningen.
Stoppa försurningen. (1) Åtgärder i Sverige (2) Handbok
för kretsarbete (3) Försurningen och framtiden (4) Folder.
1982–1984.

Riksarkivet Marieberg (Swedish National Archives), Marieberg, Sweden

SE/RA/322619/~/7. Nationalkommittén för 1972 års FN-
konferens om den mänskliga miljön. Kommittén för forskning
och andra sakfrågor, protokoll jämte bilagor nr 50–90. 1972.
SE/RA/322619/~/8. Nationalkommittén för 1972 års FN-
konferens om den mänskliga miljön. Kommittén för
forskning och andra sakfrågor, protokoll jämte bilagor nr
91–141. 1972.
SE/RA/322619/~/10. Nationalkommittén för 1972 års
FN-konferens om den mänskliga miljön. Kommittén för
forskning och andra sakfrågor, protokoll jämte bilagor nr
159–193.

SE/RA/322619/~/11. Nationalkommittén för 1972 års
FN-konferens om den mänskliga miljön. Kommittén för
forskning och andra sakfrågor, protokoll jämte bilagor nr
194–224. 1972.
SE/RA/323591. Kommittén med uppgift att samordna aktiviteter i samband med uppföljningen av Stockholmskonferensen 1972 om den mänskliga miljön. No. 1–5. 1980.

Royal Society Archives, London, United Kingdom

GB 117 repository, reference number EJD/5. Notes on staff
programmes. Acid Rain—Notes on Meetings Sweden,
Norway. 1950–1986.

Note: In addition to these documents, the author also received copies of John Mason's private papers held by Dr. Peter Collins, emeritus
director of the Royal Society, which are currently not available to the
public.

Surrey History Centre Archives, Woking, Surrey, United Kingdom

Central Electricity Research Laboratories, Leatherhead: Brochures, Technical Reports, Newsletters, Plans, and Photo-
graphs, 1965–2000.

7802/1/ Brochures, 1966–2000. No. 1–16. 1966–1990.
7802/2/ Various technical bulletins relating to topics including pollution testing, oil analysis, magnetics, and television and radio interference. No. 1–16. 1970–1989.

**United Nations Economic Commission for Europe Archives, Geneva,
Switzerland**

GX 10/6/2 Economic Commission for Europe—High-Level
Meeting on the Protection of the Environment, Geneva, 13–
16 November 1979. 1979–1980.
GX 33/1/1 Air Pollution—General—Air Pollution Arising from
Various Domestic, Commercial, and Industrial Sources.
1963–1975 .
GX 33/1/10 Air Pollution—General—SO2 Emissions from
Stationary Sources. 1974–1980.

GX 33/1/14 Air Pollution—General—Transboundary Air Pollution. 1975–1981.

GX 33/1/15 Air Pollution—General—Cooperation with Other International Organizations. 1976–1981.

GX 33/1/18 Air Pollution—General—First Meeting of the Interim Executive Body for the Convention on Long-Range Transboundary Air Pollution. 1980.

GX 33/1/19 Air Pollution—General—Working Group on Effects of Sulphur Compounds on the Environment. 1981.

GX 33/1/2 Air Pollution—General—Examination of Economic Studies in Air Pollution Prevention. 1970–1972.

GX 33/1/3 Air Pollution—General—Working Party on Pollution Problem. 1970–1981.

GX 33/1/4 Air Pollution—General—Review of Situation in ECE Countries. 1970–1981.

World Health Organization Archives, Geneva, Switzerland.

A 6: Air Pollution.

6/86/19. WMO Panel on Meteorological Aspects of Air Pollution.

6/86/22. 3 EM. Reunion d'Experts Nationaux sur la Mesure de la Pollution Atmospherique par les Composes du Soufre et les Particules en Suspension—Luxembourg, November 1973—Organisee par la Commission des Communautes Europeennes.

6/86/23. 3 EME. Reunion d'Experts Nationaux sur la Mesure de la Pollution Atmospherique par les Vechicles a Moteur—Oxydes d'Azote, Monoxyde de Carbone—Luxembourg, 14–15/1/74—Organisee par la Commission des Communautes Europeennes.

6/86/24. Expert Meeting to Establish a Cooperative Pogramme for Monitoring and Evaluation of the Transmission of Air Pollutants in Europe, Organized by the Norwegian Government, Oslo, December 1974.

E 15: Environmental Pollution

15/86/7. OECD Exploratory Meeting on National and International Efforts to Appraise Present Knowledge of

Effects of Environmental Pollutants—Paris,
April 1972.
15/86/22. Conference on the Effects of Acid Precipitation
Organized by the Norwegian Ministry of Environment
and the Norwegian Joint Research Project on the Effects of Acid Precipitation on Forest and Fish, Telemark
County, Norway, 14–19/6/1976.

Oral Histories

Note: A considerable proportion of Central Electricity Generating Board
(CEGB) documents cited in this book were collected by the author
through in-person interviews with former scientists from the organization. Although the CEGB had an extensive library during its existence,
much of these records appear to have been misplaced or destroyed with
the privatization of the power industry in the 1990s, although some can
be found in the UK National Archives. The author also obtained personal
papers concerning Norwegian research into acid rain from in-person interviews with Norwegian scientists. Thus, the following citations are followed by an "in person" or "telephone" notation to make clear which interviews resulted in the procurement of additional documentary evidence
currently in the author's possession.

Author interview with John Miller, NOAA, Stable Isotope
Lab (Institute of Arctic and Alpine Research), October 15,
2010. (telephone)
Author interview with Dr. Jerome (Nick) Heffter, NOAA,
Air Resources Laboratory (retired), formerly of the Special
Projects, Weather Bureau, October 29, 2010. (telephone)
Author interview with Anton Eliassen, director of the Nor-
wegian Meteorological Institute, May 16, 2011, Oslo,
Norway. (in person)
Author interview with Øystein Hov, research director of the
Norwegian Meteorological Institute, May 19, 2011, Oslo,
Norway. (in person)
Author interview with Gene Likens, Cary Institute for Eco-
system Studies, March 9, 2012, and August 18, 2016.
(telephone)
Author interview with Richard Skeffington, professor, Univer-
sity of Reading, former biologist at the CERL, March 15,
2013. (in person)

Author interview with Alan Webb, retired, environmental
    chemist at CERL, March 16, 2013. (in person)
Author interview with Richard (Dick) Derwent, retired profes-
    sor of Atmospheric Chemistry, Imperial College, London,
    March 18, 2013, Newbury, England. (in person)
Author interview with Anthony (Tony) Kallend, retired, head
    of environmental chemistry at CERL, March 25, 2013. (in
    person)
Author interview with Ronald Barnes, retired, former principal
    environmental adviser, Esso, March 4, 2013, Wantage,
    England. (in person)
Author interview with Bernard Fisher, principal scientist,
    United Kingdom Environment Agency, March 21, 2013,
    London, England. (in person)
Author interview with James Irwin, retired, former scientist at
    Warren Spring Laboratory and director of the Institution of
    Environmental Sciences, March 22, 2013. (in person)
Author interview with Hans Martin Seip, professor emeritus,
    University of Oslo, January 15, 2014. (telephone)
Author interview with David Fowler, professor, University of
    Edinburgh, January 15, 2014. (telephone)
Author interview with Robert Goldstein, Electric Power
    Research Institute, November 3 and November 19, 2015.
    (telephone)

Published Sources

For a sortable list of primary and secondary sources in this book, please visit press.uchicago.edu/sites/rothschild/.

Index